Pay for Performance in Health Care: Methods and Approaches

Edited by
Jerry Cromwell, Michael G. Trisolini, Gregory C. Pope,
Janet B. Mitchell, and Leslie M. Greenwald

March 2011

RTI Press

©2011 Research Triangle Institute. RTI International is a trade name of Research Triangle Institute.

All rights reserved. Please note that this document is copyrighted and credit must be provided to the authors and source of the document when you quote from it. You must not sell the document or make a profit from reproducing it.

Library of Congress Control Number: 2011921923
ISBN 978-1-934831-04-5

doi:10.3768/rtipress.2011.bk.0002.1103
www.rti.org/rtipress

Suggested Citation

Cromwell, J., Trisolini, M. G., Pope, G. C., Mitchell, J. B., and Greenwald, L. M., Eds. (2011). *Pay for Performance in Health Care: Methods and Approaches*. RTI Press publication No. BK-0002-1103. Research Triangle Park, NC: RTI Press. Retrieved [date] from http://www.rti.org/rtipress.

This publication is part of the RTI Press Book series.
RTI International
3040 Cornwallis Road, PO Box 12194, Research Triangle Park, NC 27709-2194 USA
rtipress@rti.org
www.rti.org

Contents

Acknowledgments	v
Contributors	vi
Abbreviations and Acronyms	vii

Introduction — 1
Janet B. Mitchell

Chapter 1. Introduction to Pay for Performance — 7
Michael G. Trisolini

Chapter 2. Overview of Pay for Performance Models and Issues — 33
Gregory C. Pope

Chapter 3. Theoretical Perspectives on Pay for Performance — 77
Michael G. Trisolini

Chapter 4. Quality Measures for Pay for Performance — 99
Michael G. Trisolini

Chapter 5. Incorporating Efficiency Measures into Pay for Performance — 139
John Kautter

Chapter 6. Who Gets the Payment Under Pay for Performance? — 161
Leslie M. Greenwald

Chapter 7. Attributing Patients to Physicians for Pay for Performance — 181
Gregory C. Pope

Chapter 8. Financial Gains and Risks in Pay for Performance Bonus Algorithms — 203
Jerry Cromwell

Chapter 9. Overview of Selected Medicare Pay for Performance Demonstrations — 221
Leslie M. Greenwald

Chapter 10. Evaluating Pay for Performance Interventions   267
    Jerry Cromwell and Kevin W. Smith

Chapter 11. Converting Successful Medicare Demonstrations into National Programs   315
    Leslie M. Greenwald

Chapter 12. Conclusions: Planning for Second-Generation Pay for Performance   341
    Michael G. Trisolini, Jerry Cromwell, and Gregory C. Pope

Index   371

Acknowledgments

The authors would like to thank the many editors, reviewers, and document preparation specialists who contributed to the development of this book. They include Kathleen Lohr, PhD, who was the RTI Press editor-in-chief when this book was written; she provided extensive comments on each chapter and advice on layout, cover design, and many other details of writing and production. Karen Lauterbach, RTI Press managing editor, provided detailed advice and comments on editing, layout, and production, and kept the many components of the overall process moving steadily forward. Anne Gering and Carol Offen provided detailed edits for each chapter and managed the process of moving each chapter through writing, editing, and production. Joanne Studders and Sonja Douglas provided detailed final reviews and edits for each chapter, and prepared the final layouts. We would also like to thank the many anonymous reviewers who provided valuable comments and suggestions for each of the chapters when they were in earlier versions.

Contributors

Jerry Cromwell, PhD, is an RTI Senior Fellow in Health Economics. In the past 40 years, he has participated in more than 75 federally funded evaluations and technical analyses of health care payment reforms, including Medicare's hospital prospective payment system, the physician fee schedule, anesthesia payment, disease management programs, and federal-state Medicaid cost sharing. He also is an adjunct professor in the University of Massachusetts's College of Nursing, where he teaches health economics, finance, secondary data analysis, and cost-effectiveness analysis.

Leslie M. Greenwald, PhD, is a principal scientist at RTI International. Her research interests include Medicare program policy, health care costs and payment, managed care, and health care reform. Dr. Greenwald received a BA from Dartmouth College and an MPA and PhD from the University of Virginia.

John Kautter, PhD, is a senior economist at RTI International. His research interests include the development, implementation, and evaluation of health care payment models, including payment models for fee-for-service and managed care, as well as hybrid payment models. Dr. Kautter received his doctorate in economics from the University of Illinois at Urbana-Champaign, where he specialized in health economics, industrial organization, and applied statistics.

Janet B. Mitchell, PhD, heads RTI's Social Policy, Health, and Economics Research unit. She received her doctorate from the Heller School at Brandeis University in 1976. She has studied physician payment under Medicare and Medicaid for many years and conducted the seminal work on bundling inpatient physician services (physician diagnosis-related groups).

Gregory C. Pope, MS, directs RTI's Health Care Financing and Payment program. Mr. Pope is a health economist whose primary research interest is health plan and provider payment in the US Medicare program, including pay for performance, accountable care organizations, and risk adjustment.

Kevin W. Smith, MA, is a senior health research analyst in RTI's Health Care Quality and Outcomes Group. His research interests include evaluation research, quality-of-life measurement, psychometric assessment, structural equation modeling, and survey methodology. Mr. Smith received a BA from Colgate University and an MA from Tufts University.

Michael G. Trisolini, PhD, MBA, is the director of RTI International's Heath Care Quality and Outcomes Program. Dr. Trisolini has more than 27 years of experience in health services research and management. His research focuses on quality-of-care measurement, quality improvement programs, pay for performance, value-based purchasing, and health information technology. He has a BA from Oberlin College, an MBA from Harvard University, and a PhD from Brandeis University.

Abbreviations and Acronyms

ACE	acute care episode
ACO	accountable care organization
ACSC	ambulatory care sensitive condition
ADL	activity of daily living
AHRQ	Agency for Healthcare Research and Quality (formerly known as the Agency for Health Care Policy and Research)
ALOS	average length of stay
AMA	American Medical Association
AMI	acute myocardial infarction
APR-DRG	all-payer refined diagnosis-related group
CAD	coronary artery disease
CAMC	Charleston Area Medical Center
CBO	Congressional Budget Office
CHF	congestive heart failure
CMHCB	Care Management for High-Cost Beneficiaries
CMO	care management organization
CMS	Centers for Medicare & Medicaid Services
CoE	Center of Excellence
COPD	chronic obstructive pulmonary disease
CPOE	computerized physician order entry
CPPI	California Physician Performance Initiative
CPT	Current Procedural Terminology
CPTD	Cancer Prevention and Treatment Demonstration
CQI	continuous quality improvement
CT	computed tomography
DEA	data envelopment analysis
DHHS	Department of Health and Human Services

D-in-D	differences-in-differences
DM	disease management
DRA	Deficit Reduction Act of 2005
DRG	diagnosis-related group
EDSS	Extended Disability Status Scale
EHR	electronic health record
EMR	electronic medical record
ESRD	end-stage renal disease
FFS	fee-for-service
FIM	Functional Independence Measure
GDP	gross domestic product
HAC	hospital-acquired condition
HbA1c	glycosylated hemoglobin
HCC	Hierarchical Condition Category
HEDIS	Healthcare Effectiveness Data and Information Set (as of 2007); Health Plan Employer Data and Information Set (in use 1993–2007)
HIE	health information exchange
HIPAA	Health Insurance Portability and Accountability Act
HMO	health maintenance organization
HQA	Hospital Quality Alliance
HQID	(Premier) Hospital Quality Incentive Demonstration
HVBPP	Hospital Value-Based Purchasing Program
IADL	instrumental activity of daily living
ICD-9	International Classification of Diseases, Ninth Revision
ICER	incremental cost-effectiveness ratio
ICU	intensive care unit
IDS	integrated delivery system
IHA	Integrated Healthcare Association

IHIE	Indiana Health Information Exchange
IOM	Institute of Medicine
IT	information technology
ITT	intent-to-treat
LL	lower limit
MCCD	Medicare Coordinated Care Demonstration
MCPT	maximum percentage eligible for bonus
MCS	Mental Component Summary
MedPAC	Medicare Payment Advisory Commission
MFIS	Modified Fatigue Impact Scale
MHQP	Massachusetts Health Quality Partners
MHS	Medicare Health Support
MHSO	Medicare Health Support Organization
MMA	Medicare Modernization Act
MRI	magnetic resonance imaging
MS	multiple sclerosis
MSA	metropolitan statistical area
MS-DRG	Medicare severity diagnosis-related group
MSQLI	Multiple Sclerosis Quality of Life Inventory
MSSP	Medicare Shared Savings Program
NCQA	National Committee for Quality Assurance
NICE	National Institute for Health and Clinical Excellence (United Kingdom)
NIMBY	not in my backyard
NJHA	New Jersey Hospital Association
P4P	pay for performance
PAC	post-acute care
PBPM	per beneficiary per month
PCP	primary care physician

PCS	Physical Component Summary
PDDS	Patient Determined Disease Steps
PDP	prescription drug plan
PGP	physician group practice
PHO	physician-hospital organization
PMPY	per member per year
PN	patient navigator
PO	physician organization
POA	present on admission
PPO	preferred provider organization
PQI	Prevention Quality Indicator
PRO	peer review organization
QALY	quality-adjusted life year
QOL	quality of life
RN	registered nurse
RtoM	regression to the mean
SCHIP	State Children's Health Insurance Program
SE	standard error
SFR	stochastic frontier regression
SNF	skilled nursing facility
TRHCA	Tax Relief and Health Care Act of 2006
UL	upper limit
VA	Veterans Affairs
WHO	World Health Organization

Introduction
Janet B. Mitchell

This book provides a balanced assessment of pay for performance (P4P), addressing both its promise and its shortcomings. P4P programs have become widespread in health care in just the past decade and have generated a great deal of enthusiasm in health policy circles and among legislators, despite limited evidence of their effectiveness. On a positive note, this movement has developed and tested many new types of health care payment systems and has stimulated much new thinking about how to improve quality of care and reduce the costs of health care.

The current interest in P4P echoes earlier enthusiasms in health policy—such as those for capitation and managed care in the 1990s—that failed to live up to their early promise. The fate of P4P is not yet certain, but we can learn a number of lessons from experiences with P4P to date, and ways to improve the designs of P4P programs are becoming apparent. We anticipate that a "second generation" of P4P programs can now be developed that can have greater impact and be better integrated with other interventions to improve the quality of care and reduce costs.

With the March 2010 passage of the Patient Protection and Affordable Care Act (officially, P.L. 111-148 and referred to hereafter as the "Affordable Care Act"), health care reform has moved from a much-debated policy concept to a major policy implementation challenge. The Affordable Care Act seeks to reform private health insurance regulation and practice and to extend access to private health insurance through subsidies and the creation of state-based health insurance cooperatives. Some of these provisions took effect in late 2010, and others will take effect in the following years. Although the Affordable Care Act legislation is famously detailed in many areas, the operational issues in turning policy concepts and goals into workable programs will require considerable additional effort. Making health care reform work is the next crucial step.

The Affordable Care Act seeks to reduce unsustainable US health care spending and improve health care value partly by using a wide range of demonstrations and pilot projects, many of which focus on P4P as a conceptual model. P4P models (also known as "shared savings," "accountable care organizations," or "value-based purchasing") reimburse providers based all or in part on meeting specified outcomes rather than simply paying for the services that the providers render. This change, in theory, offers providers incentives to consider the quality, value, and cost of the health care delivered and to shift away from the opposite incentives that traditional fee-for-service gives providers to increase the volume of highly profitable services.

For the past decade, multiple Medicare demonstrations and some programs in the private sector have experimented and continue to experiment with the broad notion of P4P. As of the enactment of the Affordable Care Act, many of the Medicare evaluations either had just begun or have only preliminary results available. These preliminary results are mixed at best. They suggest that P4P programs cannot guarantee improved quality of care, better health care value, or meaningful or net health care savings—or a "bending of the cost curve." Nevertheless, we can learn critical lessons from the experiences of these Medicare demonstrations and private-sector projects that will help us to plan for the considerable health care reform implementation tasks ahead.

This book identifies and evaluates the full range of issues associated with implementing P4P models. It gives policy makers and researchers thorough descriptions of alternative P4P models, examines their pros and cons, and discusses lessons learned from prior experience with these models. The authors' experience with evaluating several Medicare P4P demonstrations yields a comprehensive look at how these projects have fared in the real world. The book consists of 12 chapters, which are not necessarily intended to be read in order, although the first 3 provide valuable background and conceptual information. Readers may pick and choose among these chapters, according to their interests. The brief summaries below will help guide readers as to the content and technical detail for each chapter.

The first chapter, "Introduction to Pay for Performance," provides an overview of the historic origins of P4P and briefly describes the different forms that P4P models may take. For easy reference, this chapter also includes a table that illustrates a variety of recent and ongoing P4P projects in both the public and private sectors.

Chapter 2, "Overview of Pay for Performance Models and Issues," presents a much more detailed discussion of P4P models. We describe both alternative measures of performance and incentive schemes that payers may attach to performance measurement. Performance measurement consists of several components: defining domains of performance, selecting domains to be measured, selecting indicators to measure each domain of performance, defining the unit for performance measurement and accountability, choosing data sources for measuring performance, and deciding whether participation will be voluntary or mandatory. We also identify limitations of the P4P model.

Chapter 3, "Theoretical Perspectives on Pay for Performance," uses theoretical perspectives from economics, sociology, psychology, and organization theory to broaden our understanding of the range of factors affecting health care quality and cost outcomes, as well as the reasons a focus on economic incentives may have limited impact. We use these perspectives to describe the ways in which other factors—such as the social norms of professionalism among physicians, the range of motivational factors affecting physician behavior, and the organizational settings in which clinicians practice—affect the influence of economic incentives on the outcomes of P4P programs.

Chapter 4, "Quality Measures for Pay for Performance," describes the different types of quality measures that P4P programs can use, including structure, process, and outcome measures. We then review issues that programs should consider in selecting quality measures and comment on methods for analyzing those quality measures. We conclude by discussing public reporting of quality measures and how payers can integrate that approach to quality improvement with P4P programs.

Chapter 5, "Incorporating Efficiency Measures into Pay for Performance," is a companion to Chapter 4. We review alternative measures of provider and system efficiency and technical challenges to setting efficient and equitable P4P payment incentives. We conclude with a discussion of risk adjustment and quality in the context of efficiency measurement.

Arguably, the greatest challenge in any P4P program is whom to pay. Chapter 6, "Who Gets the Payment Under Pay for Performance?" begins with a discussion of why deciding whom to pay can be so complex technically and politically and what factors can influence this decision. We then outline which specific health care entities (e.g., hospitals, physicians, integrated delivery systems) might receive payments under P4P, depending on a patient's condition, and evaluate the respective pros and cons of the various choices. Finally, we consider the related topic of what to pay for (e.g., hospital admission, episode of care).

Chapter 7, "Attributing Patients to Physicians for Pay for Performance," discusses another complex and controversial decision that policy makers must make under P4P: how to assign physicians responsibility for a defined group of patients and their episodes of patient care. Assignment, or attribution, is necessary to reward or penalize those providers who are in the best position to manage a patient's health care needs. In this chapter, we first discuss challenges to patient attribution and give selected examples of real-world assignment strategies. We then consider basic concepts and alternatives for patient attribution in a fee-for-service context.

Chapter 8, "Financial Gains and Risks in Pay for Performance Bonus Algorithms," is one of two highly technical chapters in this book (the other is Chapter 10). Once the question of whom to pay is answered, payers must integrate quality performance measures into financial incentive schemes. First, we present a range of P4P payment models and investigate their key parameters. This includes examining the effects on bonuses (and penalties) of increasing the number of quality indicators, changing their relative weights, and using different mechanisms to set targets. We then present multiple simulations of actual quality performance against preset targets and test the sensitivity of a payer's expected bonuses and losses to different sharing arrangements and key parameters. We conclude by suggesting a few steps for payers to follow in designing P4P incentive programs that maximize the likelihood of positive responses on the part of provider organizations.

As we noted previously, recent and ongoing Medicare demonstration projects give policy makers and researchers an opportunity to observe how specific P4P pilot programs have been implemented and how successful these programs have been in raising quality while lowering costs. Chapter 9, "Overview of Selected Medicare Pay for Performance Demonstrations," provides an overview of each P4P demonstration, describes the key features of the initiative, and summarizes the current status of each project. When evaluation findings are publicly available, they are presented here as well.

Chapter 10, "Evaluating Pay for Performance Interventions," explores many of the technical challenges of deriving scientifically rigorous estimates of P4P impacts. We begin by reviewing common threats to the internal validity of demonstration findings that can introduce positive or negative bias into the quantitative estimate of P4P effects. Because most Medicare demonstrations employ quasi-experimental designs, we then introduce the theory and approaches underlying the selection of representative comparison groups that are necessary to isolate intervention effects from other confounding baseline and temporal factors. Having considered alternative ways to form the comparison group, we then investigate two external threats to valid findings that are quite common in P4P demonstrations. These threats undermine the generalizability or replicability of P4P effects to a national program. In the last section of the chapter, we summarize how evaluators of five Medicare P4P demonstrations formed their comparison groups, and we critique their success in avoiding the various threats to validity discussed earlier in the chapter.

Chapter 11, "Converting Successful Medicare Demonstrations into National Programs," examines reasons that Medicare's significant 35-year experience in conducting innovative demonstration projects has had a less-lasting impact on the current national program than might be expected. Many of the P4P projects described in this book are Medicare pilot projects, or demonstrations, which test both the administrative feasibility and success of various performance models. For both technical and political reasons, win-win initiatives that reduce costs while raising quality have been elusive. This chapter will help policy makers understand the potential barriers to turning a successful pilot project into an accountable care organization or similar entity, as the Affordable Care Act mandates.

Finally, Chapter 12, "Conclusions: Planning for Second-Generation Pay for Performance," draws on the analyses and lessons from earlier chapters and recommends steps for improving future P4P programs. We review the main problems with private markets and incentives in health care that motivated the development of P4P programs in the first place. We next summarize the major shortcomings of the first generation of P4P programs. This is followed by a set of policy and implementation recommendations to improve on current initiatives and develop more effective second-generation P4P programs. We conclude with a brief analysis of the P4P provisions in the Affordable Care Act and suggest ways that those provisions could be implemented most effectively by the Secretary of the Department of Health and Human Services, who is granted fairly wide latitude by Congress for implementing the P4P provisions of the law.

This book is the most definitive and comprehensive review to date of P4P. We believe that the many lessons learned in this book can help guide the Secretary and other policy makers in designing, implementing, and evaluating P4P programs under the Affordable Care Act. These lessons may also greatly benefit private-sector insurers as they seek to redesign their own payment and quality improvement systems.

CHAPTER 1

Introduction to Pay for Performance

Michael G. Trisolini

In just the past decade, pay for performance (P4P) programs have become widespread in health care despite a lack of rigorous evidence to support their effectiveness and a lack of consensus about how to design and implement these programs. A positive feature of this movement is that new types of health care payment systems have been developed and tested. Because of its focus on rewarding quality of care performance, P4P has also provided added momentum for improving quality in health care. The Affordable Care Act, passed in 2010, features a range of P4P initiatives and pilot programs under the closely related rubric of value-based purchasing.

The enthusiasm for P4P in health policy circles, however, echoes earlier enthusiasm for national health insurance (in the 1960s and 1970s) and for capitation and managed care (in the 1990s). Both of these policy initiatives failed to live up to their early promise. National health insurance was only partially implemented through Medicare; capitation and managed care were implemented broadly but soon scaled back. Whether P4P will prove to have more staying power than those movements is not yet clear. The more rigorous evaluations to date have shown P4P programs to have limited impact (Christianson et al., 2008). The variety of P4P programs and the organizational and health policy contexts in which they have been implemented (McDonald et al., 2009) make summary judgment difficult.

The term *pay for performance* is used in a number of different ways by different writers and practitioners. A good general definition is that P4P is an approach used to provide incentives to physicians and health care provider organizations to achieve improved performance by increasing quality of care or reducing costs. In this sense, P4P differs from the predominant fee-for-service (FFS) payment system that provides incentives for producing defined health care services (e.g., ambulatory care visits, hospital admissions). A common criticism of FFS, which P4P is intended to address, is that FFS rewards providers for producing higher volumes of health care services without direct assessment of the effect on quality of care or overall costs of the health care system. By providing direct financial incentives tied to quality of care performance measures and cost of care performance measures, P4P should provide countervailing incentives that directly promote improved quality and reduced costs.

This chapter includes five sections that provide context and background on P4P. The first section reviews the historical factors that led to the current policy interest in P4P. The second describes the different types of P4P programs currently active, including private sector, public sector, and international examples. The third section discusses the roles that physicians can take in the implementation of P4P. The fourth section compares P4P with public reporting of quality measures, another increasingly popular policy option for promoting quality improvement. The fifth section summarizes the challenges and promise of P4P.

Why Pay for Performance?

Health policy has traditionally focused on the usually competing goals of increasing access, containing costs, and improving quality. P4P has become prominent primarily as a means to improve quality of care, and sometimes for improving efficiency or reducing costs as well. At the same time, P4P has its roots in health sector policies and problems that developed from earlier efforts to expand access and contain costs. For most of the past 50 years, US federal and state health policy initiatives have focused primarily on increasing access and containing costs. This section reviews several key points in the history of health policy that provide context for P4P and clarify why interest in the P4P concept has increased so much in recent years.

Historical Policy Focus on Access and Cost

The passage of Medicare and Medicaid legislation in 1965 was a landmark accomplishment that increased access to health care for millions of elderly, low-income, and other Americans who did not have health insurance through employer-based or commercial plans. In 1973, Congress extended Medicare eligibility to people with disabilities and those with end-stage renal disease. At the time, those initiatives were expected to lead to universal access through national health insurance. Several national health insurance proposals were introduced in Congress in the 1970s, but none were ultimately passed into law (Starr, 1982).

The 1970s was also a period of new awareness of health care cost escalation and concerns for its containment. One result of expanding access to third-party insurance coverage was increased costs, especially in the contexts of primarily FFS reimbursement for physicians, cost-based reimbursement for hospitals, and rapid innovation in health care technology. As a result, federal health policy began to turn from emphasizing access to a new focus on cost containment. New initiatives in the 1980s included the development of Medicare's prospective payment system for hospitals using diagnosis-related groups (DRGs) and development of the resource-based relative value scale for physician fees (Altman & Wallack, 1996). In the 1990s, initiatives included expansion of capitated reimbursement options and enrollment for Medicare and Medicaid insurance plans (Hurley & Somers, 2001; Zarabozo & LeMasurier, 2001).

In the 1980s and 1990s, private health insurance plans also faced cost-containment concerns from employers, who demanded reductions in high rates of health care cost inflation. In the context of increasing international competition, such inflation often adversely affected their business prospects. In response, private plans turned increasingly to capitated reimbursement and followed Medicare's lead by implementing prospective payment systems for hospitals and fee schedules for physicians. Many employers liked capitation because it sets a fixed annual limit on per capita health care costs, unlike FFS, which allows open-ended per capita costs.

Capitation also has two theoretical advantages for quality improvement. First, it allows health care providers and clinicians to be more flexible in tailoring treatment to individual patients' needs, without being restricted by a fee structure that may limit the types of interventions that are reimbursed. Second, capitation provides more incentives for preventive care than FFS does because insurance plans can benefit financially if patients have fewer future illnesses. When enrolled patients have fewer illnesses, health plans pay less to health care providers for medical treatments and thus incur lower costs in the context of fixed annual revenue per person.

These incentives can include both primary and secondary prevention. Primary prevention involves reducing risk factors, such as cholesterol levels, before physicians diagnose disease. Secondary prevention involves early detection or diagnosis of disease so that physicians can apply early interventions, which usually cost less. Some large health plans, such as Kaiser Permanente, that had long periods of continuity with enrollees took advantage of these incentives to improve both primary and secondary prevention for enrollees for selected higher cost chronic diseases, such as kidney failure (Tompkins et al., 1999). However, the quality improvement incentives of capitation often were limited in practice because enrollees in most health insurance plans switched plans too frequently for any one plan to reap the cost savings rewards from more effective preventive care (fewer future illnesses) or early intervention (fewer complications).

Capitation also has two key weaknesses, however, and these eventually led to a public backlash and forced insurance plans to cut back on capitated reimbursement. First, capitation gives providers and health plans incentives to profit by selecting healthier patients (with lower costs) rather than by improving the quality of care. Second, capitation gives providers and payers incentives to increase profits by undertreating patients once health plans receive the up-front capitated revenues. Although some capitated health plans avoided these temptations and used incentives to improve care in creative ways, enough insurance plans focused on patient selection or undertreatment for short-term profits to erode public confidence in capitation by the end of the 1990s, and capitation's promise as an alternative to FFS faded. This led to a search for new policy initiatives that could provide alternatives to FFS, which contributed to the recent surge of interest in P4P.

Quality and Value Rise to the Forefront

Up until the 1990s, the task of ensuring health care quality was left largely to the medical profession and hospital accreditation organizations. Government agencies and private health insurance companies shied away from intruding on what they viewed as the professional domain of physicians. Medical associations successfully established and defended that professional autonomy throughout most of the twentieth century, enabling physicians to earn high salaries and enjoy high status in US society (Starr, 1982). As recently as the mid-1990s, Congress almost withdrew funding for the US Agency for Health Care Policy and Research (AHCPR; now the Agency for Healthcare Research and Quality, AHRQ) because of lobbying by orthopedic surgeons. The surgeons were upset by AHCPR publication of clinical guidelines that cast doubt on the value of some orthopedic surgical procedures for low back pain.

Nonetheless, starting in the 1990s, several developments led to increasing policy concerns about quality of care. A health policy movement aimed at value-based purchasing introduced quality of care into health care payment proposals in the 1990s. In this context, "value" is usually defined as *focusing on both quality and cost at the same time in purchasing and delivering health care* (Thomas & Caldis, 2007). The goal is to organize health care purchasing efforts and incentives to achieve either higher quality care for the same cost, or the same quality care for lower cost, or possibly even higher quality care for lower cost. As with P4P, value-based purchasing contrasts with the prevailing FFS reimbursement system, where the incentives encourage higher utilization of health care services, which does not necessarily raise quality and often raises costs. Value-based purchasing did not catch on in the 1990s because concerns about quality of care were not as strong at the time (Meyer et al., 1997). However, in the following decade quality became a much larger focus in health policy initiatives as several notable studies highlighted inconsistencies in the quality of care.

Recent studies have found large and unexplained variations in rates of health care utilization and clinical outcomes across geographic areas, questioning the traditional reliance on the medical profession to ensure the uniform delivery of high-quality care (Davis & Guterman, 2007; Wennberg et al., 2002). Since 1999, several landmark publications, most notably from the Institute of Medicine (IOM) and RAND, have highlighted widespread problems with patient safety and quality of care (IOM Board on Health Care Services, 2000; 2001; McGlynn et al., 2003). These studies have helped to galvanize federal and state governments, private employers, and private health insurance plans to focus their policy, regulatory, and management interventions more directly on measuring and improving the quality of care.

Policy makers' frustration with the lack of success of cost-containment initiatives has also led to a renewed focus on value in health care in recent years. If high costs are inevitable in the high-technology environment of US health care, then the quality-of-care benefits should also be high. However, several studies of variations in health care spending from high-cost to low-cost regions did not find any evidence that patients in high-cost regions received a higher quality of care (Davis & Guterman, 2007; Fisher et al., 2003a, 2003b).

Comparisons with health care systems in other countries have also highlighted the poor value Americans receive for the high cost of US health care. The United States spends far more than any other high-income country on health care, both as a percentage of gross domestic product and on a per capita basis. At the same time, available measures of health care outcomes such as infant mortality, child mortality, maternal mortality, and life expectancy in the United States are poor compared with those of other industrialized countries (IOM Board on Health Care Services, 2007). Moreover, most other industrialized countries have national health insurance covering all or most of their citizens, which may explain some of the differences in outcomes. Many countries that spend much less on health care perform much better than the United States on these outcome measures. Even some developing countries, such as China and Costa Rica, spend far less on health care per capita and have outcomes similar to those in the United States.

Promise of Pay for Performance

The increased interest in P4P programs is based on the belief that the health care payment system can be designed to offer incentives to improve the quality of care provided in multiple settings, including physicians' offices, hospitals, and other types of provider organizations. This is intended to ensure that patients and payers receive good value for high levels of spending on health care. Moreover, many economists have supported the idea of linking payment and quality, based on their traditional focus on using pricing signals to produce internally motivated changes in supplier (physician or health care provider organization) behavior rather than relying on more cumbersome regulatory mechanisms that try to impose external rules, reporting requirements, and other structures that suppliers often evade.

Traditional FFS reimbursement in health care has been useful in improving access to care, but it lacks incentives for improving quality. In a sense, FFS reimbursement was originally viewed as paying for quality, because it enabled formerly uninsured people to have much better access to licensed doctors and hospitals, who were assumed to provide high-quality care because of formal medical training, professional ethics, and accreditation status.

P4P is intended to bring incentives for improving quality of care directly into the payment system. By paying for specified standards of quality care, P4P may help equalize quality across different regions of the country and among different providers in the same region. P4P can also include explicit incentives for other goals, such as reducing costs or improving coordination of care among providers.

Up until the 1990s, quality assurance in health care focused mainly on inputs or structural factors, such as physicians being licensed after receiving degrees from accredited medical schools, and hospitals receiving accreditation based on evaluations of staffing, facilities, equipment, administrative procedures, and related measures. In contrast, most P4P programs include a focus on process factors that assess quality of care through the ways in which doctors and hospitals provide medical care to patients. Process measures of quality scrutinize the tests and procedures administered to patients with particular diseases, as well as pharmaceuticals and other interventions used in treatment, explicitly to check for errors or missed tests or treatments—for example, whether people with diabetes have at least annual tests to check on their disease, and whether people with heart disease are avoiding high blood pressure levels.

P4P enables quality assurance and quality improvement programs to move beyond information sharing and managerial sanctions to disbursing payments based on process measures of quality of care. As recently as the 1980s, such second-guessing of medical treatment was largely unknown. P4P programs sometimes include structural measures of quality for performance assessment, but process measures have been the main focus. P4P programs focus mainly on providing financial incentives, but linking them to nonfinancial, systems interventions for improving processes of care is another approach that could be tried in the future—for example, linking P4P process of care incentives to point-of-care decision support and collaborative care models (Bufalino et al., 2006).

P4P programs could also include a broader focus on health care outcomes as the basis of payment for quality. Outcomes include reducing morbidity and mortality and improving quality of life and patient satisfaction. P4P programs are beginning to include some types of outcome measures of performance (e.g., with patient satisfaction surveys), although they are using them much less frequently than process measures of care. Process measures are usually easier than outcomes to measure and are considered to be more closely related to clinician or provider organization performance (given that other factors besides medical care can affect patient outcomes). However, exploring ways to expand the use of outcome measures is one potential area for future development of P4P programs. The Centers for Medicare & Medicaid Services (CMS), on its Hospital Compare Web site, has made initial efforts for measurement and public reporting of outcomes measures for hospitals, which could lay the groundwork for including more outcomes measures in hospital P4P programs.

Varieties of Pay for Performance

P4P can mean a number of different things in both concept and practice. The field is still young and evolving, with new programs being designed and tested every year. Because of the pace of innovation, the terminology for describing P4P programs is not yet standardized. The primary variation in defining pay for performance is in the definition of *performance*, which varies by the aspects of care or results being rewarded. The main definitions of P4P include the following:

- **Pay for quality.** These programs can assess quality in several ways, using structure, process, outcome, or coordination of care measures. Such programs may also use composite measures to quantitatively combine multiple quality indicators into a single metric.

- **Pay for reporting.** Often termed P4R, pay for reporting focuses on provider reporting of quality-related data. These programs usually intend to develop into pay for quality once providers become more comfortable with the validity and reliability of the quality measures and data collection procedures.

- **Pay for efficiency.** Paying for efficiency generally means rewarding cost reduction or cost containment. Cost measures may include annual expenditures for patients with chronic diseases or episode-based spending measures for patients with acute illnesses. Alternately, efficiency-based programs may use health care utilization measures that focus on the number of physician visits or hospital days per patient per year. Some payers have also developed composite measures or indexes of efficiency to profile and compare provider performance.

- **Pay for value.** This approach combines quality and cost measures. For example, a pay for value program may reward providers for improving quality while keeping cost constant or reducing cost while maintaining or improving quality. Payers may give providers simultaneous incentives for increasing quality and containing costs and then allow the providers to sort out the best approaches for responding to both incentives. The Affordable Care Act health reform legislation took this approach with hospital P4P in its Hospital Value-Based Purchasing Program (HVBPP), in which cost savings are guaranteed through across-the-board reductions in hospital reimbursement; hospitals are then able to earn back a portion of the lost reimbursement through performance on quality measures. As a result, both cost and quality factors are included in the HVBPP.

The Leapfrog Group and Med-Vantage, Inc., have conducted nationwide surveys of P4P programs in recent years to provide a more comprehensive picture of the range and scope of the programs in operation or being developed across the country. The Leapfrog Group is a coalition of employers working to improve health care quality and affordability; Med-Vantage is a company that conducts surveys and provides services related to health care quality and cost performance analysis. Their surveys on P4P included programs sponsored by payers and health plans serving enrollees covered by private health insurance, Medicare, and Medicaid. They identified 148 organizations that were P4P program sponsors in 2006; 62 percent of these were commercial payers, 21 percent were government sponsors, 10 percent were coalitions or employers, and 7 percent were still in the process of development (Baker & Delbanco, 2007; Med-Vantage, 2006–2007). Moreover, the 148 program sponsors sometimes provided multiple programs; as a result, the survey found a total of 258 P4P programs, with 130 targeted at primary care physicians, 72 for specialist physicians, and 56 for hospitals or other health care facilities. In addition, these surveys have tracked growth in the number of P4P programs, from 52 in the 2003 survey to 120 in 2004, 220 in 2005, and 258 in 2006.

Table 1-1 includes 15 examples of P4P programs: 4 from the private sector, 10 from the public sector, and 1 international program from the United Kingdom. Table 1-1 illustrates the broad range of P4P program designs that payers use. The table compares programs across four design factors: (1) types of providers targeted, (2) performance measures used, (3) types of performance targets, and (4) the size of the financial incentives. This table provides descriptions of the P4P programs discussed in the following chapters, and thus provides reference summaries of them.

The providers targeted in the P4P programs in Table 1-1 include individual physicians, physician groups, disease management organizations, and hospitals. P4P can include other types of health care providers, but these types are the ones most widely involved to date.

The performance measures included in the programs in Table 1-1 focus mainly on clinical process measures of quality, but some also include other measures. Several programs include structural measures of information technology (IT) investment, use of electronic medical records, and organization of care. Outcome measures are included in some programs through patient satisfaction indicators. Cost or resource utilization measures are sometimes included through assessment of drug utilization, annual cost per patient or per beneficiary, or cost per patient per month.

Performance targets in Table 1-1 focus mainly on preset thresholds but also include examples of improvement-over-time targets and rankings of providers against one another. A number of variations are also found (e.g., using tiered thresholds to provide increasing rewards for increasing levels of performance).

The size of P4P incentives has typically been modest in US programs; those included in Table 1-1 reflect this pattern. P4P incentives in the United

Table 1-1: Comparison of Selected Pay for Performance Programs

Pay for Performance Program	Providers Targeted	Performance Measures
1. Private Sector		
Integrated Healthcare Association[a]	• Physician organizations in California serving enrollees of 7 large health plans	• Clinical quality • Patient satisfaction • Information technology investment • Measures and weighting vary by year and by health plan
Bridges to Excellence[b]	• Physicians and physician organizations in Albany, Boston, Cincinnati, and Louisville	• Diabetes care measures • Heart/stroke care • Physician office care—implementing information management systems
Hawaii Medical Service Association (Blue Cross Blue Shield of Hawaii)[c]	• Physicians treating preferred provider organization plan enrollees in Hawaii	• Clinical performance • Patient satisfaction • Use of electronic records • Medical and drug utilization
Blue Cross Blue Shield of Michigan Rewarding Results[d]	• Hospitals in Michigan	• Joint Commission on the Accreditation of Healthcare Organizations measures • Medication safety measures • Community health • Efficient utilization
2. Public Sector		
Medicare Physician Group Practice Demonstration[e]	• Large multispecialty physician groups—10 groups each with at least 200 physicians, located in 10 different states	• Annual cost per beneficiary • 32 ambulatory care quality measures for diabetes, heart failure, coronary artery disease, hypertension, and preventive care
Medicare Health Support Pilot Program[f]	• Private disease management companies	• Cost per beneficiary per month, includes beneficiaries with diabetes or heart failure

States generally range up to only about 5–10 percent of baseline provider reimbursement. The UK program is notable because it includes much larger incentives relative to baseline reimbursement, with a goal of increasing family practitioners' income by 25 percent.

Performance Targets	Size of Financial Incentives
• Thresholds (1 health plan) • Relative rankings (6 health plans), physician groups in most plans in the 50th to 100th percentile paid on a sliding scale • Public reporting of performance included as a nonfinancial incentive	• About 1.5% of physician group compensation (2004 average) Goal of increasing to 10% of compensation
• Per member per year (PMPY) bonus for meeting requirements for certification in physician recognition programs in each measure category	• $80–$100 PMPY for diabetes patients • $50 average PMPY for meeting physician office criteria
• Rankings of physicians relative to scores of other practitioners	• Ranged from 1% to 7.5% of physicians' base professional fees in 2003 • Average total payment of $4,785 per physician in 2003
• Thresholds	• Up to 4% increase in diagnosis-related group (DRG) payments per admission
• Percentage reduction in cost more than 2% greater than comparison group • Quality targets with both fixed thresholds and Improvement over time	• Up to 5% of combined Part A and Part B expenditures for assigned beneficiaries, depending on both cost and quality performance
• Threshold of 5% cost savings, compared to a randomized control group	• Up-front management fees paid to each company, but none achieved the 5% savings required to retain at least some fee revenue

(continued)

Table 1-1: Comparison of Selected Pay for Performance Programs *(continued)*

Pay for Performance Program	Providers Targeted	Performance Measures
Premier Hospital Quality Incentive Demonstration[g]	• Hospitals—250 throughout the United States	• 35 inpatient process quality measures for heart failure, acute myocardial infarction, pneumonia, coronary artery bypass graft, surgery, and hip and knee replacement; one outcome measure for mortality • Composite quality measure scores calculated to determine incentives for each condition
Care Management for High-Cost Beneficiaries Demonstration[h]	• Care management organizations—6 total, in different regions of the country	• Cost per beneficiary, including beneficiaries with one or more chronic diseases and high costs or high-risk status
Medicare Participating Heart Bypass Center Demonstration[i]	• Seven hospitals with affiliated physician groups • Sites selected for demonstrated quality of care, high volumes of the selected surgical procedure, and willingness to offer CMS a discount on the average combined FFS payments to hospitals and physicians for the selected procedures	• CMS paid single negotiated global rate for all Parts A and B inpatient hospital and physician care associated with heart bypass surgery (DRGs 106 and 107)
Medicare Acute Care Episode (ACE) Demonstration[j]	• Five hospitals with affiliated physician groups • Sites selected for demonstrated quality of care, high volumes of the selected surgical procedure, and willingness to offer CMS a discount on the average combined FFS payments to hospitals and physicians for the selected procedures	• CMS paid single negotiated global rate for both Part A and Part B services for selected cardiac and orthopedic surgical services and procedures
Medicare Physician-Hospital Collaboration Demonstration[k] (Another very similar demonstration is the Medicare Hospital Gainsharing Demonstration)	• Integrated Care Consortium • Focus on gainsharing between hospitals and physicians based on Medicare reimbursement for episodes of care, including both acute and long-term care	• Hospitals make payments to physicians based on achieved net savings over episodes of care, where payments are based on improvements in quality or efficiency resulting in savings

Performance Targets	Size of Financial Incentives
• Competition against other hospitals in each module; top decile received 2% bonus payment for each clinical condition module, second decile gets 1% bonus • Penalties of 2% and 1% for bottom deciles in third year	• Average bonus was $71,960 per year; range of $914 to $847,227
• Cost savings per beneficiary for Medicare	• Up-front monthly fees paid to each care management organization; demonstrated Medicare savings required to retain the management fee revenue
• Cost savings below the negotiated global payment rate • Hospitals also compete to be admitted to the program to gain marketing benefits from recognition as a Medicare Participating Heart Bypass Center	• Hospitals shared global payments with surgeons and cardiologists based on cost savings • Participating hospitals allowed to market a demonstration imprimatur as a "Medicare Participating Heart Bypass Center"
• Cost savings below the negotiated global payment rate • Hospitals also compete to be admitted to the program to gain marketing benefits from recognition as a Value-Based Care Center	• Hospitals shared global payments with surgeons and physicians • Participating hospitals allowed to market a demonstration imprimatur as a "Value-Based Care Center"
• Focus on net savings, with quality performance targets required for physicians to be eligible for incentive payments	• Physician payments limited to 25 percent of Medicare payments made to physicians for similar cases

(continued)

Table 1-1: Comparison of Selected Pay for Performance Programs *(continued)*

Pay for Performance Program	Providers Targeted	Performance Measures
CMS Cancer Prevention and Treatment Demonstration for Ethnic and Racial Minorities[l]	• Six cancer centers	• Implementation of patient navigator programs to reduce disparities in cancer care for racial and ethnic minorities
Medicare Coordinated Care Demonstration[m]	• Disease management organizations, including 5 commercial disease management firms, 3 academic medical centers, 3 community hospitals, 1 integrated delivery system, 1 long-term care facility, and 1 retirement community	• Cost per beneficiary per month, including beneficiaries with diabetes, heart failure, coronary artery disease, chronic obstructive pulmonary disease, and other chronic conditions • Quality measures included for evaluation, but were not used to determine incentive payments
Local Initiative Rewarding Results Demonstration[n]	• Physicians and physician groups serving Medicaid-focused health plans in California	• Well-child, well-adolescent, and Health Plan Employer Data and Information Set quality measures
3. Other Countries		
British National Health Service[o]	• Family practitioners (primary care physicians) throughout the United Kingdom	• 146 indicators, including clinical quality measures for 10 chronic diseases, organization of care, and patient experience

CMS = Centers for Medicare & Medicaid Services; DRG = diagnosis-related group; FFS = fee-for-service; PMPY = per member per year.

[a] Folsom et al., 2008; IOM Board on Health Care Services, 2007; Lempert & Yanagihara, 2006; Young et al., 2007.
[b] Bridges to Excellence, 2008; Folsom et al., 2008; IOM Board on Health Care Services, 2007; Young et al., 2007.
[c] Gilmore et al., 2007; IOM Board on Health Care Services, 2007.
[d] Folsom et al., 2008; Young et al., 2007.
[e] Kautter et al., 2007; Trisolini et al., 2008.
[f] Cromwell et al., 2008.
[g] CMS, 2009b; Davidson et al., 2007; Glickman et al., 2007; Grossbart, 2006; Lindenauer et al., 2007.
[h] CMS, 2005, 2009a.
[i] CMS, 1998a; 1998b.
[j] CMS, 2009b.
[k] CMS, 2007.
[l] CMS, 2008.
[m] Peikes et al., 2009.
[n] Felt-Lisk et al., 2007; Folsom et al., 2008; Young et al., 2007.
[o] Campbell et al., 2007; Doran et al., 2006; Epstein, 2006, 2007.

Performance Targets	Size of Financial Incentives
• Enrollment of patients in the program for care navigator services	• Variable by site, includes start-up payments of $50,000 per site, payments for surveys administered per patient, and capitation payments to sites depending on the cost of patient navigator services
• Programs at financial risk if savings on Medicare outlays on intervention beneficiaries were less per month than the monthly management fee paid to the programs by CMS	• Up-front management fees ranging from $80 to $444 per beneficiary paid to each program, but none achieved cost savings for Medicare net of the management fees
• Varied by plan, with focus on thresholds for performance-based risk pools, capitation increases, and bonus payments	• Varied by plan (e.g., bonus payments ranged from $50 possible per child to $200 per child)
• Sliding scale of thresholds with points awarded for achieving several different tiers for each measure, up to a maximum of 1,050 points overall per practice	• Goal of increasing family practitioners' income by 25% • Payments were $133 per point ($139,650 maximum per year) in 2004–2005 and $218 per point ($228,900 maximum) in 2005–2006 and beyond

Private Sector Pay for Performance Programs

The private sector programs in Table 1-1 have several noteworthy features. The Integrated Healthcare Association (IHA) program includes multiple types of quality measures, including structure (IT investment), process (clinical quality), and outcome (patient satisfaction) measures. IHA is the largest P4P program in the United States, covering 8 million health plan members (Folsom et al., 2008; IOM Board on Health Care Services, 2007; Lempert & Yanagihara, 2006; Young et al., 2007). IHA also emphasizes public reporting of performance results through a commitment to transparency for its P4P program, which is not the case for most other P4P programs.

A coalition of large employers developed the Bridges to Excellence program. It focuses on recognizing physicians for achieving high-quality care (Bridges to Excellence, 2008; Folsom et al., 2008; IOM Board on Health Care Services, 2007; Young et al., 2007). Bridges to Excellence implemented four original regional programs (Albany, Boston, Cincinnati, and Louisville) and later expanded to include additional regions and clinical conditions.

The Hawaii Medical Service Association is a local health insurance organization affiliated with Blue Cross Blue Shield. The Hawaii Medical Service Association started its P4P program in 1999, making it one of the longest running programs in the US (Gilmore et al., 2007; IOM Board on Health Care Services, 2007). It provides some of the largest incentive payments in the United States, up to 7.5 percent of baseline provider reimbursement.

The Blue Cross Blue Shield of Michigan program is an example of private-sector P4P that focuses on hospitals (Folsom et al., 2008; Young et al., 2007). It includes patient safety performance measures that other P4P programs have not widely applied.

Public Sector Pay for Performance Programs

Medicare is the largest public-sector sponsor of P4P programs to date, as reflected in the examples provided in Table 1-1; it sponsors most of these programs. A more detailed description of many of these Medicare P4P pilot programs can be found in Chapter 9. The role Medicare plays in sponsoring and championing P4P programs will only grow in coming years as a result of the Affordable Care Act health care reform legislation Congress passed in March 2010. That legislation mandates several new or expanded Medicare P4P programs and also provides funding for new pilot programs that will be largely Medicare-focused as well. Because Medicare is the largest payer for health care in the United States, many commentators have called for it to lead the way in designing and implementing P4P programs, with the goal of providing precedents for private-sector payers, as it did in the 1980s, when Medicare led development of prospective payment for hospitals, and the private sector soon followed suit.

The Medicare Physician Group Practice Demonstration includes both cost and quality performance measures; it expects participating groups to respond to both incentives at the same time (Kautter et al., 2007; Trisolini et al., 2008). In order to provide incentives to providers at varying initial levels of measured quality performance, the demonstration includes both threshold and improvement-over-time targets for quality measures.

The Medicare Health Support Pilot Program targets private disease management companies (Cromwell et al., 2008). It focuses on P4P incentives for cost containment, but also includes quality-of-care measures to enable a more global evaluation of performance. This program includes a randomized evaluation design; this rigorous approach has not been widely used to study the impacts of P4P.

The Premier Hospital Quality Incentive Demonstration is a public sector example of hospital P4P (CMS, 2010; Davidson et al., 2007; Glickman et al., 2007; Grossbart, 2006; Lindenauer et al., 2007). CMS awarded more than $24 million to participating hospitals in the first 3 years of this demonstration. It also includes payment penalties on lower performing hospitals starting in the third year of the demonstration; this disincentive complements the bonus payments made to higher performing hospitals. This approach differs from that of most P4P programs, which reward positive performance but do not impose penalties for poor performance.

The Care Management for High-Cost Beneficiaries Demonstration is an FFS demonstration that focuses on providing incentives for cost containment (CMS, 2005, 2009a). The participating beneficiaries have one or more chronic diseases and either high-cost or high-risk status. Care management organizations that participate in the demonstration receive up-front fees as incentives but must demonstrate Medicare savings to retain the fee revenue.

The Medicare Participating Heart Bypass Center Demonstration ran from 1991 to 1996 and thus is an earlier example of P4P than the other programs included in Table 1-1 (CMS, 1998a, 1998b). It was a bundled payment demonstration, in which Medicare paid hospitals and physicians a combined rate for all inpatient Part A and Part B services for coronary bypass surgery DRGs. (Medicare pays reimbursements for hospitals and physicians separately under traditional FFS.) Bundling reimbursements provided an incentive for hospitals and physicians to work together to reduce overall inpatient costs because, under the demonstration, they could share any savings achieved if their combined costs were lower than the combined payment rate. Quality of care performance was assessed in the application process, and participating hospitals were allowed to market a demonstration imprimatur as a "Medicare Participating Heart Bypass Center." This approach differs from most P4P programs that measure quality performance after the program begins operations. This approach is termed a Centers of Excellence (CoE) model.

A more recent CoE model P4P program is the Medicare Acute Care Episode Demonstration (CMS, 2009b). This demonstration was implemented in 2009 and includes a bundled payment for both Part A and Part B services provided during an inpatient stay. This demonstration includes a range of both cardiac and orthopedic procedures. As in the Participating Heart Bypass Center Demonstration, quality of care will be assessed in an application process and approved centers will be able to market themselves as "Value-Based Care Centers."

Gainsharing is the focus of the Medicare Physician-Hospital Collaboration Demonstration (CMS, 2007). It is intended to use incentive payments from hospitals to physicians to align their financial incentives under Medicare reimbursement, where hospitals can benefit financially from lower costs of care in relation to their fixed DRG reimbursement, but physicians have countervailing incentives to increase volumes of care to increase their reimbursement. Under this demonstration, integrated delivery systems that include hospitals can provide incentive payments to physicians for up to 25 percent of the Medicare payments the physicians would receive for similar cases. However, the payments must be linked to net savings that result from improvements in quality and efficiency over espisodes of care, and not based on increases in volumes of patients or other factors.

P4P programs have sometimes been criticized for providing incentives that could increase disparities in care, but the CMS Cancer Prevention and Treatment Demonstration for Ethnic and Racial Minorities includes payments for programs specifically intended to reduce disparities (CMS, 2008). This demonstration is based on a structure measure of quality—enrollment of patients in programs that have patient navigators, who are staff who help minorities to gain better access to preventive care and cancer treatment care. Payments are made to the participating programs based on the number of patients enrolled in these programs.

The Medicare Coordinated Care Demonstration included a diverse set of 15 disease management organizations based at academic medical centers, community hospitals, an integrated delivery system, a long-term care facility, a retirement community, and for-profit disease management companies (Peikes et al., 2009). The program paid the disease management organizations' monthly management fees, averaging $235 per beneficiary, to improve coordination of care for chronic diseases, reduce costs, and maintain or improve quality of care. In addition to other interventions, all of the disease management organizations assigned enrollees to nurse care coordinators. However, an evaluation study found that none of the programs produced statistically significant cost savings relative to a control group.

The Local Initiative Rewarding Results Demonstration focuses on providers that treat Medicaid enrollees (Felt-Lisk et al., 2007; Folsom et al., 2008; Young et al., 2007). Unlike the other P4P programs profiled in Table 1-1, this program emphasizes health care services for children. To date, most P4P programs have focused on clinicians and provider organizations that treat adults.

Pay for Performance Programs in Other Countries

The United Kingdom, through the British National Health Service, has implemented the largest P4P program (Campbell et al., 2007; Doran et al., 2006; Epstein, 2006, 2007). It is noteworthy for its nationwide scope, very large number of quality measures (146 measures that cover 10 clinical conditions, organization of care, and patient experience), and large financial incentives for providers (which can be 25 percent or more of family practitioners' incomes). By comparison, the P4P programs implemented in the United States to date are much less ambitious.

P4P programs with published documentation have yet to develop in additional countries. It will be interesting to see in coming years if other countries follow the examples of the United States and United Kingdom by developing P4P programs, and what types of program designs they may pursue.

The Role of Providers in Pay for Performance Implementation

The potential conflict between the financial incentives included in P4P programs and physicians' interest in maintaining their professional autonomy has raised concerns that physicians should be involved from the outset in designing and implementing P4P programs. Although both public-sector and private-sector P4P initiatives have stressed the importance of this approach, the best way to organize physicians' participation in P4P programs may vary widely across different regions, communities, and provider organizations.

Payers may supply providers with periodic feedback through performance reports that anticipate future P4P performance assessments and bonus payment calculations. The frequency of reporting and the amount of detail in these reports can be organized at many levels, however, and it is still unclear what is the best approach. Lag time between clinical activity and receipt of feedback reports is a common concern in that the lag may lessen the value of reports to providers. Some providers have emphasized the need for real-time information from electronic medical records or other on-site information systems, to provide prompts to physicians during patient visits to alert them about tests or preventive treatments that a patient may need and that will affect their quality performance scores.

The Question of Public Reporting

Public reporting of quality measure results for health care providers is another quality improvement strategy that has gained popularity among policy makers in recent years. For example, Medicare recently began reporting a series of quality measures for individual hospitals on its Hospital Compare tool within its public Web site, www.medicare.gov. The goal is to provide the public with better information on how hospital quality of care can be measured objectively, and to enable consumers to compare the quality performance of individual hospitals. P4P and public reporting of quality performance are not necessarily linked, but some payers, notably the Integrated Healthcare Association in California, have developed both in tandem. The IHA views public reporting as important for promoting transparency in the quality performance results used to determine the financial incentives paid to health care provider organizations under P4P programs (Lempert & Yanagihara, 2006). Congress also linked P4P and public reporting in the Hospital Value-Based Purchasing Program included in the Affordable Care Act health reform legislation.

However, other payers often choose to keep P4P performance data confidential to enhance physician cooperation and buy-in to P4P programs. Physicians may view P4P quality measures as limited to a subset of overall clinical performance issues (some of which may be hard to measure quantitatively) and vulnerable to overemphasis if payers make results public.

Public reporting also requires that results be presented in formats that consumers who lack clinical or statistical expertise can easily understand.

If sophisticated statistical analysis is part of the P4P methodology, then consumers may be misled about the significance of results.

In addition, some P4P methodologies may not lend themselves to public reporting. For example, rankings of providers and payment of P4P incentives for the top one or two deciles can mask absolute levels of high-quality performance for the third or fourth deciles. As a result, provider rankings based on quality measures may sometimes indicate only very small differences in actual quality measure performance.

Conclusion

P4P encompasses a broad range of interventions and programs, and we are only beginning to discover its potential. A number of program design options have yet to be explored, and several types of existing programs, particularly pay for efficiency and pay for value, warrant more extensive testing. The Affordable Care Act is expected to facilitate testing of new P4P models in coming years. To date, P4P program results have not lived up to the original expectations, but evaluation studies indicate that impacts are possible and that policy, organizational, and professional culture contexts may be intervening variables that affect the success of P4P programs.

The challenge for the future is to identify ways to design P4P programs that are better aligned with other interventions at the individual physician, practice site, group practice, hospital, delivery system, community, and policy levels. Policy makers need to address numerous practical and policy problems to make P4P more effective—for example, how to avoid or mitigate incentives for physicians to select more affluent patients under P4P, which might increase their measured quality and increase existing disparities in care. A related issue is ensuring that facilities that serve higher numbers of lower-income patients receive sufficient funding so they can compete effectively for P4P incentive payments.

Subsequent chapters of this book explore the range of theoretical, design, implementation, and evaluation issues related to P4P programs, and review how these programs can be improved for greater impact. Existing programs have focused on relatively simple theoretical models that assumed straightforward effects of financial incentives on quality and cost outcomes. In the future, payers and policy makers need to test more sophisticated models and programs that may be termed second-generation P4P, in which P4P is one element of broader health policy and health care delivery interventions.

These ideas are discussed further in Chapter 12, the concluding chapter. Second-generation P4P should reinforce the financial incentives of P4P with other types of quality improvement and efficiency improvement initiatives implemented at multiple levels of the health care system, rather than relying on financial incentives alone.

References

Altman, S., & Wallack, S. (1996). Health care spending: can the United States control it? In S. Altman & U. Reinhardt (Eds.), *Strategic choices for a changing health care system* (chap. 1). Chicago: Health Administration Press.

Baker, G., & Delbanco, S. (2007). *Pay for performance: National perspective, 2006 longitudinal survey results with 2007 market update*. Retrieved November 9, 2009, from http://www.medvantage.com

Bridges to Excellence. (2008). *BTE fifth anniversary report—Five years on: Bridges built, bridges to build*. Retrieved January 26, 2011, from http://www.bridgestoexcellence.org

Bufalino, V., Peterson, E., Burke, G., LaBresh, K., Jones, D., Faxon, D., et al. (2006). *Payment for quality: Guiding principles and recommendations. Circulation, 113*, 1151–1154.

Campbell, S., Reeves, D., Kontopantelis, E., Middleton, E., Sibbald, B., & Roland, M. (2007). Quality of primary care in England with the introduction of pay for performance. *New England Journal of Medicine, 357*(2), 181–190.

Centers for Medicare & Medicaid Services. (1998a). *Medicare Participating Heart Bypass Center Demonstration: Summary*. Retrieved June 17, 2009, from http://www.cms.hhs.gov/DemoProjectsEvalRpts/downloads/Medicare_Heart_Bypass_Summary.pdf

Centers for Medicare & Medicaid Services. (1998b). *Medicare Participating Heart Bypass Center Demonstration: Final report*. Retrieved May 14, 2010, from http://www.cms.hhs.gov/demoprojectsevalrpts/MD/ItemDetail.asp?ItemID=CMS063472

Centers for Medicare & Medicaid Services. (2005, July 1). *Medicare to award contracts for demonstration projects to improve care for beneficiaries with high medical costs*. Press release for award of Care Management for High Cost Beneficiaries Demonstration. Retrieved June 17, 2009, from http://www.cms.hhs.gov/DemoProjectsEvalRpts/downloads /CMHCB_Press_Release.pdf

Centers for Medicare & Medicaid Services (2007). *Physician-Hospital Collaboration Demonstration solicitation*. Retrieved March 9, 2010, from http://www.cms.hhs.gov/DemoProjectsEvalRpts/downloads/PHCD_646_ Solicitation.pdf

Centers for Medicare & Medicaid Services. (2008). *Cancer Prevention and Treatment Demonstration for Ethnic and Racial Minorities*. Fact sheet. Retrieved November 12, 2009, from http://www.cms.hhs.gov/ DemoProjectsEvalRpts/downloads/CPTD_FactSheet.pdf

Centers for Medicare & Medicaid Services (2009a). *Care Management for High Cost Beneficiaries Demonstration: Summary – February 2009*. Retrieved June 17, 2009, from http://www.cms.hhs.gov/DemoProjectsEvalRpts //downloads/CMHCB_summary.pdf

Centers for Medicare & Medicaid Services. (2009b). *Acute Care Episode Demonstration*. Retrieved March 9, 2010, from http://www.cms.hhs.gov/ DemoProjectsEvalRpts/downloads/ACEFactSheet.pdf

Centers for Medicare & Medicaid Services. (2010). *Premier Hospital Quality Incentive Demonstration: Fact sheet*. Retrieved January 26, 2011, from https://www.cms.gov/HospitalQualityInits/35_HospitalPremier.asp

Christianson, J. B., Leatherman, S., & Sutherland, K. (2008). Lessons from evaluations of purchaser pay-for-performance programs: A review of the evidence. *Medical Care Research and Review, 65*(6 Suppl), 5S–35S.

Cromwell, J., McCall, N., & Burton, J. (2008). Evaluation of Medicare Health Support Chronic Disease Pilot Program: 6-month findings. *Health Care Financing Review, 30* (1), 47–60.

Davidson, G., Moscovice, I., & Remus, D. (2007). Hospital size, uncertainty, and pay-for-performance. *Health Care Financing Review, 29*(1), 45–57.

Davis, K., & Guterman, S. (2007). Rewarding excellence and efficiency in Medicare payments. *Milbank Quarterly, 85*(3), 449–468.

Doran, T., Fullwood, C., Gravelle, H., Reeves, D., Kontopantelis, E., Hiroeh, U., et al. (2006). Pay-for-performance programs in family practices in the United Kingdom. New England Journal of Medicine, 355(4), 375–384.

Epstein, A. M. (2006). Paying for performance in the United States and abroad. *New England Journal of Medicine, 355*(4), 406–408.

Epstein, A. M. (2007). Pay for performance at the tipping point. *New England Journal of Medicine, 356*(5), 515–517.

Felt-Lisk, S., Gimm, G., & Peterson, S. (2007). Making pay-for-performance work in Medicaid. *Health Affairs (Millwood), 26*(4), w516–w527.

Fisher, E. S., Wennberg, D. E., Stukel, T. A., Gottlieb, D. J., Lucas, F. L., & Pinder, E. L. (2003a). The implications of regional variations in Medicare spending. Part 1: The content, quality, and accessibility of care. *Annals of Internal Medicine, 138*(4), 273–287.

Fisher, E. S., Wennberg, D. E., Stukel, T. A., Gottlieb, D. J., Lucas, F. L., & Pinder, E. L. (2003b). The implications of regional variations in Medicare spending. Part 2: Health outcomes and satisfaction with care. *Annals of Internal Medicine, 138*(4), 288–298.

Folsom, A., Demchak, C., & Arnold, S. B. (2008). *Rewarding Results pay-for-performance: Lessons for Medicare*. Washington, DC: AcademyHealth.

Gilmore, A. S., Zhao, Y., Kang, N., Ryskina, K. L., Legorreta, A. P., Taira, D. A., et al. (2007). Patient outcomes and evidence-based medicine in a preferred provider organization setting: A six-year evaluation of a physician pay-for-performance program. *Health Services Research, 42*(6 Pt 1), 2140–2159; discussion 2294–2323.

Glickman, S. W., Ou, F. S., DeLong, E. R., Roe, M. T., Lytle, B. L., Mulgund, J., et al. (2007). Pay for performance, quality of care, and outcomes in acute myocardial infarction. *JAMA, 297*(21), 2373–2380.

Grossbart, S. R. (2006). What's the return? Assessing the effect of "pay-for-performance" initiatives on the quality of care delivery. *Medical Care Research and Review, 63*(1 Suppl), 29S–48S.

Hurley, R., & Somers, S. (2001). Medicaid managed care. In P. Kongstvedt (Ed.), *The managed health care handbook* (chap. 57). Gaithersburg, MD: Aspen Publishers.

Institute of Medicine, Board on Health Care Services (2000). *To err is human: Building a safer health system.* Washington, DC: National Academies Press.

Institute of Medicine, Board on Health Care Services (2001). *Crossing the quality chasm: A new health system for the 21st century.* Washington, DC: National Academies Press.

Institute of Medicine, Board on Health Care Services (2007). *Rewarding provider performance: Aligning incentives in Medicare.* Washington, DC: National Academies Press.

Kautter, J., Pope, G. C., Trisolini, M., & Grund, S. (2007). Medicare Physician Group Practice Demonstration design: Quality and efficiency pay-for-performance. *Health Care Financing Review, 29*(1), 15–29.

Lempert, L., & Yanagihara, D. (Contrib. Eds.). (2006). Advancing quality through collaboration: The California Pay for Performance program. S. McDermott and T. Williams (Eds.). Retrieved May 14, 2010, from http://www.iha.org/pdfs_documents/p4p_california/P4PWhitePaper1_February2009.pdf

Lindenauer, P. K., Remus, D., Roman, S., Rothberg, M. B., Benjamin, E. M., Ma, A., et al. (2007). Public reporting and pay for performance in hospital quality improvement. *New England Journal of Medicine, 356*(5), 486–496.

McDonald R., White J., & Marmor, T. (2009). Paying for performance in primary medical care: learning about and from "success" and "failure" in England and California. *Journal of Health Politics, Policy and Law, 34*(5), 747–776.

McGlynn, E. A., Asch, S. M., Adams, J., Keesey, J., Hicks, J., DeCristofaro, A., et al. (2003). The quality of health care delivered to adults in the United States. *New England Journal of Medicine, 348*(26), 2635–2645.

Med-Vantage. (2006–2007). *2006 National Pay for Performance Study results with '07 projections.* Retrieved May 14, 2010, from http://www.medvantage.com/Pdf/National%20List%20of%2020062007P4PProgramswithEnrollment123107.pdf

Meyer, J., Rybowski, L., & Eichler, R. (1997). *Theory and reality of value-based purchasing: Lessons from the pioneers.* Available from http://www.ahrq.gov

Peikes, D., Chen, A., Schore, J., & Brown, R. (2009). Effects of care coordination on hospitalization, quality of care, and health care expenditures among Medicare beneficiaries: 15 randomized trials. *JAMA, 301*(6), 603–618.

Starr, P. (1982). *The social transformation of American medicine.* New York: Basic Books.

Thomas, F. G., & Caldis, T. (2007). Emerging issues of pay-for-performance in health care. *Health Care Financing Review, 29*(1), 1–4.

Tompkins, C. P., Bhalotra, S., Trisolini, M., Wallack, S. S., Rasgon, S., & Yeoh, H. (1999). Applying disease management strategies to Medicare. *Milbank Quarterly, 77*(4), i–ii, 461–484.

Trisolini, M., Aggarwal, J., Leung, M., Pope, G. C., & Kautter, J. (2008). The Medicare Physician Group Practice Demonstration: Lessons learned on improving quality and efficiency in health care. Commonwealth Fund, 84, 1–46.

Wennberg, J. E., Fisher, E. S., & Skinner, J. S. (2002). Geography and the debate over Medicare reform. *Health Affairs (Millwood)*, Web Exclusives, W96–W114.

Young, G. J., Burgess, J. F., Jr., & White, B. (2007). Pioneering pay-for-quality: Lessons from the rewarding results demonstrations. *Health Care Financing Review, 29*(1), 59–70.

Zarabozo, C., & LeMasurier, J. (2001). Medicare and managed care. In P. Kongstvedt (Ed.), *The managed health care handbook* (chap. 55). Gaithersburg, MD: Aspen Publishers.

CHAPTER 2

Overview of Pay for Performance Models and Issues

Gregory C. Pope

For the purposes of this chapter, we define "pay for performance" (P4P) as *a set of performance indicators linked to an incentive scheme*. The performance indicators are the *performance* component of P4P, and the incentive scheme is the *pay* component. In health care, P4P contrasts with traditional fee-for-service (FFS) payment, which pays for quantity of services without regard to performance.

This chapter considers the elements that go into designing P4P systems. A very large number of specific P4P schemes can be formed from various combinations of the elements described in this chapter. Given the lack of compelling evidence for particular approaches, payers have experimented with many different approaches. P4P encompasses a large range of real-world programs that have not yet coalesced into a small number of accepted models. All P4P programs, however, are based on decisions about a common set of design elements.

This chapter presents measures of performance and the incentive schemes that payers (e.g., health plans or government programs such as Medicare or Medicaid) may attach to performance measurement. We identify the limits of the P4P model and offer alternative ways to reach the same goals. For concreteness and simplicity, throughout this chapter we focus mostly on situations in which payers apply incentives to health care provider organizations (including group practices, hospitals, and integrated delivery systems) and physicians or other clinicians. Payers, health plan sponsors, and policy makers can apply many of the same principles and even specific approaches in other situations (e.g., employers or the government giving incentives to health plans).

Measuring Performance

P4P systems attempt to reward explicitly measured dimensions of performance. Performance measurement consists of several components: defining domains of performance, selecting domains to be measured, selecting indicators to measure each domain of performance, defining the unit for performance measurement and accountability, choosing data sources for measuring performance, and deciding whether participation will be voluntary or mandatory.

Defining Domains of Performance

The first crucial step in designing a P4P system is defining the domains or dimensions of performance that the program might reward. In health care, performance domains might include clinical outcomes, clinical process quality, patient safety, access to and availability of care, service quality, patient experience or satisfaction, cost efficiency or cost of care, cost-effectiveness, adherence to evidence-based medical practice, productivity, administrative efficiency and compliance, adoption of information technology, reporting of performance indicators, and participation in performance-enhancing activities. We discuss these in turn below.

Clinical outcomes. The ultimate goal of health care is to maintain or improve patient health status. Clinical outcomes are, therefore, a desired performance domain. Outcome measures include mortality, morbidity, functional status, quality of life and quality-adjusted life years (QALYs), and avoidance of acute exacerbations of chronic conditions. However, using outcomes to measure quality faces challenges (Eddy, 1998). Some outcomes, such as mortality, are rare or observed only with a long time lag. Outcomes such as functional status can be expensive to measure in large populations. Also, outcomes can be influenced by many factors, and some important ones (e.g., patient adherence to recommended care) may be outside of physicians' control.

Clinical process quality. Given the limitations in using clinical outcomes to judge performance, process measures are currently the most widespread method that evaluators use to assess clinical quality. Examples of process measures include eye examinations, lipid tests for patients with diabetes, and mammograms for women in certain age groups. Compared with outcomes measures, process measures are often frequent and controllable. In recent years, the efforts of several national bodies—including the US Agency for Healthcare Research and Quality, the National Committee for Quality Assurance, the American Medical Association, and the National Quality Forum—have substantially increased the number of available clinical guidelines and detailed quality process measure specifications. However, quality measurement in health care is not as straightforward as one might hope. Professional organizations, policy makers, and regulating bodies often base clinical guidelines and quality measures more on expert opinions than on the results of randomized controlled trials. While process quality measures may still be appropriate in many cases, there are not always well-established linkages between process quality measures and final outcomes of interest (see Chapter 4 for more on quality measures).

Patient safety. Reports of the large numbers of patients injured by medical care have stimulated interest in improving patient safety by reducing medical errors (Kohn et al., 1999). An example of a patient safety performance measure is the rate of hand washing among hospital patient care employees (a higher rate of washing reduces the rate of patient infections).

Access and availability of care. Measuring enrollee access to care may be especially important in settings, such as capitated payment systems, that have incentives to withhold services. The health plan, which controls the benefit design and provider network, is often a natural unit for measuring access.

Service quality. "Service quality" refers to nonclinical aspects of the patient experience that may be valuable to patients. Service quality can include such factors as patient waiting time to see physicians, patient telephone or e-mail access to provider organizations, convenience and length of office hours, and so forth.

Patient experience or satisfaction. Patient reports, which researchers usually obtain from patient surveys, provide evidence of provider organization or physician performance from the point of view of the patients who receive medical care. Typical domains include how individual physicians are rated for attributes such as communication; whether patients have difficulty getting referrals, tests, or care; whether patients receive needed care; whether patients receive care quickly; how well physicians communicate; how good physicians' customer service is; and how provider organizations and physicians submit and process claims. The number of existing patients who have changed doctors or new patients who have selected doctors can also be used to infer patient experience or preferences.

Cost efficiency or cost of care. Cost efficiency refers to the cost of providing a given level of quality of care or health outcome. Together with the quality of care, cost efficiency defines the value of care (see Chapter 5 for more about efficiency). An example of a cost efficiency measure is the cost of producing an extra QALY. Cost of care is the cost of producing an intermediate health care services output. Examples of cost of care measures include the rate of prescribing generic drugs by a physician or within a health plan, hospital days per 1,000 health plan enrollees, case mix–adjusted hospital average length of stay, and cost per episode of care. Because cost of care measures are much easier to quantify than cost efficiency measures, they tend to be much more prevalent than the latter measures (Hussey et al., 2009).

Cost-effectiveness. Cost-effectiveness refers to the relative cost of alternative interventions that produce desired outcomes such as improvement in health (e.g., QALYs). To reduce continued increases in health care costs, P4P programs may provide incentives for more cost-effective medical treatment patterns. For example, P4P might reward physicians who order fewer expensive diagnostic imaging tests that are not considered medically necessary by clinical practice guidelines.

Adherence to evidence-based medical practice. Medical practice encompasses many practice styles, some of which do not rely on evidence-based standards of care (Wennberg et al., 2004). Adhering to evidence-based standards of care may enhance physicians' quality and efficiency. P4P may reward physicians for following clinical practice guidelines in their treatment of patients (e.g., following an evidence-based decision algorithm when deciding on ordering advanced imaging tests for low-back pain).

Productivity. Productivity refers to the amount of output per unit input. Payers may wish to measure and explicitly reward productivity in situations in which base compensation for physicians or provider organizations is not tightly tied to work effort and generated output. For example, if physicians are salaried, a payer may want to find a way to reward productivity to stimulate work effort, efficiency, and provided services.

Administrative efficiency and compliance. Administrative compliance refers to performance outside the clinical and patient domains on indicators that may be relevant to payers. For example, a health plan might want to reward provider organizations and physicians based on their electronic submission of claims (invoices) for medical treatment, timely submission of claims, and low error rates in claims submission.

Adoption of information technology. Most payers consider measuring information technology (IT) critical to improving the coordination, quality, and efficiency of care. For example, payers might reward organizations and physicians based on physicians' use of electronic software to order prescriptions for their patients. This use may both lower costs and improve quality by reducing medication errors.

Reporting of performance indicators. Especially early in the implementation of a P4P system, complete reporting of requested performance indicators may be an important measure of performance. For P4P to be comprehensive, fair, and equitable, provider organizations and physicians must report performance indicators frequently and accurately. "Pay for reporting" is a first step toward improving the data to which payers apply incentives.

Participation in performance-enhancing activities. Payers may provide incentives for physicians to participate in performance-enhancing activities. Participation in such activities could include attending collaborative quality-improvement workgroup meetings and developing quality improvement action plans. The limitation of this "pay for participation" is that payers can measure the fact that participation occurred but not the performance outcomes of participation.

Selecting Performance Domains for Measurement

Some P4P systems may be comprehensive and include many domains of performance; others may focus on only a single domain. Payers may implement systems in stages, starting with a single domain and gradually adding others. Payers may determine domains, specific performance indicators, and the relative size of rewards by considering numerous variables: the importance of individual domains; the goals of the program; the availability of meaningful measures; the potential for clinical improvement; existing problem areas; and cost, burden, and data availability (Dudley & Rosenthal, 2006; Sorbero et al., 2006).

Importance and goals. Some domains may be more important to the priorities of the sponsor of the P4P system (e.g., a health plan) or to its members or clients (e.g., enrollees) than others. Many P4P programs focus on clinical quality of care. For example, California's Integrated Healthcare Association (IHA) P4P program weights clinical quality at 50 percent of total performance (McDermott & Williams, 2006). As another case in point: in six Rewarding Results demonstration sites, the weight on clinical quality ranges from 40 to 100 percent (Young et al., 2007). Patient satisfaction is also often weighted heavily; for example, the IHA program weights it at 30 percent. Early in the implementation phase, to facilitate implementation of the system, programs may place more weight on adopting IT and reporting systems and on reporting performance indicators. The California IHA program weights IT at 20 percent.

Availability of meaningful performance measures. The availability of meaningful (reliable, valid, and significant) performance measures varies across domains. Performance in domains for which a larger number of meaningful measures is available is likely to be assessed more accurately, facilitating inclusion of these domains in P4P programs.

Potential for improvement. Payers have fewer reasons to focus on domains in which performance is difficult to improve (e.g., domains in which performance is already high) than on domains in which the need and potential for improvement are substantial. A wide range of performance in a domain may indicate that it has considerable potential for improvement.

Current problems or areas of poor performance. P4P programs may emphasize areas in which current performance is poor or needs improvement. Focusing measurement and incentives on problem areas can lead to improvements in these areas.

Cost, burden, and data availability. P4P programs are more likely to include domains for which data are available or can be generated at low cost without undue burden on providers or health plan enrollees. For example, domains with measures for which programs can obtain data from existing computerized administrative data systems or health insurance claims are typically easier to implement than domains with measures that require new methods of medical chart abstraction or patient surveys.

Selecting Performance Indicators for Measured Domains

Once P4P programs choose the domains they will include in their systems, they need to specify indicators of performance for each of these domains. Good performance indicators should be valid, reliable, important, relevant, specific, controllable, actionable, efficient, and cost-effective.

Validity. The indicators should be valid indicators of the performance dimension that they purport to measure. Programs may choose indicators that have been peer reviewed and endorsed by a national accreditation organization (for example, the National Quality Forum). If programs use process-of-care indicators, the indicators should be linked with the ultimate outcome of interest (e.g., patient mortality, morbidity, or functional status).

Reliability. The indicators should be reported as consistently as possible across participants and across time. The sample size of patients that the indicators use should be large enough for statistically reliable calculation of rates. The data underlying the measurement process should be reliable. Physicians may dislike P4P programs that they feel do not measure their performance accurately.

Importance and relevance. Indicators should measure an important or relevant aspect of the performance domain to which they correspond. An outcome indicator, for example, should measure a significant aspect of patient health, such as mortality or functional status, and there should be evidence that physicians' actions can appreciably affect it. A process indicator should measure a process that has a demonstrable link to health outcomes of interest and that is under the control of physicians.

Specificity and controllability. The indicators should be specific to the performance domains they measure. They also should be specific to factors under the control of the entity whose performance is being measured. Indicators should match accountability with control. For provider organizations and physicians, one advantage of process measures over outcome measures is that process measures often measure factors under the direct control of provider organizations and physicians, whereas patient and other characteristics may affect outcomes measures in ways that are difficult to adjust for.

Actionability. The indicators should provide information that provider organizations and physicians can act upon to improve performance.

Efficiency. The indicator set should be the smallest possible that is still broad enough to cover the performance domain. Too many quality measures may impose excessive data collection costs on provider organizations and physicians, and the sheer number of measures may cause a lack of focus in quality improvement activities. On the one hand, with many indicators, the potential reward from improving performance for any one indicator may be too small to justify the investment in doing so. On the other hand, having too few performance measures creates the risk that provider organizations or physicians will focus too narrowly on the selected measures while ignoring other dimensions that are important for overall performance in a domain.

Cost-effectiveness and cost benefit. P4P programs prefer indicators that have greater expected benefit of improved performance relative to their costs of collection and compliance. It should be possible to improve, collect, and report indicators in a cost-effective manner. The data needed to calculate the indicator should be available at a reasonable cost. The cost of complying with and reporting a performance indicator should correspond with the expected benefits of improved performance on the measure.

The availability of indicators that score highly on these criteria may vary greatly across performance domains and across particular settings. Hence, at the current time, implementing P4P programs that emphasize certain domains (e.g., clinical quality) rather than others (e.g., cost efficiency) may be more feasible.

Defining the Unit for Performance Measurement and Accountability

P4P systems differ in whose performance is measured. Performance may be measured for any or all of the following: provider organizations and physicians, disease management companies and other third-party care management organizations, and health plans. We discuss the issues for each of these target units of analysis below.

Provider organizations and physicians. Most commonly, health care P4P systems apply directly to provider organizations and physicians. These entities directly deliver services; therefore, they have the most direct control over important aspects of performance such as clinical quality. The provider organizations and physician entities that are held accountable in P4P programs may be classified into three broad categories.

- **Institutional providers.** Institutional providers include hospitals, nursing facilities, and home health agencies. Institutions are important targets of P4P for several reasons. First, a large percentage of health care spending occurs in institutions. Second, institutions are often large organizations with considerable resources. They are more likely to have sophisticated information systems that can capture and report performance measures. Also, they are more likely to have the management systems and organizational structures to respond to incentives to improve performance. Third, institutions typically treat large patient populations. Thus, events (e.g., treated patients) that are eligible for performance measurement occur frequently and allow statistically reliable and valid measurement of performance. Fourth, institutions facilitate attributing responsibility for care. For example, one and only one hospital is responsible for a given hospital stay.[1] An example of a P4P program in which institutions are the unit of accountability is Medicare's Premier Hospital Quality Incentive Demonstration, which rewards or penalizes hospitals for their performance on selected inpatient quality measures.

- **Physicians and other clinicians.** Clinicians, particularly physicians, control most health care spending because they make the decisions about whether to order or authorize care. For this reason, P4P programs tend

---

[1] This assertion assumes that a patient is not transferred from one hospital to another. Quality measures often exclude such cases.

to measure the performance of physicians and other clinicians. P4P programs often focus on the performance of primary care physicians (PCPs) because the PCPs may be responsible for managing patients' overall care. The United Kingdom's General Medical Services Contract, for example, rewards PCPs for their performance on 146 performance measures (Doran et al., 2006). Although fewer performance measures exist for specialist physicians, those physicians are also important because they control a considerable portion of health care spending, including many high-cost and possibly discretionary services. Payers sometimes hold specialists accountable for episodes of specialty care beginning with primary care referral or first contact with a patient.

Patients often receive treatment from several different physicians. Assigning responsibility to particular physicians is a problem for open access insurance arrangements that do not require enrollees to select primary care gatekeepers. Some P4P systems allow multiple physicians to earn incentive payments (e.g., all physicians who provided at least one or two primary care visits for the patient). These are often termed the *one-touch* or *two-touch* rules for assignment. Other P4P systems require a plurality of primary care visits to determine which provider is assigned performance accountability. Managed care systems or medical home systems that require patients to select an accountable PCP at the time of enrollment avoid this problem, at least on the primary care level. See Chapter 7 for more on patient attribution to physicians or organizations.

Ideally, programs should measure the performance of the individual physicians who provide care to particular patients. Alternatively, programs can measure the performance of the physician group. The group has the advantages of larger sample size and greater statistical reliability for performance measurement, and it may also have organizational mechanisms to provide feedback to individual physicians. Further, rewarding groups of providers—including support staff— emphasizes interdependence and team delivery of health care (Young & Conrad, 2007). Measuring the physician group also reduces concerns about determining accountability among multiple physicians who may be treating a patient because some or all of those physicians may practice in the same group.

For these reasons, P4P programs often focus on physician groups for patient assignment and performance accountability measures (Christianson et al., 2006; Rosenthal et al., 2006). For example, the

Medicare Physician Group Practice Demonstration requires a plurality of visits for assignment, but it assigns patients to groups, not to individual physicians (Kautter et al., 2007). Physician groups may be traditional integrated group practices, other physician organizations such as independent practice associations, or virtual groups (e.g., hospital medical staff or all physicians practicing in a geographic area) established for the explicit purpose of performance measurement.

- **Integrated delivery systems and other combinations of providers.** Payers may evaluate integrated delivery systems (IDSs), physician-hospital organizations, or other organizational forms that combine provider types on both professional and institutional components of performance. Measurement at the level of the IDS allows payers to attribute larger bundles of care, such as episodes, to the provider units they are profiling. Moreover, measurement at this level recognizes and incentivizes the coordination of care across multiple provider types.

Disease management companies and other third-party care management organizations. In some P4P models, payers may hold a third party, outside of provider organizations, responsible for performance (e.g., the quality or efficiency of care). For example, the Medicare Health Support Pilot Program holds third-party disease management organizations (e.g., for-profit organizations that payers hire to monitor patients' chronic conditions) accountable for aspects of the care provided to Medicare beneficiaries enrolled in the traditional FFS program. Third-party organizations have certain advantages over provider organizations in achieving performance objectives. They can exploit economies of scale in developing and implementing specialized disease management programs; because they serve large populations for relatively little cost, third-party organizations perform well on cost-effectiveness and cost efficiency. Also, unlike health care providers, third-party organizations do not face the disincentive of foregone revenues when they reduce their clients' use of health care services. However, because third-party organizations do not provide care directly, they must establish mechanisms to gain the cooperation of and influence the behavior of patients and physicians.

Health plans. Health insurance plans are a natural unit for performance measurement because they are responsible for arranging all care for covered conditions and services for enrolled members. Because members are enrolled in health plans, the health plans are clearly responsible for their care.

Many individuals have a choice of multiple health plans. The availability of comparative information about the quality and efficiency of health plans may aid individuals in choosing health plans. Employers, governments, or other health plan sponsors can establish incentives at the health plan level and then let plans be responsible for transmitting the incentives to downstream provider organizations.

The health plan is an aggregated level of measurement that is far removed from the individual physicians who treat patients. Given heterogeneity among physicians in a plan network, an individual who enrolls in a plan that has a certain rated performance may receive care that deviates substantially from this average, depending on the particular physician who supplies treatment. Hence, performance measurement at the health plan level does not obviate the need for measurement at disaggregated levels (e.g., at the level of the individual physician).

Data Sources for Measuring Performance

Depending on the domain or specific indicator, a variety of data sources may be used to measure performance in P4P programs. The central clinical quality-of-care domain is typically measured by one or more of three data sources: administrative claims, medical records, and patient surveys. Claims data are useful for some types of quality measures that are consistently and reliably recorded in those data and that are used primarily for billing by provider organizations reimbursed through FFS. For example, one measure that payers commonly derive from claims data is whether patients with diabetes have had annual HbA1c testing. An important benefit of claims data is that no additional data collection burden is placed on provider organizations and physicians because the data have already been submitted to payers for billing purposes. This benefit does not exist for providers who are reimbursed by capitation, however, because they often do not submit claims for individual visits, hospital admissions, or other types of medical services. Claims may also have reporting lags (e.g., when pharmacy data are held by contracted pharmacy benefit managers and the data are not easily available to health plans or payers) (Young & Conrad, 2007). Another limitation of claims data is that they contain a restricted range of clinical information.

Medical records are generally superior to claims for determining more clinically detailed quality measures. The high cost of manual medical records data collection is often viewed as a barrier, however. Widespread adoption of electronic medical records (EMRs) might mitigate this concern, although the dissemination and use of EMRs remains limited in medical practice. The lack of standardization across medical records is also an issue in judging performance.

Patient surveys are useful for some types of data, such as patient satisfaction or patient experience of care measures that cannot be collected from other sources. However, provider organizations and physicians may be concerned about whether patients are able to report accurately on technical aspects of medical care. Also, patient surveys can be expensive and a burden to patients, and they can suffer from low response rates and nonresponse bias.

Mandatory Versus Voluntary Participation in Pay for Performance

Voluntary programs are easier to implement than mandatory programs, and provider organizations and physicians are less likely to resist the implementation of voluntary programs than required programs. Moreover, providers who expect to do well are more likely to participate in voluntary programs than providers who do not expect to do well. The lack of participation by poor performers may limit the ability of voluntary programs to improve overall system performance.

In programs that offer bonuses for good performance, voluntary participation may lead to the same results as a mandatory program because provider organizations and physicians may choose to participate based on the likelihood of earning a bonus. Thus, if a program is voluntary, the sponsor must offer incentives that lead at least some provider organizations and physicians to want to participate because they expect that the rewards they can achieve under the program will exceed the costs of participating. P4P programs that are less favorable to providers (e.g., those that involve penalties for poor performance or downside financial risk) may need to be mandatory. A strategy that payers may use is to start with a voluntary program to demonstrate feasibility and work out operational problems and then gradually increase the penalties for nonparticipation or eventually mandate participation as a condition of eligibility for receiving any reimbursement.

Incentive Schemes to Reward Performance

Given a measurement of performance, an incentive scheme to reward good performance (or penalize bad performance) is the second crucial ingredient of a P4P program. This section discusses the elements of P4P incentive schemes. We consider how to fund incentive payments and to structure financial incentives. As well as discussing direct incentives to provider organizations and physicians, we also address financial and nonfinancial incentives that programs can offer patients for using high-performing providers.

Funding of Performance Payments

P4P systems must identify a source of funding for the performance incentives. According to the Institute of Medicine Board on Health Care Services (2007), three possibilities are existing payments, generated savings, and new money

Existing payments. Redistributing existing payments is attractive to payers because they do not have to add money to the system. One justification for using existing payments to reward quality is that payers already expect high-quality care, and they should not have to add new money to payments to expect provider organizations and physicians to supply high-quality care. However, this approach inevitably means that low-quality providers will receive lower payments than before.

Generated savings. Generated savings are those produced by high performance. Payers often claim that improving quality of care (e.g., reducing medical errors and complications of care) will generate savings. Generated savings are also attractive to payers because they do not require new money and because savings are a prerequisite for any performance payments. Basing performance payments on generated savings favors efficiency improvements and quality enhancements that generate savings.

New money. Provider organizations and physicians may justifiably argue that performance payments should be funded out of new money if improving and reporting performance requires new investments and higher costs on their part. For example, improving and reporting performance may require providers to invest in expensive IT equipment and training and to hire additional support personnel (e.g., nurse case managers and IT support workers). However, adding new money to the system raises the question of how cost-effective the program is. The question that arises is whether the performance gains that will result from the system are worth its extra cost.

Performance Benchmarks

P4P programs must establish benchmarks against which performance is judged and that will trigger performance payments. The benchmarks that programs choose can significantly affect the amount of P4P performance payments and the extent to which P4P schemes reward high quality or improvements in quality. The choice of a benchmark is thus a critical decision that each P4P program should tailor to its goals (Werner & Dudley, 2009). P4P programs have three possible benchmarks for rewarding performance—absolute performance, improved performance, and relative performance—discussed below.

Absolute performance (target attainment). Some performance indicators may have natural benchmarks. For example, for clinical process-of-care indicators and practice guidelines, payers expect that every patient satisfying the relevant eligibility criteria should receive the indicated service (and/or not receive an obsolete or contraindicated service). The natural benchmark and goal for such indicators is 100 percent compliance or performance. In real-world situations, 100 percent compliance is unlikely because of patient refusal and other factors, but payers may establish a high absolute threshold or target for rewarding performance (e.g., 90 percent). A target provides a clear, simple, direct standard of expected performance.

Target attainment tends to reward existing high performance, not necessarily improvement. Provider organizations and physicians that exceed the target at baseline can enjoy performance payments without improving their performance, although they must maintain it. Conversely, providers with low performance may see high targets as unattainable and may not attempt to improve. Therefore, if the goal of P4P is to improve overall system performance, not merely to reward current high performers, absolute thresholds have drawbacks (Rosenthal et al., 2005).

Improved performance. Explicitly rewarding improved performance focuses P4P on improving overall system performance, rather than just rewarding existing high performers. Both low and high performers are rewarded only if they improve compared with their past performance. However, if payers only reward improvement, then low-performing provider organizations and physicians may find it easier than their high-performing counterparts to earn performance payments because improving from a low rather than a high starting point is easier. Giving greater rewards to low-performing provider organizations and physicians, even if they are improving, may lack face validity and appear inequitable to high performers. One way of ameliorating these concerns would be to phase out rewards based on improvement after some period of time (Institute of Medicine [IOM] Board on Health Care Services, 2007), under the logic that providers should be able to transition to a high absolute level of performance within a limited period of time.

Relative performance. A third approach is to reward relative performance. In this approach, payers identify a comparison group for the participating provider organizations and physicians. An advantage of rewarding relative performance is that the comparison group defines the performance benchmark, relieving the program designers of the need to choose a particular reward threshold and adjust it over time. If the general level of performance improves over time, the performance benchmark automatically adjusts upward. Moreover, payers can define regional or local comparison groups, customizing the benchmark to local conditions and baseline performance. Payers can risk adjust comparisons among groups to standardize for differences in group composition in at least two ways.

First, payers can use a usual care comparison group. One variant of relative incentives defines a comparison group of provider organizations and physicians who are not participating in the P4P program. For example, several of Medicare's FFS P4P demonstrations (discussed in detail in Chapter 9) compare the performance of participating providers to that of nonparticipating providers as representatives of the usual standard of care. An advantage of a usual-care comparison group is that the payer can potentially reward all participating provider organizations and physicians, if they all exceed the performance of their (nonparticipating) usual-care comparison group. In a relative ranking approach (discussed next), payers reward only the top performers among the participants. If P4P participation is voluntary, payers may benefit from creating the potential for all participating providers to earn a reward. A usual-care performance standard presents a more feasible improvement target for low-performing providers than high absolute and relative performance criteria do, but such a standard does not reward below-average performance. Identifying a nonparticipating comparison group may not be feasible in all situations. If a program is extended to all providers to maximize its impact, no nonparticipating providers will exist. Then the only feasible comparison group is other participating providers.

Second, payers can use an approach based on a relative ranking of provider organizations. Sometimes called the "tournament" approach or "yardstick competition," this variant ranks participating provider organizations and physicians and rewards only those in the top ranks. Medicare's Premier Hospital Quality Incentive Demonstration uses this method, rewarding only hospitals in the top two deciles of quality performance (2 percent payment bonus for the top decile and 1 percent bonus for the second decile). This approach does not consider absolute performance. Thus, high-ranked provider organizations and physicians will be rewarded, even if their absolute performance is poor, and low-ranked providers will not be rewarded, even if their absolute performance is good. The tournament approach provides the greatest incentive for improvement to the provider organizations and physicians who are near the threshold that defines the top-ranked providers. Provider organizations and physicians who are already top performers or poor performers have less incentive to improve. Top performers can simply maintain their current relative performance, and poor performers may have difficulty substantially improving their current rankings. Penalties for poor performance can be added to spur quality improvement among relatively poor performers.

This is a competitive approach that will not foster collaboration among providers. Competitive ranking requires provider organizations and physicians to outperform others in order to earn a P4P bonus payment. This can stimulate higher levels of quality improvement because no one knows in advance how high performance needs to be to earn the bonus payments. However, sample size can sometimes be an issue in differentiating providers' ranks because random variation may affect the measured performance results and hence the levels of P4P bonus payments. For example, one study found that smaller hospitals had a greater risk for misclassification in rankings than larger hospitals when this type of target-setting was simulated (Davidson et al., 2007). Also, relative ranks may not distinguish substantively different performance, and ties (identical scores) may be problematic.

Combined benchmark approaches. The various approaches to establishing performance benchmarks are not mutually exclusive. The different incentives and distribution of rewards established by alternative benchmarks may be used in combination. For example, Medicare's Premier Hospital Quality Incentive Demonstration combines all the following four elements: a target attainment award for hospitals exceeding median performance; a top performer award for the top-ranked 20 percent of hospitals; an improvement award for hospitals that attain targets and are in the top 20 percent of improvement; and a threshold penalty for hospitals scoring below the ninth decile of performance (Centers for Medicare & Medicaid Services [CMS], 2009b). In Medicare's Physician Group Practice Demonstration, participating physician group practices can satisfy quality performance standards by exceeding either (1) an absolute threshold (75 percent compliance on process quality measures), (2) an improvement threshold (reducing the gap between baseline performance and 100 percent attainment by 10 percent or more), or (3) an external relative target (established with reference to the performance of Medicare private health plans) (Kautter et al., 2007).

Graduated or tiered rewards. Rewards based on achieving a single performance benchmark using any of these approaches—absolute, improved, or relative performance—have the disadvantage of not giving incentives for improvement along the entire spectrum of performance. This limitation can be addressed by a system of graduated or tiered rewards that increase as the level of performance rises. For example, in the absolute approach, a payer may give a reward to provider organizations or physicians for exceeding 70 percent compliance on a process quality measure, a larger reward for exceeding 80 percent compliance, and a still larger reward for exceeding 90 percent compliance. Improvement rewards can also be graduated, with payers giving larger rewards for greater improvements. A graduated relative reward system might give a reward to provider organizations and physicians in the top 50 percent, a larger reward for those in the top 25 percent, and a penalty for those in the bottom 10 percent. All rewards systems are likely to have a minimum performance threshold below which no rewards are given.

Continuous rewards (percentage of patients receiving recommended care). An alternative approach is not to rely on specific thresholds of performance at all but to pay provider organizations and physicians more for each appropriately managed patient, episode, or recommended service. For example, a PCP could be paid more for each patient in her panel who had diabetes and had received clinically recommended eye and foot examinations. Under this model, physicians at any level of performance will always do better by achieving recommended care processes for more patients.

Rebasing benchmarks. Over time, the general level of performance may improve. Approaches that use absolute thresholds or improvement from baseline to reward performance should eventually rebase to a higher level of expected performance. Payers must find a balance between not rebasing too often, which gives provider organizations and physicians too little reward for performing well or improving performance, and rebasing too infrequently, which gives provider organizations and physicians too little incentive to continue improving performance. In the context of cost efficiency, the payer can financially capture the initial efficiency gain by rebasing (i.e., by lowering) provider payment rates, and thereby give the provider an incentive to achieve further efficiency improvements from the new, higher-efficiency baseline.

Implementing Financial Incentives

P4P programs may implement financial incentives in a wide variety of ways. Several typical approaches to distributing incentive payments are discussed below.

Bonus or withhold. One common approach to distributing reward payments is through a bonus pool, which is disbursed at the end of the measurement period (e.g., annually) and is contingent on performance. A bonus pool can be funded either by using new money or by withholding a portion of regular payments throughout the year. For physicians, payers might withhold 5 or 10 percent of physicians' fees or employers might withhold a small percentage of premiums paid to health plans. The Excellus/Rochester (New York) Individual Practice Association Rewarding Results demonstration project returned to individual physicians 50 to 150 percent of a 10 percent withhold based on relative performance. The Blue Cross of California Preferred Provider Organization Rewarding Results demonstration made available a bonus of up to $5,000 to physicians, based on their performance on selected clinical indicators (Young et al., 2007). In the context of health plans, some employers have put a percentage of health plan premiums at risk, with payments contingent on performance on administrative services measures (e.g., percentage of claims processed accurately), clinical quality, member access to services, and data reporting (Bailit & Kokenyesi, 2002). About 2 percent of the premium is typically put at risk.

Penalties. Payers may reduce payments to provider organizations and physicians who do not achieve an acceptable level or improvement of performance. For example, in year 3 of Medicare's Premier Hospital Quality Incentive Demonstration, participating hospitals faced a 1 percent payment reduction if they scored below the 9th decile baseline quality level and a 2 percent reduction if they scored below the 10th decile baseline level (CMS, 2009b).

Fee schedule adjustment. In FFS environments, payers may adjust fee schedule payments up or down, depending on performance, by adjusting the fee schedule conversion factor that translates fee schedule relative value units per service into dollar payments. For example, a PCP might be paid 105 percent of an insurer's base fee schedule if he or she ranked in the top 25 percent of network PCPs on performance measures. The Blue Cross Blue Shield of Michigan Rewarding Results demonstration allows participating hospitals to earn up to a 4 percent diagnosis-related group fee enhancement for meeting absolute thresholds of performance on selected quality measures (Young et al., 2007).

Per-member payment. In capitated environments, or plans in which patients are enrolled with PCPs, a health plan might pay providers an additional or incremental per member per month or per member per year payment that is contingent on measured performance. For example, the Bridges to Excellence Rewarding Results demonstration pays a per patient per year bonus of $100 for diabetes care and $160 for cardiac care based on National Committee for Quality Assurance performance recognition (Young et al., 2007).

Differential payment update. Payers can reward provider organizations and physicians that perform well with a update factor to their payments that is higher than those given to provider organizations and physicians that perform poorly. For example, under the Medicare Reporting Hospital Quality Data for Annual Payment Update program, hospitals that did not report designated quality measures received a 0.4 percent reduction (later raised to a 2 percent reduction) in their annual payment update (CMS, 2009a).

Payment for provision of a service. A payer can establish payment, or enhanced payment, for services that further the goals of the P4P program. For example, if raising the rate of mammography screening is a quality goal of a P4P program, then the payer can increase the provider payment for mammography. Payers could also institute payments for activities involving coordinating and managing patient care. These might include completing an annual patient health-risk assessment and action plan or performing patient education activities.

Payment for participation or payment for reporting. Programs might pay provider organizations and physicians to engage in performance-enhancing activities, such as developing quality improvement action plans, attending continuing education programs, or implementing computerized physician order entry. Alternately, payers might pay provider organizations and physicians for reporting performance measures, as in Medicare's Physician Quality Reporting Initiative, which pays successfully reporting physicians 2 percent of their Medicare covered allowed charges.

Lack of payment for poor performance. Payers can deny payment for services that appear to be ineffective, harmful, or inefficient. Notably, payers may deny payment for preventable medical errors or their sequelae, including performing surgery on the wrong patient or body part, leaving a foreign object in a patient during surgery, or wrongly prescribing or incorrectly administering drugs. Since October 1, 2008, Medicare no longer pays for extra costs associated with eight preventable occurrences, including transfusion with the wrong blood type, pressure ulcers, and certain hospital-acquired infections.

Shared savings. Payers can give providers incentives to improve efficiency and generate savings by allowing them to share in the realized savings. For example, in Medicare's Physician Group Practice Demonstration, Medicare retains 20 percent of annual measured savings and shares up to 80 percent with participating provider groups, depending on the quality performance (Kautter et al., 2007).

Quality grants or loans. A provider could apply to a payer for a grant to implement quality-enhancing infrastructure changes, such as an EMR or patient registry. Payers could commit to invest or lend capital to high-performing providers to build their delivery systems.

Single versus multiple reward pools. Payers can set up multiple reward pools to reward performance in the services supplied by each different type of provider organization or physician. For example, one reward pool might focus on hospital services, a second on PCP services, and a third on specialist physician services. With multiple pools, payers can attribute accountability more easily, but a smaller number of pools or linked pools increase the incentives for coordination of care and overall efficiency. For example, given the primary role of physicians in hospitalization, payers might partly fund physician performance payments out of the hospital pool if one of the performance goals is to keep enrollees out of the hospital. In the long run, as provider organizations that can take responsibility for entire episodes of care evolve, consolidating multiple pools into a single pool can establish better incentives for overall efficiency.

Magnitude and Risk of Financial Incentives

Several important characteristics of financial incentive schemes will affect provider response to them. Among the more important are the magnitude of the incentive and the financial risk to which programs subject the provider organizations and physicians. Payer design choices affect these characteristics. Payment frequency is also an important implementation issue.

Magnitude of incentives. Payers must decide on the magnitude of performance incentives that they will offer to providers. The necessary incentive will depend on the cost to the provider of the intervention that the payer is rewarding. Prescribing more generic drugs may be relatively costless, but coordinating a patient's care through a nurse case manager is not. The incentive per provider depends on the total payout and the proportion of providers who will receive the incentive. Extending incentive payments to more providers will involve a higher proportion of providers in the incentive scheme but will lower the incentive payment per provider, holding total payments constant. If a payer has a small market share, then to represent a meaningful incentive to provider organizations and physicians it may have to offer a larger incentive per member than payers with larger market shares would need to offer.

Most P4P systems have started out with incentives of limited size, although the United Kingdom's program is an exception. Reasons for limiting the size of incentives include concerns about the validity and reliability of quality measurement and data collection, the controversy created by payment disparities between providers, and provider market power to resist P4P programs. Many P4P systems in the United States provide incentives of below 5 percent of providers' total FFS incomes, although this amount may grow over time (Dudley & Rosenthal, 2006).

In this context, studies and reviews of the scientific literature on P4P have reported only limited evidence of its impact (Mullen et al., 2010; Petersen et al., 2006; Rosenthal & Frank, 2006; Sorbero et al., 2006). However, given the limited size of the incentives implemented in P4P to date, one can ask whether evaluation results showing no impact or limited impact of P4P are a fair test of this new approach to provider payment. Indeed, recent evidence has shown more positive effects of P4P, although studies have found that the effect size remains modest in most cases and the largest effects are often for provider organizations and physicians that have started at lower levels of performance (Campbell et al., 2007; Felt-Lisk et al., 2007; Gilmore et al., 2007; Glickman et al., 2007; Golden & Sloan, 2008; Grossbart, 2006; Lindenauer et al., 2007). One review suggests that incentives of about 5 percent of total physician earnings are large enough to attract "meaningful attention" from physicians (Young et al., 2007).

Nonfinancial factors may either enhance or dilute the effects of financial payments under P4P; they may certainly affect the size of the incentive payments that are needed to improve performance. Incentive payments that payers make to organizations such as hospitals or physician groups may have diluted (or enhanced) effects in relation to the individual physicians working in those organizations (Christianson et al., 2006; Young & Conrad, 2007). The organizations may or may not transmit the incentive payments directly to the physicians. Conversely, some physicians in group practices may free-ride on the efforts of their colleagues. Organizations and payers may also support P4P programs in complementary ways, with investments in electronic health records, public reporting of performance, patient incentives for adherence to care, education of boards of directors, feedback reports to providers, and staff support for case management and care coordination (IOM Board on Health Care Services, 2007). Senior staff may work actively to promote an organizational culture that fosters quality improvement and collaboration among staff. Large incentive payments may generate more quality-maximizing behavior but may also break down the norms of clinical teamwork that are needed to improve quality. Large incentive payments may also lead to gaming or manipulation of measurement systems that could defeat the purpose of P4P.

Risk to providers. An important aspect of a P4P financial incentive program is the financial risk to which programs subject provider organizations and physicians. Different designs of P4P programs may greatly affect the amount and type of risk that participating providers face. We discuss different aspects of provider risk below.

- **Upside versus downside risk ("carrots versus sticks").** Shared-savings incentives involve only upside bonus risk. If provider organizations and physicians generate savings, then they benefit by sharing in those savings. If provider organizations and physicians do not realize savings, the status quo ante is maintained, so provider organizations and physicians face no downside risk. A withhold, however, involves downside risk because the payer will not return the withhold to the provider unless the provider has met performance objectives. If participation in a P4P program is voluntary, positive incentives will be necessary to induce providers to participate.

- **Limitations on risk. Provider risks in P4P systems are typically capped.** For example, withholds are limited to 5 or 10 percent of provider payments, which is the largest amount that provider organizations and physicians can lose because of poor performance. Upside risk is also typically limited. For example, in Medicare's Physician Group Practice Demonstration, the maximum performance payment that participating providers may earn is 5 percent of the target expenditure amount.

- **Additional versus foregone revenues.** Process quality measures may involve the provision of additional services, which are separately reimbursed under FFS payment. Because provider organizations' and physicians' costs of meeting the performance objective are entirely or largely covered, the risk that they incur by meeting the performance objective is low. The necessary incremental P4P incentive may be small.

Other quality interventions reduce needed services, thus reducing provider organizations' or physicians' revenues under FFS payment and thereby creating a foregone revenue risk. For example, better ambulatory management of care may reduce hospital admissions, which would lower inpatient revenues for IDSs. In this case, the P4P incentive may need to be larger to offset the foregone revenues (e.g., the provider organization could share in generated savings). Alternatively, the provider organization may realize substantial cost savings because of reduced utilization, or it may be operating at capacity and can replace lost utilization from the queue of patients waiting to use its services. To avoid the disincentive of foregone revenues, a payer may give the performance incentive to an entity that does not forego revenue, such as a physician group without an affiliated hospital or a third party such as an independent disease management organization.

- **Business risk of performance-enhancing investments.** Improving performance typically requires providers to make investments in systems and processes to improve and report their performance. Provider organizations and physicians incur business risk in making these investments because there is usually no guarantee that investments will lead to performance payments. The larger the required investments and the greater their perceived risk, the less likely provider organizations and physicians are to make them. One aspect of risk is the certainty of reward. Absolute thresholds or improvement targets have greater certainty than relative rewards, which depend on the performance of other provider organizations and physicians.

An approach that payers can take to reduce the business investment risk is to pay an upfront fee, either a lump sum "grant" or a periodic per-member payment, that finances a provider organization's or physician's performance-enhancing investments. This ameliorates the provider organization's or physician's cash flow concerns, given the lag between the necessary investments and the realization of performance payments. The greatest reduction in provider risk occurs if the upfront fee does not depend on ultimate performance. Alternatively, the upfront fee can be used only as an advance on ultimate performance payments. In this case, the provider is at risk for the fee and ultimately for the investments it supports.

Payment frequency. The frequency of P4P payments may also be an issue (Young et al., 2005). Annual payments are common, but more frequent payments may provide more visibility for P4P programs and have more impact on provider behavior. However, more frequent payments will necessarily be smaller and thus may dilute a behavioral response. More frequent payments may also raise administrative burden and cost.

Nonfinancial Incentives

P4P programs may also use nonfinancial incentives. Nonfinancial incentives may require less investment on the part of payers and may be less threatening to providers, whose income is not directly affected. Nonfinancial incentives include performance profiling, public recognition, technical assistance, practice sanctions, reduced administrative requirements, and automatic assignment of patients (Llanos & Rothstein, 2007).

In performance profiling, payers provide confidential feedback to providers on their performance. Public recognition, discussed in greater detail in the next section, publicizes provider performance and recognizes high-performing provider organizations and physicians. Technical assistance might occur when the payer provides help to providers in improving, for example, their achievement rates related to process-of-care criteria. A practice sanction might involve an insurer's excluding provider organizations and physicians from the provider network until they meet a threshold level of quality or efficiency performance. Reduced administrative requirements could involve quality audits of provider organizations and physicians every other year instead of annually if they meet specified performance thresholds.

In the Medicaid context, enrollees may be automatically assigned to health plans, provider organizations, or physicians (Kuhmerker & Hartman, 2007). In other contexts, payers may also automatically assign patients who fail to choose their own plans, provider organizations, or physicians. The payer or sponsor managing the enrollment process (e.g., state government or employer) can direct more patients to higher-performing plans, provider organizations, or physicians by automatically assigning more patients to them. Providers not achieving a minimum level of performance may not be assigned any patients. Assuming they are not already at capacity, automatic assignment provides an incentive for plans or providers to perform better because they will receive more enrollees or patients without incurring marketing or other acquisition costs to enroll them.

Incentives for Patients to Use High-Performing Providers

Steering patients to high-performing provider organizations and physicians creates an indirect but potentially powerful incentive for providers to improve their performance. Even a small proportion of patients changing their providers based on these incentives could represent significant revenue risk for provider organizations and physicians. From a patient's point of view, these incentives maintain freedom of provider choice but create consciousness of quality and cost differences among provider organizations and physicians. For example, a patient can continue to patronize a high-cost provider organization or physician if he or she chooses to do so, but the patient will have to pay more for this choice and must weigh the perceived quality or other advantages of the provider against the higher cost. The ability to use financial incentives may be limited for some populations, such as Medicaid enrollees, who may not be able to afford significant out of pocket payments.

Nonfinancial incentives. Nonfinancial incentives may be used prior to, in lieu of, or together with financial incentives. They provide a means of introducing the concept of performance measurement with less controversy than with financial incentives. Nonfinancial incentives create reputational effects that may drive referrals and patient choice.

- **Public reporting or report cards.** The mildest form of patient incentive is public reporting of quality and cost efficiency information for provider organizations and physicians. These reports are sometimes known as report cards. Public reporting arms patients with information that may help them choose provider organizations and physicians, although there is no clear reason why patients should prefer to go to more efficient provider organizations or physicians unless these providers offer lower patient out-of-pocket costs. Provider organizations and physicians may feel peer pressure, or pressure from their own internal norms of professionalism, competence, or competition, to improve their performance scores. Public reporting may also be a first stage in the introduction of patient incentives, to vet the performance measures and work out the kinks in the system.

- **Designation of high-performing provider organizations/Centers of Excellence.** The next step may be to designate certain provider organizations as superior in some way, perhaps on both cost and quality.

For example, payers may designate certain provider organizations as Centers of Excellence, giving those groups an imprimatur of quality. This designation is designed to steer patients to these groups even in the absence of financial incentives to use these providers. Provider organizations may be willing to accept a discounted payment from the payer to achieve this designation.

Financial Incentives. To create stronger patient incentives to use high-performing provider organizations, payers may introduce financial incentives. Payers may implement financial incentives and categorize provider performance in several ways.

- **Differential premiums.** Health plans may require their members to choose a health care system from which to receive their care or a PCP to direct and authorize their care. At the point of annual enrollment, lower health plan premiums may be charged to members who choose higher-performing health care systems.

- **Differential cost sharing.** Payers may impose financial incentives at the point of service rather than at the annual premium stage. The basic procedure is to create differential cost sharing based on the measured performance of provider organizations so that patients pay less to use higher-performing providers. Payers may charge lower copayments, coinsurance, and/or deductibles for higher-performing provider organizations and physicians.

- **Provider tiering.** Payers may classify provider organizations and physicians into tiers based on cost and quality performance. The FFS rates (e.g., discount off Medicare payment rates) that the provider offers the payer may measure cost performance. Alternatively, cost performance could be measured by case-mix–adjusted episode or per-patient-per-month expenditures that the provider organization or physician incurs for episodes or patients attributed to it. These latter approaches measure performance in controlling use as well as price charged per service. Payers may employ standard approaches to measuring process, outcome, or structural quality. Payers charge patients lower premiums or cost sharing for using providers in the higher-performance tiers.

- **Centers of Excellence.** In this approach, payers do not rank all provider organizations into tiers. Rather, payers award a smaller number of provider organizations the special designation of Centers of Excellence for specified procedures or episodes of care, such as expensive organ transplants, or heart or orthopedic surgeries, based on their charge to the payer and the quality of care that they provide for these episodes. Patients may be required to use these centers or receive lower cost sharing or premium reductions for receiving care for these procedures at the designated centers. The goal of this approach is to consolidate volume into these centers and to use the resulting higher volume to drive quality improvement and cost reductions.

- **Other methods.** Other ways of structuring consumer incentives also exist. For example, one of the Medicaid health plans in the Local Initiative Rewarding Results P4P demonstration tried providing low-income parents with gift certificates as incentives to bring children in consistently for well-child visits, but the effort was unsuccessful in that case. Only about 3 percent of the parents sent in cards to document well-child visits to receive their gift certificates (Felt-Lisk et al., 2007).

Health plan provider network designation. For certain types of health plans, plan sponsors create networks of providers. In health maintenance organizations (HMOs), enrollees generally have no coverage for out-of-network providers. In preferred provider organizations (PPOs), enrollees have some out-of-network coverage but face higher cost sharing if they use out-of-network providers. Health plans may base their provider network selection largely on the cost and quality performance of provider organizations and physicians. Cost is typically the payment rate that the provider organization or physician is willing to accept from the health plan, and quality may be based on simple credentialing or more sophisticated quality indicators. Network-based health plans provide a means of translating provider performance into differential premiums or cost sharing for patients.

Limitations of Pay for Performance

Although P4P is currently a powerful movement in health care, payers and policymakers should recognize its limitations. This section discusses those limitations and approaches to dealing with them. The next section identifies some alternatives and complements to P4P.

Lack of Valid, Reliable, and Important Performance Indicators

Measuring performance in health care can be quite difficult. Quality of care, for instance, is influenced by many physician, patient, health care system, and environmental factors. Determining the marginal contribution of a provider organization or physician to a given process or outcome is often challenging (Hahn, 2006). Also, many areas of medical practice suffer from large uncertainties about the best approaches. The relationship between many health care processes and outcomes is difficult to discern. The number of times that payers can observe recommended processes and especially outcomes for individual physicians is often small, leading to concern about the statistical reliability of performance measurement.

Health care expenditures, often used to measure efficiency, are subject to enormous variation because of patient case mix, which is unrelated to efficiency. It is challenging to hold constant or adjust for this underlying patient health status variation so that payers can distinguish differences of a few percentage points in the efficiency of provider organizations. Large sample sizes—a minimum of 10,000 to 15,000 patients per profiled provider unit—and powerful risk adjustment methodologies that provider organizations cannot manipulate are necessary but often not available (Kautter et al., 2007; Nicholson et al., 2008).

In these circumstances, reliably isolating, measuring, and attributing the incremental contributions of individual provider organizations, physicians, or even health plans to quality or efficiency is difficult. Available performance indicators are often driven by the data (e.g., administrative billing data) that are available at reasonable cost and that have usually been collected for purposes other than measuring performance. Administrative data have limitations, however; for example, they may lack the clinical detail necessary to measure quality of care adequately. The result is that payers may base P4P programs on available performance indicators rather than important or optimally measured performance indicators.

Provider organizations are understandably concerned about having their performance judged on measures that may not be valid, reliable, or important. Approaches to improving the value of a P4P program include focusing on areas that have a high degree of consensus about appropriate medical practice, that amass accurate data and sample sizes sufficient to measure performance reliably, that represent important areas of medical practice in terms of quality or cost, and adjusting for as many noncontrollable factors as possible. Using a transparent and relatively simple performance assessment and reward system can also promote understanding and acceptance by provider organizations and physicians (Folsom et al., 2008).

Lack of Comprehensive Performance Indicators

Even if performance can be gauged accurately in some areas, comprehensive performance measurement may not be possible or may be too costly to obtain. If such assessments are not comprehensive, provider organizations and physicians may focus on improving their performance in the areas that can be measured and neglect areas that are not examined or rewarded. Performance could actually deteriorate in unmeasured areas, and this unintended consequence may be more important to ultimate outcomes than measured areas. One solution to this problem is to rotate measures among multiple areas across performance periods (e.g., across years), so that provider organizations and physicians cannot consistently do well by focusing on only one performance domain or narrow set of indicators.

Prescriptiveness or Lack of Flexibility of Performance Measures

Process measures may use considerable detail to specify how patients should be treated in specific circumstances. If the goal of a P4P program is to promote the adoption of certain evidence-based care processes, this level of detail may be an advantage, especially when exceptions processes are available to excuse justified noncompliance with recommended care. Alternatively, the performance measurement approach may be overly prescriptive and may intrude on provider organizations' and physicians' autonomy, flexibility, and ability to use professional judgment to decide the best course of care in particular situations (Epstein et al., 2004). Measuring and rewarding the ultimate outcomes of interest rather than detailed intermediate care processes allows provider organizations and physicians to have the autonomy and flexibility to determine the best means to achieve ultimate outcomes.

Lack of Cost-Effectiveness

Implementing and administering a P4P program may be quite costly. P4P programs may impose large costs on provider organizations. Simply reporting performance measures may be quite expensive for provider organizations, especially for solo practices, smaller groups, or institutions with limited resources and when reporting requires large fixed investments. Such providers may need to purchase and implement complex information systems and collect and validate expensive data. The investments required to improve performance may also be costly, depending on the performance measure. Provider organizations may have to hire additional staff to manage patient care (e.g., to document which patients are not receiving recommended care and to convince them to receive it), purchase information technology systems, and allocate portions of individual physicians' time to complying with performance indicators.

Payers also have burdens for administering P4P programs. They must define performance measures, collect and process the necessary data, evaluate performance, disseminate results, and implement incentives. Payers must involve, educate, assist, and adjudicate appeals from provider organizations and physicians. Organizations and physicians are likely to demand higher reimbursement from payers to defray their costs of reporting performance measures and otherwise participating in a P4P program.

The value of the performance gains that one can reasonably expect from a P4P system may not clearly justify the large costs that the system imposes on both providers and payers. The business case for P4P may be especially hard to make when the financial gains from improved performance are likely to mostly accrue in the future—as may be true with better management of chronic disease—but the costs are immediate. The imbalance of short-run costs and long-run savings is especially difficult to justify in settings, such as in employer-based health insurance, that have a high turnover of plan members; the principal reason is that the gains are not likely to accrue to the same health plan or even to the same employer as the one that incurs the initial cost.

Given these considerations, sponsors of P4P programs should evaluate all performance indicators for cost-effectiveness. For example, insurers with a high rate of patient turnover may wish to focus on measures that have short-term payback. The Medicare program, or the government more generally, has more reason to establish P4P efforts that can invest in improving longer-term performance.

Unintended consequences. P4P incentive payments may have unintended consequences that could be detrimental in several ways. One concern is that physicians may begin to avoid taking on more "difficult" patients so that they can avoid scoring poorly on quality or efficiency (Epstein et al., 2004). However, difficult patients—such as those with multiple chronic diseases or low socioeconomic status—may need high-quality, coordinated health care more than other patients.

Performance measures may be risk adjusted for the characteristics of patients to reduce or eliminate providers' disincentive to accept high-risk patients; that is, programs might disburse higher payments to providers for taking on more difficult patients (IOM Board on Health Care Services, 2007). Risk adjustment is complex and controversial, however. Adjusting quality indicators for patient characteristics may implicitly create a lower benchmark standard for the care of high-risk patients, who often have lower socioeconomic status, are minorities, and have worse health than other patients. Moreover, whether (or to what extent) provider organizations and physicians trust the currently available risk adjustment systems to protect them from the potential for negative performance assessments in P4P programs is not clear (Dudley & Rosenthal, 2006). If provider organizations and physicians believe that P4P payments put a substantial portion of their compensation at financial risk, they may demand as a condition of participating a risk premium (higher payments) from payers to compensate them for this risk (Nicholson et al., 2008).

Another potential unintended consequence of P4P would be exacerbated disparities in care (IOM Board on Health Care Services, 2007; Karve et al., 2008). For example, provider organizations in high-income communities might be able to fund IT and quality improvement systems at a higher level than organizations in low-income communities, thereby earning a larger share of P4P bonus payments. A study of P4P in Medicaid health plans found that provider organizations often reported lacking the office staff and systems needed to respond to the quality improvement incentives, and they did not have the financial resources needed to hire more staff and install better information systems (Felt-Lisk et al., 2007).

In general, P4P incentives have the potential to either narrow or widen disparities in health care (Chien et al., 2007). P4P programs can be designed to reduce disparities in care, if that is established as an explicit goal; however, very few P4P programs to date have been designed explicitly either to limit their impact on disparities or to actively reduce them. One program that has is the CMS Cancer Prevention and Treatment Demonstration for Ethnic and Racial Minorities, a randomized controlled trial that uses patient navigators to reduce racial and ethnic disparities in cancer screening and treatment (Mitchell et al., 2008). Demonstration sites receive monthly capitation payments to provide navigation services for beneficiaries enrolled in the intervention arm of the study.

Payers may also be vulnerable to unintended consequences. In the United Kingdom, doctors initially met more targets than projected, resulting in much larger payouts than the government had expected (Epstein, 2006). This produced a larger deficit for the National Health Service than the government had anticipated. Advocates of P4P programs may believe that they will reduce costs, but an exclusive focus on improving quality, especially service underuse as indicated by process-of-care measures, may or not may not reduce long-run costs. In the short run, under FFS payment, payer costs are likely to rise as additional services are provided (Hahn, 2006).

Difficulty of Patient Attribution

In situations in which many provider organizations or physicians treat individual patients without coordinated care, attributing care to individual provider organizations or physicians may be difficult; similarly, ascertaining which provider organizations or physicians are responsible for the observed processes or outcomes is challenging. Analysts may use various attribution rules, but none of these may be ideal (see Chapter 7 for more on patient attribution to physicians or organizations). An alternative approach to using attribution rules is to institute a voluntary or mandatory system in which patients choose a provider organization or physician (e.g., a primary care gatekeeper or "medical home") that is assigned overall responsibility for managing the patient's care. However, making such assignments mandatory may conflict with patient freedom of choice. Inherent conflicts may exist between complete patient freedom of provider choice and provider responsibility in P4P programs. Organizational changes that clarify provider responsibility—changes that to some degree may compromise patient freedom of choice—may be necessary precursors to effective P4P programs.

Multiple Payers with Inconsistent Programs

Most provider organizations and physicians treat enrollees who, together, are covered by multiple private insurers and government programs. If each payer implements its own P4P program with different performance measures, reporting requirements, and incentive schemes, the costs to provider organizations and physicians of participation or compliance will be much higher than with a single, coordinated P4P program. Working with multiple programs is likely to result in confusion and to dilute the impact of P4P (Hahn, 2006). The obvious solution is for payers to coordinate their programs, which they have done in some cases, such as the Integrated Healthcare Initiative in California. However, coordinating programs may be costly and difficult, especially among competing private insurers, when P4P programs are an aspect of competitive advantage or coordination may be subject to antitrust restrictions.

Alternatives and Complements to Pay for Performance
Payers and policymakers can consider using several alternative approaches instead of or in combination with P4P to further the goals of improving the quality and efficiency of medical care. The main ones are provider reimbursement; professionalism/provider education; quality regulation and accreditation; malpractice insurance; market competition, reputation, and public reporting; and incentives to patients or enrollees.

Provider Reimbursement
The basic system that payers use to reimburse provider organizations and physicians embodies incentives for quality and efficiency. FFS and capitation are the two canonical reimbursement systems. FFS rewards the provision of extra services; thus, it tends to perform well on access but poorly on cost efficiency. Capitation rewards withholding services; therefore, it tends to perform well on cost efficiency but poorly on access. A payer that is particularly concerned about its members' access to care may find using FFS provider reimbursement more effective than adopting a P4P program that rewards good performance on access measures. Similarly, a payer particularly concerned about cost efficiency and controlling costs might want to use capitated reimbursement rather than pay for performing well on cost efficiency indicators.

P4P programs can be added to the underlying reimbursement system to reinforce its incentives or to provide incentives for performance that the reimbursement system does not. In the latter circumstance, if a P4P program is going to be incremental to FFS provider reimbursement, the program logically should include a focus on cost efficiency. A P4P program that is added to capitated reimbursement would logically incorporate a focus on access.

In terms of incentives for quality, if higher quality is associated with the provision of more services, FFS promotes higher quality. FFS contains no incentives to achieve the ultimate outcome (i.e., good health) and may be inimical to it if achieving good health involves providing fewer services (e.g., avoiding medical mistakes that require additional treatment). Capitation provides incentives to avoid mistakes and invest in cost-effective quality enhancements that reduce long-run costs, but a basic incentive of capitation is to provide fewer services, especially expensive ones, which may be inconsistent with high-quality care.

Given these considerations, a P4P program that has the goal of improving the quality of care may be a useful supplement to either FFS or capitated provider reimbursement. The form of the P4P program should complement the underlying reimbursement system. For example, an extra per member per month payment for good quality performance is feasible in a capitated environment, while a higher fee schedule conversion factor for high-quality provider organizations and physicians could only be used in a FFS environment.

Professionalism/Provider Education

Payers may rely on providers' sense of professionalism to promote good care. That is, provider organizations and physicians may provide good care because that is "the right thing to do," because they are being paid to care for patients and want to do their job well. For physicians, the power of professional training, ethics, and norms of patient-centered behavior are important factors that may limit the impact of financial incentives from P4P (Golden & Sloan, 2008). Intrinsic motivation—the internal reward of performing a task for its own sake—may be as important for some providers as extrinsic motivation from P4P incentive payments, peer and community recognition, and other external factors.

The statistics that show that the current health care system provides recommended care inconsistently (McGlynn et al., 2003) have somewhat undermined this professionalism argument and have abetted the rise of P4P programs with explicit financial incentives for high-quality care. Professionalism may not be enough to ensure high-quality care, but it is an important adjunct to financial incentives. Provider profiling, feedback, and education fit in with the professionalism approach. Programs may attain performance improvements, the argument goes, by educating provider organizations and physicians about their performance and relying on their sense of professionalism to improve, even without public reporting and financial incentives.

Quality Regulation and Accreditation

An alternative to offering financial incentives is regulating quality. Payers and accreditation bodies may regulate provider organizations and physicians with minimum quality standards. For example, government programs such as Medicare and Medicaid and private accreditation organizations such as the Joint Commission review the credentials, eligibility, and suitability of provider organizations and individual physicians to provide care paid for by their programs or member organizations. State agencies may license organizations and physicians to allow them to operate legally. These licensure, program approval, and accreditation assessments may review structural and other quality indicators as well as legal compliance, malpractice actions, and other factors. If provider organizations or physicians do not satisfy these organizations' quality and other standards, payers can deny payment for provided care and regulating bodies can prohibit provider organizations and physicians from practicing medicine.

Quality regulation can ensure that all provider organizations and physicians meet a minimum threshold of measured quality, but achieving improvements in quality above the minimum threshold may be cumbersome and expensive. If something is so critical to quality that a payer feels that all organizations and physicians that it pays must have it, then mandating it as a condition of eligibility for the payer's reimbursement can be effective. For aspects of quality in which a higher rate of performance is desirable but 100 percent compliance is infeasible, prohibitively expensive, or not critical, it makes more sense for a P4P program to offer incentives for compliance rather than to mandate it.

Malpractice Insurance

Legal actions against provider organizations and physicians by patients who believe they have suffered adverse outcomes of care create an incentive for providers to avoid medical mistakes and furnish, if not high-quality care, at least the usual standard of care. Physicians and provider organizations purchase malpractice insurance against lawsuits, but if their policy premiums are experience rated, those who suffer more adverse malpractice judgments will pay higher insurance premiums.

However, the legal system is a limited mechanism for maintaining and improving the quality of medical care. Only a small portion of adverse medical events result in malpractice lawsuits, and many malpractice claims are unrelated to physician negligence (Weiler et al., 1993). P4P programs can provide a much more comprehensive and systematic measurement of the quality performance of provider organizations and physicians than the occasional malpractice lawsuit. Another problem with relying on the legal system is that malpractice lawsuits are very expensive (e.g., lawyer's fees) and time consuming. A P4P program may be a much more cost-effective and timely means of improving quality than the legal process. The legal process does have the advantage of compensating some victims of poor medical care, however, which is not a feature of P4P programs.

Market Competition, Reputation, and Public Reporting

In typical markets, competition among sellers (provider organizations and physicians) and seller reputation are important forces for maintaining quality. If sellers do not maintain their quality and reputation for quality, buyers (patients) will buy from other sellers or lower the price that they are willing to pay to low-quality sellers. Competition on quality can be enhanced through credible third-party measurement and reporting of seller quality (Nicholson et al., 2008).

The health care market is different from typical markets in several important ways. One is the presence of insurance, which means that payment for medical services is mostly made by the insurer, not out of pocket by the patient. A second is that quality is difficult to measure and judge in health care, especially for many patients. These factors create a strong role for the payer to ensure and promote quality in health care markets. Competition on and reputation for quality are important in health care, just as they are in other industries. One way in which the health care payer can promote quality is by measuring and publicly reporting the quality performance of provider organizations and physicians. Moreover, the payer's role in purchasing care means that it can also create payment incentives around the quality of care, as in a P4P program.

Incentives to Patients or Enrollees

P4P tends to focus on provider organizations and physicians rather than patients, although, as mentioned earlier, payers may use provider performance measurement to create incentives for patients to patronize high-performing providers. An alternative or complement to provider P4P is patient or enrollee P4P. Rather than giving provider organizations and physicians an incentive for the proportion of their patients complying with a process quality measure, payers could give patients direct incentives to comply, either through a direct payment or lower insurance cost-sharing for the service in question (so-called value-based insurance design). For example, payers could give patients with diabetes an incentive payment to keep their blood sugar under control, or they could reduce these patients' cost sharing for annual eye examinations. Some companies give employees incentives for lifestyle changes (e.g., lower health insurance premiums for nonsmokers, bonuses for losing weight or participating in fitness programs). Enabling patients to benefit from P4P payments is an explicit acknowledgement that they are part of what economists call the health care production function. However, only a limited number of programs include patients in P4P incentive payments today.

Patient incentives can complement provider incentives. Patient incentives can address concerns that provider organizations and physicians have sometimes expressed about being held accountable for quality performance for patients who do not adhere to prescribed tests and treatments. For example, physicians may prescribe angiotensin-converting enzyme inhibitors or angiotensin II receptor blockers for patients with heart failure, but physicians cannot force the patients to fill the prescriptions or take the medications as prescribed. Similarly, a study of P4P in Medicaid health plans found that low-income parents often lacked time and transportation needed to bring children in consistently for well-child visits that were included in P4P assessments (Felt-Lisk et al., 2007).

Clearly, direct patient incentives are not appropriate for some of the more technical aspects of medical care that are not under patient control (e.g., avoiding surgical mistakes). Patient and provider incentives may be more complements than substitutes, but for performance measures that are ultimately under patient control, such as lifestyle, it is an empirical question whether provider or patient incentives are more effective in improving performance. On the efficiency side, consumer-directed health plans put consumers in charge of managing their own health care and focus financial incentives on the consumer. Demand side (consumer, patient) efficiency incentives are an important complement to supply side (provider organization, physician) efficiency incentives.

Concluding Comments

P4P has substantial conceptual appeal. It seems logical that payment should be related to demonstrated performance on the objectives established by the payer. However, P4P is a general framework for payment, not a specific method that can be applied in every situation. As this chapter illustrates, a very large variety of performance measurement and payment schemes can fall under the rubric of P4P. The success or failure of P4P in particular applications depends on how payers evaluate performance and structure incentives. As is often true, the devil is in the details.

Because P4P is a general conceptual framework, considerable experimentation and evaluation is likely to continue for some time. We are unlikely to conclude that P4P universally fails or succeeds. Over time, payers and policy makers will discover and disseminate the successful elements of P4P and discard the unsuccessful elements. Payers and policy makers will incorporate the successful elements of P4P into other big conceptual frameworks, such as managed care.

A major limitation of P4P is that implementing it well in practice is often difficult. Achieving a valid, reliable, and comprehensive measurement of performance in an area as complex as medical care is extremely challenging. Structuring financial incentives to achieve the intended goals while avoiding unintended consequences can also be difficult. The theory of optimal incentive contracts shows that when available performance measures are "noisy" (imprecise in their relation to the outcomes of ultimate interest) and "distorted" (improving the measure does not necessarily improve the outcome of ultimate interest), the proportion of compensation that should be based on them is lower (Baker, 2002). Thus, payers should be cautious about tying a large proportion of physician and other provider reimbursement to incomplete and flawed performance measures. P4P may prove most useful in specific, narrow applications in which an accurate assessment of performance can be obtained.

Because of the diversity of P4P programs and their contexts and environments, evaluating and generalizing individual programs is hard, too. Whether evaluation results from one P4P program will apply to other programs is rarely, if ever, clear (Hahn, 2006). Rigorous evaluation evidence to support P4P's impact on quality of care and other performance metrics remains limited (IOM Board on Health Care Services, 2007; Christianson et al., 2008; Damberg et al., 2009; Epstein, 2007).

P4P is not a panacea for improving health care (Sorbero et al., 2006). We need to consider it as part of a set of complementary and substitutable strategies to achieve payer objectives, such as those discussed in this chapter. P4P is not necessarily the best strategy, or even appropriate, in all situations. Nicholson et al. (2008) discuss circumstances under which P4P is more or less useful. An important contribution of the P4P movement, however, is payers' increased emphasis on holding provider organizations and physicians accountable for the value of their health care, rather than simply paying for the volume of care. This orientation, the P4P framework, has the potential to eventually contribute significantly to improving the quality and efficiency of health care.

References

Bailit, M., & Kokenyesi, C. (2002). *Financial performance incentives for quality: The state of the art* (National Health Care Purchasing Institute Executive Brief). Retrieved January 13, 2011, from http://www.bailit-health.com/articles/NHCPI-brief1.pdf

Baker, G. (2002). Distortion and risk in optimal incentive contracts. *The Journal of Human Resources, 37*(4), 728-751.

Campbell, S., Reeves, D., Kontopantelis, E., Middleton, E., Sibbald, B., & Roland, M. (2007). Quality of primary care in England with the introduction of pay for performance. *New England Journal of Medicine, 357*(2), 181–190.

Centers for Medicare & Medicaid Services. (2009a, May 1). *Hospital quality initiatives: Reporting hospital quality data for annual payment update.* Retrieved June 18, 2009, from http://www.cms.hhs.gov/HospitalQualityInits/08_HospitalRHQDAPU.asp

Centers for Medicare & Medicaid Services. (2009b, July). *Premier Hospital Quality Incentive Demonstration: Fact sheet.* Retrieved February 10, 2010, from http://www.cms.hhs.gov/HospitalQualityInits/downloads/HospitalPremierFactSheet200907.pdf

Chien, A. T., Chin, M. H., Davis, A. M., & Casalino, L. P. (2007). Pay for performance, public reporting, and racial disparities in health care: How are programs being designed? *Medical Care Research and Review, 64*(5 Suppl), 283S–304S.

Christianson, J. B., Knutson, D. J., & Mazze, R. S. (2006). Physician pay-for-performance. Implementation and research issues. *Journal of General Internal Medicine, 21*(Suppl 2), S9–S13.

Christianson, J. B., Leatherman, S., & Sutherland, K. (2008). Lessons from evaluations of purchaser pay-for-performance programs: a review of the evidence. *Medical Care Research and Review, 65*(6 Suppl), 5S–35S.

Damberg, C. L., Raube, K., Teleki, S. S., & Dela Cruz, E. (2009). Taking stock of pay-for-performance: A candid assessment from the front lines. *Health Affairs (Millwood), 28*(2), 517–525.

Davidson, G., Moscovice, I., & Remus, D. (2007). Hospital size, uncertainty, and pay-for-performance. *Health Care Financing Review, 29*(1), 45–57.

Doran, T., Fullwood, C., Gravelle, H., Reeves, D., Kontopantelis, E., Hiroeh, U., et al. (2006). Pay-for-performance programs in family practices in the United Kingdom. *New England Journal of Medicine, 355*(4), 375–384.

Dudley, R. A., & Rosenthal, M. B. (2006). Pay for performance: A decision guide for purchasers (final contract report). Retrieved January 12, 2011, from http://www.ahrq.gov/qual/p4pguide.htm

Eddy, D. M. (1998). Performance measurement: Problems and solutions. *Health Affairs (Millwood), 17*(4), 7–25.

Epstein, A. M. (2006). Paying for performance in the United States and abroad. New England Journal of Medicine, 355(4), 406–408.

Epstein, A. M. (2007). Pay for performance at the tipping point. *New England Journal of Medicine, 356*(5), 515–517.

Epstein, A. M., Lee, T. H., & Hamel, M. B. (2004). Paying physicians for high-quality care. *New England Journal of Medicine, 350*(4), 406–410.

Felt-Lisk, S., Gimm, G., & Peterson, S. (2007). Making pay-for-performance work in Medicaid. *Health Affairs (Millwood), 26*(4), w516–w527.

Folsom, A., Demchak, C., & Arnold, S. B. (2007). *Rewarding results pay-for-performance: Lessons for Medicare.* Washington, DC: AcademyHealth, Changes in Health Care Financing and Organization Program of the Robert Wood Johnson Foundation. Retrieved January 12, 2011, from http://www.hcfo.org/files/hcfo/HCFOmonograph0308.pdf

Gilmore, A. S., Zhao, Y., Kang, N., Ryskina, K. L., Legorreta, A. P., Taira, D. A., et al. (2007). Patient outcomes and evidence-based medicine in a preferred provider organization setting: A six-year evaluation of a physician pay-for-performance program. *Health Services Research, 42*(6 Pt 1), 2140–2159; discussion 2294–2323.

Glickman, S. W., Ou, F. S., DeLong, E. R., Roe, M. T., Lytle, B. L., Mulgund, J., et al. (2007). Pay for performance, quality of care, and outcomes in acute myocardial infarction. *JAMA, 297*(21), 2373–2380.

Golden, B., & Sloan, F. (2008). Physician pay for performance: Alternative perspectives. In F. Sloan & H. Kasper (Eds.), *Incentives and choice in health care.* Cambridge, MA: MIT Press.

Grossbart, S. R. (2006). What's the return? Assessing the effect of "pay-for-performance" initiatives on the quality of care delivery. *Medical Care Research and Review, 63*(1 Suppl), 29S–48S.

Hahn, J. (2006). *Pay-for-performance in health care.* Congressional Research Service Report for Congress, Order Code RL33713. Retrieved January 13, 2011, from http://www.allhealth.org/briefingmaterials/CRSReportingforCongress-Pay-for-PerformanceinHealthCare-501.pdf

Hussey, P. S., de Vries, H., Romley, J., Wang, H. C., Chen, S. S., Shekelle, P. G., et al. (2009). A systematic review of health care efficiency measures. *Health Services Research, 44*(3), 784–805.

Institute of Medicine, Board on Health Care Services. (2007). *Rewarding provider performance: Aligning incentives in Medicare.* Washington, DC: National Academies Press.

Karve, A. M., Ou, F. S., Lytle, B. L., & Peterson, E. D. (2008). Potential unintended financial consequences of pay-for-performance on the quality of care for minority patients. *American Heart Journal, 155*(3), 571–576.

Kautter, J., Pope, G. C., Trisolini, M., & Grund, S. (2007). Medicare Physician Group Practice Demonstration design: Quality and efficiency pay-for-performance. *Health Care Financing Review, 29*(1), 15–29.

Kohn, L. T., Corrigan, J. M., & Donaldson, M. S. (Eds.). (1999). *To err is human: Building a safer health system.* Washington, DC: Institute of Medicine, National Academies Press.

Kuhmerker, K., & Hartman, T. (2007). P*ay-for-performance in state Medicaid programs: A survey of state Medicaid directors and programs.* Retrieved January 13, 2011, from http://www.commonwealthfund.org/usr_doc/Kuhmerker_P4PstateMedicaidprogs_1018.pdf?section=4039

Lindenauer, P. K., Remus, D., Roman, S., Rothberg, M. B., Benjamin, E. M., Ma, A., et al. (2007). Public reporting and pay for performance in hospital quality improvement. *New England Journal of Medicine, 356*(5), 486–496.

Llanos, K., & Rothstein, J. (2007). *Physician pay-for-performance in Medicaid: A guide for states.* Report prepared for the Commonwealth Fund and the Robert Wood Johnson Foundation. Hamilton, NJ: Center for Health Care Strategies, Inc.

McDermott, S., & Williams, T. (2006). Advancing quality through collaboration: The California Pay for Performance Program: A report on the first five years and a strategic plan for the next five years. Retrieved January 13, 2011, from http://www.iha.org/pdfs_documents/p4p_california/P4PWhitePaper1_February2009.pdf

McGlynn, E. A., Asch, S. M., Adams, J., Keesey, J., Hicks, J., DeCristofaro, A., et al. (2003). The quality of health care delivered to adults in the United States. *New England Journal of Medicine, 348*(26), 2635–2645.

Mitchell, J. B., Holden, D. J., & Hoover, S. (2008). Evaluation of the Cancer Prevention and Treatment Demonstration for Racial and Ethnic Minorities: Report to Congress. Available from http://www.cms.hhs.gov/reports/downloads/Mitchell_CPTD.pdf

Mullen, K. J., Frank, R. G., & Rosenthal, M. B. (2010). Can you get what you pay for? Pay-for-performance and the quality of healthcare providers. *RAND Journal of Economics, 41*(1), 64–91.

Nicholson, S., Pauly, M. V., Wu, A. Y., Murray, J. F., Teutsch, S. M., & Berger, M. L. (2008). Getting real performance out of pay-for-performance. *Milbank Quarterly, 86*(3), 435–457.

Petersen, L. A., Woodard, L. D., Urech, T., Daw, C., & Sookanan, S. (2006). Does pay-for-performance improve the quality of health care? *Annals of Internal Medicine, 145*(4), 265–272.

Rosenthal, M. B., & Frank, R. G. (2006). What is the empirical basis for paying for quality in health care? *Medical Care Research and Review, 63*(2), 135–157.

Rosenthal, M. B., Frank, R. G., Li, Z., & Epstein, A. M. (2005). Early experience with pay-for-performance: from concept to practice. *JAMA, 294*(14), 1788–1793.

Rosenthal, M. B., Landon, B. E., Normand, S. L., Frank, R. G., & Epstein, A. M. (2006). Pay for performance in commercial HMOs. *New England Journal of Medicine, 355*(18), 1895–1902.

Sorbero, M., Damberg, C. L., Shaw, R., Teleki, S., Lovejoy, S., DeCristofaro, A., et al. (2006). Assessment of pay-for-performance options for Medicare physician services: Final report. Available from http://aspe.hhs.gov/health/reports/06/physician/report.pdf

Weiler, P. C., Hiatt, H. H., Newhouse, J. P., Johnson, W. G., Brennan, T. A., & Leape, L. L. (1993). A measure of malpractice: Medical injury, malpractice litigation, and patient compensation. Cambridge, MA: Harvard University Press.

Wennberg, J. E., Fisher, E. S., Stukel, T. A., & Sharp, S. M. (2004). Use of Medicare claims data to monitor provider-specific performance among patients with severe chronic illness. *Health Affairs (Millwood), Supplemental Web Exclusives,* VAR5–VAR18.

Werner, R. M., & Dudley, R. A. (2009). Making the "pay" matter in pay-for-performance: implications for payment strategies. *Health Affairs (Millwood), 28*(5), 1498–1508.

Young, G. J., Burgess, J. F., Jr., & White, B. (2007). Pioneering pay-for-quality: Lessons from the rewarding results demonstrations. *Health Care Financing Review, 29*(1), 59–70.

Young, G. J., & Conrad, D. A. (2007). Practical issues in the design and implementation of pay-for-quality programs. *Journal of Healthcare Management, 52*(1), 10–18; discussion 18–19.

Young, G. J., White, B., Burgess, J. F., Jr., Berlowitz, D., Meterko, M., Guldin, M. R., et al. (2005). Conceptual issues in the design and implementation of pay-for-quality programs. *American Journal of Medical Quality, 20*(3), 144–150.

CHAPTER 3

Theoretical Perspectives on Pay for Performance

Michael G. Trisolini

The widespread enthusiasm for pay for performance (P4P) in recent years reflects an underlying theory that we can improve the quality and efficiency of medical care by focusing on economic incentives. By paying more for evidence-based preventive care services and denying payment for preventable complications, to cite two examples, we can provide financial incentives that we expect will encourage physicians and health care provider organizations to improve the quality of care. Similarly, by paying bonuses for efficiency improvements, such as reducing hospital admissions per 1,000 chronic disease patients, we expect to motivate reductions in utilization of care and overall costs. However, the documented impacts of P4P to date have not lived up to expectations.

This chapter examines theoretical perspectives from economics, sociology, psychology, and organization theory to broaden our understanding of the range of factors affecting health care quality and cost outcomes and better understand why the focus of P4P on economic incentives has had limited impact. These theoretical perspectives describe the ways in which other factors—such as the social norms of professionalism among physicians, the range of motivational factors affecting physician behavior, and the organizational settings in which clinicians practice—affect the influence of economic incentives on the outcomes of P4P programs.

For example, we can view basic concepts in sociology and economics as presenting contrasting theories of physician behavior (Gray, 2004). The sociological perspective emphasizes physicians' extensive training and socialization, and the way in which that context leads them to provide good-quality care except in cases in which negative financial incentives disrupt their efforts. The economic perspective argues that financial rewards are important in motivating physicians (and workers of all types), and thus we need to implement financial incentives that focus specifically on quality of care. This approach will ensure that physicians do not neglect quality in favor of other goals—such as the volume of care provided—that may be more remunerative in some situations. In reality, both perspectives have merit, so we should view neither in isolation. P4P programs, however, may need to emphasize either approach, depending on the range of policy, technology, organizational, motivational, and patient factors present in a particular medical practice setting.

The high levels of complexity in today's health care sector mean that focusing solely on economic incentives may have unintended consequences. For example, despite the recent advances in medical technology, physicians still must often make high-stakes diagnoses and treatment decisions under conditions of uncertainty and ambiguity (Town et al., 2004). Scientific data from randomized controlled trials, systematic evidence reviews, and other products of evidence-based medicine may be available for only a minority of a physician's patients. Patients with multiple chronic diseases may present clinical challenges for which few scientific guidelines are available; the full range of interactions between different diseases and treatments may be unknown. Patients' actions, which physicians cannot always anticipate, can also enhance or hinder the effects of treatments. Physicians may practice in multiple settings and treat patients covered by a range of different health insurance plans. Moreover, the legal system impinges on health professionals and provider organizations through the threat of malpractice suits and myriad regulatory requirements. As a result, physicians and provider organizations may respond in unexpected ways to the economic incentives of P4P programs because the incentives are operating in the context of these other forces that are also at work at the same time in the health sector.

The next section of this chapter reviews theoretical perspectives from economics, sociology, psychology, and organization theory, with a focus on the ways in which they all can have implications for P4P. The final section of the chapter discusses the need for a multidisciplinary, composite model that includes the broad range of factors affecting the behavior of physicians and health care organizations. It also reviews how policy makers can use a broader model of that type to improve the design of P4P programs and increase their impact on health care outcomes.

Theoretical Perspectives on Health Care

Economics

Market mechanisms that make economic incentives effective for price setting and cost-control in other industries often weaken or fail the health care sector for two main reasons: (1) insurance payment for medical services, and (2) lack of consumer knowledge regarding the desired attributes of medical care. Health insurance lowers the net price of care to consumers, resulting in higher utilization at lower marginal value. Economists term this tendency the "moral hazard": patients who have health insurance often consume more health care than they would otherwise (and raise the overall costs of health care), because they are not paying out of their own pockets. Consumers may also overvalue or undervalue a broad array of medical services by lacking knowledge regarding exactly how these services contribute to quality of care in terms of accurately diagnosing and treating their diseases and symptoms. As a result, consumers delegate most medical diagnosis and treatment decisions to professional experts, most prominently physicians.

Arrow (1963), in his foundational article on health economics, recognized the asymmetry of information between patients and physicians. A decade later, economists began developing new theories of the value of information—for example, in analysis of used car "lemons"—that provided analytic approaches to address the problem of physicians as imperfect "agents" for their patients (Akerlof, 1970). These new approaches focus on "agency theory," which examines optimal contracts and payment systems between principals (patients, insurers) and agents (physicians) under conditions of uncertainty and information asymmetry (Christianson et al., 2006; Eisenhardt, 1989; Golden & Sloan, 2008; Robinson, 2001; Town et al., 2004).

Agents can take advantage of information asymmetry to increase earnings, reduce work hours, or increase their prestige with colleagues. Physicians may spend less time ensuring that they correctly diagnose one patient's condition so that they can see other patients and gain additional revenue. After making the diagnosis, physicians may choose a more expensive course of treatment to increase their own billings or those of colleagues whom they expect to reciprocate with future referrals. Physicians can also earn supplemental income if they hold equity ownership in facilities used to test or treat patients (e.g., ambulatory surgery centers, laboratories, imaging centers). Clinical uncertainty can exacerbate this situation. When clinical guidelines do not provide specific guidance on treatment protocols, as is often the case, physicians may have more latitude regarding ordering tests and treatments.

All these factors could compromise quality of care in subtle, hard-to-measure ways. As in most principal-agent problems, the principals have difficulty monitoring the quality of the work the agents provide. Even if it were technically feasible, the cost of monitoring quality may be prohibitive for individual patients. As a result, health care consumers cannot make optimal purchasing decisions, unlike those in other sectors of the economy, where quality and price information is more evenly distributed between consumers and producers.

To make matters worse, most physicians in the United States are reimbursed on a fee-for-service (FFS) basis, which the business sector calls "piece-rate" compensation. Economists generally view piece-rate compensation as a poor solution to the principal-agent problem in that it encourages exploitation of information asymmetries (Robinson, 2001). Piece-rate compensation gives physicians financial incentives to increase the quantity of services provided at the expense of quality when the deficiencies in quality are difficult to detect.

P4P is intended to address these principal-agent problems in health care in two ways: (1) by providing objective quality measures and (2) by linking payment to improvements in performance. First, evidence reviews or physician consensus panels develop clinical guidelines that are used to develop quality measures. National groups such as the National Committee for Quality Assurance and the National Quality Forum oversee development and dissemination of these measures. Patients and their insurers can rely on these organizations to help them in their roles as principals, by reducing their information asymmetries with physicians and hospitals. Second, by linking some portion of physician or hospital payment to improvements in these objective measures of quality performance, P4P provides economic incentives for improving quality rather than for increasing the quantity of services provided, as is the case under FFS.

The P4P economic incentives for improving quality can be effective, but countervailing economic incentives are often strong and the design of P4P programs sometimes underestimates them. Two large and countervailing economic factors are the much larger size of FFS reimbursement compared with P4P payments and the threat of malpractice lawsuits that encourages the practice of defensive medicine. Both of these forces provide strong economic incentives to increase use of health care services without necessarily focusing on those that increase quality of care. This could be one explanation for the limited impact of P4P programs, when examining only other economic factors, even before considering the sociological, psychological, and organizational factors discussed later in this chapter.

The business sector's experience with P4P provides additional perspectives regarding the economic incentives often promoted in the health policy debate over P4P. The business sector uses a different terminology for P4P, calling it "variable pay," a category that includes piece-rate payment, merit-based pay, bonuses, profit-sharing plans, gainsharing, and employee stock ownership plans (Robbins & Judge, 2009). In the business sector, the goal is to move away from basing pay increases on time on the job or seniority, as has been traditional in some industries, and instead shifting to a system in which at least a portion of an employee's pay is based on an individual or organizational measure of performance. However, contrary to many health sector policy makers' impression of the success of P4P economic incentives in the business sector, research has shown only mixed results from variable pay systems in business settings.

P4P programs in health care are similar to the business sector model known as merit-based pay, in which performance appraisal ratings drive pay increases. Research in business organizations has shown that if merit pay systems are well designed, and if employees perceive a strong relationship between performance and rewards, they can succeed in improving employees' motivation (Robbins & Judge, 2009).

However, business researchers have also found that, in practice, merit pay systems have at least five types of limitations (Robbins & Judge, 2009; Packwood, 2008). First, the merit pay is only as valid as the performance ratings on which it is based, and both workers and managers often perceive the ratings as problematic. For example, the impact of merit pay on the volume of production may be larger because it is easier to measure volume than quality in most industries. Second, the amounts available for pay raises may fluctuate based on economic conditions unrelated to an employee's performance, so good performance may sometimes result only in small rewards. Third, organized groups of workers, such as unions, may resist attempts to institute individual rewards for individual performance that may undermine group cohesion. Fourth, individual rewards provide disincentives for cooperation and collaboration among employees. Fifth, both employees and managers express frustration about the time and effort required for the performance review process, which often fails to achieve genuine pay for performance. Similar concerns have also emerged in regard to health-sector P4P efforts.

Rynes and colleagues (2005) reviewed the management literature and found little evidence regarding the impact of merit pay systems, which they found surprising in the context of their widespread use as P4P programs in the business sector. Available studies they reviewed showed mixed impacts of merit pay, some positive and some null. They noted that the difficulties of clearly linking pay to performance and challenges in developing credible measures of performance impeded rigorous research on this topic. Jenkins and colleagues (1998) conducted a meta-analysis of 39 studies to examine the quantitative impact of business-sector P4P programs. They found a positive relationship between financial incentives and performance quantity, but no relationship with performance quality. Packwood (2008) found that no available studies provide conclusive proof of positive impacts of variable pay plans on business results.

In sum, although economic incentives are important, they may not be sufficient alone to ensure that P4P programs are effective, in either the health sector or the business sector. Policy makers must also consider additional factors and incorporate them into the design of health care P4P programs.

Sociology

Medical education provides one of the most intensive technical training and professional socialization processes of any occupation (Town et al., 2004). The technical training is long, including 4 years of medical school and 3 or more years of residency. The training is also rigorous: extensive memorization of anatomy and physiology; detailed practice in analytical reasoning for diagnosis and treatment; extensive review of the range of available diagnostic tests, therapeutic procedures, and pharmaceutical treatments; detailed practice in the use of technologies; and training for the emotional detachment and confidence needed to conduct often painful and invasive procedures on patients. The socialization that accompanies this technical training in medicine has several common features:

- commitment to taking strong personal responsibility for patients;
- high degree of dependability when working in medical teams;
- confidence in knowledge and skills as a medical professional;
- commitment to patient care decisions based on scientific judgment when possible, but under uncertainty when necessary;

- emotional detachment from processes and outcomes;
- strong peer orientation toward physician colleagues;
- rigid lines of authority and decision hierarchies; and
- commitment to long, hard hours of work in a high-technology and high-risk environment.

Medical training teaches physicians to take personal responsibility for their patients and to be highly dependable. In the operating room and at the bedside, physicians must exude confidence in their ability to diagnose and recommend when and how to treat. Whatever doubts they may have must be quickly cleared up (e.g., with another test) or sublimated when interacting with patients and families. Because physicians may make life and death decisions, medical training teaches them the limits of their knowledge and the truism that some patients simply respond differently from everyone else to treatment. They often seek out specialized expertise from their physician colleagues who may be able to help avert mistakes and who understand these issues as few others do. At the same time, physicians learn that they often need to proceed with a treatment in situations of clinical uncertainty, which occur much more frequently than the general public realizes. This leads to an emotional detachment from their patients that is necessary in order to be able to return to work the next hour or the next day after an experience of failure (Kirk, 2007).

Since the 1920s, medicine has met all of the sociological characteristics of a profession, in being a service occupation supported by prolonged training and specialized knowledge that determines its own standards of education and training. It successfully recruits the best and the brightest students, controls its own licensing boards, influences legislation to advance its own interests, and, at least historically, has remained mostly free of formal lay evaluation and control (Cockerham, 2007). Ultimately, clinicians become different from most other people in ways that are key to understanding how best to reward them (or not) for their services under P4P.

In their training, physicians become accustomed to hierarchical arrangements as they move from student to resident to attending physician. In addition, given the downside risks from incompetence, merit and scientific qualifications necessarily play a prominent role in career progression. Consequently, physicians often have greater difficulty than nonphysicians in accepting direction from those with less training in their field (e.g., health insurance company staff sending them P4P quality performance reports with highlighted areas for improvement or hospital business managers pressuring them to change practice patterns to reduce costs). They will not "suffer fools gladly" if a P4P approach is inconsistent with their perception of what constitutes a necessary and effective course of care.

At the same time, in recent years the cumulative effect of written guidelines, second opinion requirements, documentation requirements, and regulatory intrusions into their practice has touched off a process in medicine that sociologists term "deprofessionalization" (Cockerham, 2007). Medical work, no longer the sole purview of physicians, is now under greater scrutiny by patients, health care provider organizations, health insurance organizations, business corporations, and government agencies. Health care purchasers want to know more about what exactly they are getting for their money. Ironically, medicine's technical capability to diagnose and treat diseases has steadily been increasing during this time, over the past several decades, just as the medical profession's autonomy has been diminishing.

Studies have found that physicians often have difficulty living up to the public tenets of medical professionalism, and this has eroded their public support. Core tenets such as always providing the highest quality care for patients, putting patients' interests ahead of the physician's own career or financial interests, and commitment to science, are ideals—but hard to fulfill in the realities of practice with heavy workloads and uncertain reimbursement (Wynia, 2009). For example, physicians are often unwilling to criticize one another in public for fear of reprisals and in recognition of common interests (Cockerham, 2007). In a physician survey of attitudes and behaviors toward professionalism, Campbell et al. (2007) found that

- 85 percent believed that physicians should disclose all medical errors to affected patients,
- 77 percent believed physicians should undergo periodic recertification,
- 46 percent had personal knowledge of one or more serious medical errors and did not report them to the hospital or other relevant authorities in every case,
- 45 percent had encountered impaired or incompetent colleagues and had not reported them,
- 36 percent would order an unneeded magnetic resonance imaging (MRI) scan for low back pain if a patient requested it,
- 31 percent were not accepting uninsured patients who were unable to pay, and

- 24 percent would refer patients to an imaging facility in which they had an investment and would not inform the patient of that investment.

These results indicate that the ethical and professional standards highlighted, and perhaps idealized, during a physician's professional training have been difficult to sustain.

P4P can cut two ways in response to physicians' concerns about deprofessionalization. If external government agencies or insurance organizations impose P4P, physicians may perceive the move as contributing to deprofessionalization. On the other hand, if physician groups themselves organize P4P programs, then this approach could reinforce physicians' leadership in quality of care measurement. It could also provide additional payment for services that often go unreimbursed under FFS, such as case management and patient and family education, thereby helping physicians to improve quality of care (Wynia, 2009). In these ways, the influence of concerns regarding professionalism on physicians' responses to the economic incentives of P4P could be either positive or negative, and they could enhance or inhibit the impact of P4P.

Psychology

We can also apply psychological theories and concepts to understand physician behavior for analysis and design of P4P programs. Herzberg's two-factor theory postulates two types of factors that affect workers' motivation in many industries and organizational settings: (1) motivators that encourage productive work and (2) dissatisfiers (Herzberg, 1966; Shortell & Kaluzny, 2006). Golden and Sloan (2008) similarly categorized motivators as extrinsic and intrinsic. Table 3-1 includes examples of extrinsic and intrinsic motivators for physicians.

Table 3-1. Extrinsic and intrinsic work motivators for physicians

Extrinsic Motivators	Intrinsic Motivators
Money, fringe benefits, perquisites (discretionary fringe benefits)	Accomplishment of difficult tasks, correct diagnoses, effective treatments
Workload, working conditions	Learning new skills
Avoiding paperwork, bureaucracy	Link between effort and successful outcomes
Extent and nature of job hierarchy	Autonomy, flexibility
Recognition, status	Collegial relationships with peers
Patients' appreciation	Contributing to the community and the profession

Motivators external to the person include pay, fringe benefits, vacation time, large offices with windows, reserved parking spaces, and first-class travel. Job conditions such as burdensome workloads and poor working conditions are external dissatisfiers that discourage productivity. Most people rebel against paperwork that takes time away from accomplishing tasks and against illogical bureaucracy that frustrates performance, autonomy, and flexibility. By contrast, professionals generally accept a supervisory hierarchy in the workplace if it is based on objective criteria (e.g., competence, experience, education). Most people appreciate external recognition or praise by their supervisors, peers, and clients, especially if it leads to enhanced status, higher pay, and more control over decisions affecting their work and performance.

Intrinsic, self-motivating factors include a person's satisfaction in accomplishing a challenging task for its own sake and the satisfaction derived from learning new skills or knowledge. The closer one's own effort can link to success, the more internal motivation workers may have to make the extra effort. Most people prefer more control over their work environment and support staff, which is closely associated with power over production activities. Most professionals prefer a collegial work environment, interacting with peers in solving problems. The following discussion reviews the ways in which physicians often react to extrinsic and intrinsic motivators.

Money is one of the main motivators for most people. When physicians rank their priorities, money is in the top five, although not always number one (Shortell & Kaluzny, 2006). Physicians make an enormous investment of time and money in their training, and they usually view this as requiring financial returns from high salaries or private-practice income. Increasingly, this encourages medical students and residents to pursue training in higher-paying medical and surgical specialties. As a result, we can expect the economic incentives of P4P to have a significant influence on physician behavior that may encourage improvements in quality of care (if other factors also support that goal).

Heavy workloads and time pressures, however, can negatively affect physicians' ability and willingness to adhere to clinical guidelines and quality measures based on those guidelines (Mechanic, 2008). Long lists of guidelines for good medical practice, each reasonable on its own, often overwhelm physicians. Primary care physicians often view patient visit times as being unduly shortened and expected patient workloads as too high; they increasingly experience high levels of stress and burnout. As a result, their willingness to respond to quality and cost-control measures included in P4P programs can sometimes be limited.

Most physicians value recognition and praise from their peers and patients. The profession places much emphasis on local community and national recognition that comes through research publications, conference presentations, medical professional society awards, government testimony, and the media. Recognition can result from developing novel clinical procedures, conducting groundbreaking research studies, spearheading new quality improvement innovations, or leading health policy making efforts. P4P programs that include recognition for quality-improvement accomplishments will likely achieve better support from physicians.

Intrinsic rewards are another powerful motivator for the medical profession. Physicians train intensively to perform complex tasks that require them to marshal other doctors, nurses, technicians, drugs, and devices in the care of both routine and potentially life-threatening problems. Completing these tasks successfully, caring for patients often over many years, and sometimes saving lives or curing diseases, provides psychological rewards unmatched in most other occupations.

Wynia (2009) reviewed evidence that indicates financial incentives can damage intrinsic motivation. He noted that the work of physicians, with its cognitive sophistication, open-ended thinking, and professional ethos, is exactly the type for which financial rewards may have negative impacts on intrinsic motivation. He warned that P4P could have unintended negative effects on quality (contrary to the economic perspective, which holds that explicit payment should improve quality) if not carefully designed to avoid this pitfall. For example, P4P programs may have fewer negative impacts on intrinsic motivation if (1) rewards focus on the group or team level instead of the individual physician, (2) physicians are able to retain a sense of professional control through designing the ways certain types of atypical patients can be excluded from quality measurement, and (3) physicians are involved in the efforts toward developing the quality measures themselves.

Physicians highly prize the acquisition of new skills in a rapidly changing technological environment. For the primary care and medical specialist, the choice of new drugs provides increasing challenges and rewards. For the surgical specialist, endoscopic, robotic, and minimally invasive procedures offer similar challenges and rewards. Rapid change in medical technologies brings with it rapid skill obsolescence, however. Maintaining competence is complicated by the need to keep abreast of the rapidly growing body of medical research. The number of journal articles reporting on randomized clinical trials alone reached 30,000 in 2005 (Mechanic, 2008). Risk of mistakes and professional embarrassment or failure rises with the rate of skill obsolescence, undermining physician confidence and adding to the overall time pressures of the medical profession. P4P can support acquisition of new skills and use of new technologies by updating quality measures frequently to incorporate new clinical guidelines and new types of treatments.

Organization Theory

Economic agency theory focuses on the simple example of an individual physician as the agent treating a single patient as the principal. However, the individual physician may not only be an agent for the patient, but also a principal for his or her physician group. The physician group, in turn, may be negotiating fees with health insurance organizations as an agent on behalf of all physicians in the group as principals. The multidisciplinary teams of primary care physicians, specialist physicians, surgeons, nurses, technicians, and other health care professionals that are usually needed to provide health care further complicate the principal-agent relationships.

Because P4P programs commonly apply to provider organizations such as physician groups, hospitals, or integrated delivery systems (IDSs)—and not to individual physicians—we can expect organizational structures, processes, and cultures to affect the impact of P4P in both positive and negative ways. Indeed, organizational theorists often view improving quality of care as an organizational problem (Kimberly & Minvielle, 2003). Four strands of organization theory can shed light on potential P4P program impacts: (1) ownership, (2) institutional layers, (3) cultures, and (4) change management and quality improvement.

Ownership. Economic studies of payment effects on organizations often assume that the affected individuals are employees or owners but not both (Town et al., 2004). However, physician group practices are better characterized as worker-owned firms (Robinson, 2001). Hospitals and integrated delivery systems are often nonprofit organizations, with employees and oversight from community-based boards of directors, but not owners who have a claim on profits. Salaried physicians employed in large provider organizations and sole proprietorship in solo physician practices represent two ends of a spectrum of organizational complexity. In practice, clinicians experience a wide array of middle-ground ownership approaches; one commonly found in physician groups bases physician compensation on a mix of salary and productivity standards based on relative value units such as weighted numbers of visits provided per month. Notably, this approach can accommodate P4P fairly easily by adding either groupwide or individual physician quality-of-care measures to the productivity measures for determining physician compensation.

Ownership can include partnerships, stock options, and numerous other arrangements that tie pay to financial performance in varying ways. Because physician-owners share in the financial returns from capital investments in buildings and equipment, they naturally respond to payment systems in ways different from physicians who are strictly on salary, with no vested interest in recommending more tests, procedures, or hospital admissions. Benefits of worker-ownership include an increased willingness to take risks that may translate into greater clinical and organizational elasticity in response to P4P incentives. A downside of worker-ownership can be an excessive focus on maximizing revenue.

Institutional layers. Health care is unusual in that lower levels of institutions are often not completely part of higher ones. In this situation, we can view health care organizations as an "incompletely contained hierarchical nest" (Town et al., 2004, p. 104S). Patients often see more than one physician. Physicians, in turn, often work in more than one clinical group or department. Physician groups usually contract with multiple health insurance organizations. A practicing physician can work and interact with at least five different organizational layers: (1) other physicians, (2) multispecialty groups, (3) multigroup provider organizations (e.g., independent practice associations, physician hospital organizations, IDSs), (4) multiple health insurance plans, and (5) varying consumer health plan choices within insurance plans (e.g., health maintenance organizations, preferred provider organizations, point of service plans) (Landon et al., 1998).

Moreover, each of these five layers may implement programs or systems aimed at influencing medical practice and health care quality in different ways, such as selecting or profiling physicians, promoting or discouraging particular types of services, implementing incentives though P4P, and implementing constraints through utilization review or limited investment in medical technologies. All of the influence strategies need to be aligned with P4P programs if P4P incentives are to be effective. If the other strategies are working at cross purposes, then the impact of P4P will likely be blunted. A case in point might be conducting a stringent review of "unnecessary" services (such as preventable hospital admissions) and making some P4P bonus payments based on that measure, on the one hand, while at the same time paying most of physicians' compensation according to their revenue productivity in terms of FFS billings or relative value units, on the other hand.

Still unclear is how physician groups respond to multiple, sometimes conflicting, payment arrangements that can range from FFS to capitation. Physicians in a group may see some patients with health insurance plans that reimburse using FFS (so higher utilization of care means higher reimbursement for the physician group), and then see other patients, even on the same day, with insurance plans that are capitated (so higher utilization means lower profit margins for the physician group, because reimbursement is fixed in advance and higher utilization means higher costs). Physicians in a group may treat patients differently depending on insurance coverage, or physicians may be blinded to the varying financial incentives. P4P incentives can add to that mix of broader payment incentives, but the overall impact of P4P may be hard to predict in the context of this already complex mix of incentives that often have much larger financial impacts on the group or the individual physician than those included in P4P programs.

Organizational culture. Physician groups and other health care organizations vary widely in their cultures. Some emphasize cooperation among physicians and other staff and free flow of information, whereas others emphasize competition among physicians, which can result in hoarding of information (Town et al., 2004). One study found collegiality, innovativeness, and autonomy to be negatively related to quality of care, whereas organizational trust/identity and emphasis on information flow were positively associated with quality (Smalarz, 2006). Many so-called integrated provider organizations exhibit multiculturalism by combining under one corporate umbrella different medical professions, divisions, departments, and teams that compete with one another more than they cooperate (Ferlie & Shortell, 2001).

A clash of cultures is often even more pronounced between physicians and health care managers (Shortell & Kaluzny, 2006). Physician culture is based on socialization from medical school, biological cause-effect relationships, short time frames for action, and responsibility and autonomy in caring for one's own patients. Managerial culture, by contrast, is grounded in the social sciences and business schools, and emphasizes less-clear-cut cause-effect relationships, longer time horizons, population averages, teamwork, and financial performance. Physicians sometimes resist managers' efforts to standardize clinical practices to improve organizational performance on quality measures included in P4P programs. Alternatively, physicians may be more inclined to support efforts to develop clinical guidelines and quality measures spurred by medical professional societies and termed "evidence-based medicine."

Change management and quality improvement. The ability of an organization to implement changes in medical care practices can also influence its ability to improve quality of care. The organization literature in health care identifies six main characteristics associated with organizational change in health care: (1) leadership (commitment to both quality and efficiency for financial success); (2) a culture of learning (willingness to acknowledge and correct mistakes and utilize evidence-based care); (3) working in teams across professions and clinical and functional departments; (4) effectively using health information technology; (5) care coordination across sites and services; and (6) patient-centered medicine (involving patients as active managers of their own care) (Institute of Medicine [IOM] Board on Health Care Services, 2001; Christianson et al., 2006; Ferlie & Shortell, 2001; Grol et al., 2007; Klein & Sorra, 1996; Lukas et al., 2007; Town et al., 2004; Wang et al., 2006).

The Institute of Medicine (IOM Board on Health Care Services, 2001) has identified four stages of development that health care organizations need to move through to achieve high-quality care. These stages, presented in Table 3-2, also reflect the six characteristics associated with organizational change identified above. We can identify many health care organizations operating at Stage 2 or 3 already; few have achieved Stage 4. From this perspective, most health care organizations need to implement additional organizational changes to move to Stage 4 to achieve the highest quality of care possible.

Stage 4 organizations may be more responsive to P4P and better able to benefit from its incentives. However, if they have already achieved high levels of teamwork, patient involvement, and integration of information technologies, they also may not need external P4P programs to improve quality as much as other providers do. As a result, provider organizations that are actively working to move across these stages of development may actually show the largest measured impact of P4P programs on quality if the financial incentives help to facilitate the organization's advancement to a higher stage.

One of the lessons learned from total quality management programs is that quality improvement is hard to accomplish when financial incentives are

Table 3-2. Four stages of organizational development in health care

Stage	Description
1. Traditional private practice	• Fragmented delivery system
	• Physicians work independently; rely on journals, conferences, and peers to stay current
	• Information technology absent in most settings
	• Minimal use of allied health personnel
	• Passive patients
2. Limited coordination of care	• Well-defined referral networks
	• Continued specialty-oriented care
	• Limited evidence-based practice
	• Minimal information technology
	• Increased patient information and informal involvement in care
3. Team-based care	• Team-based clinical care common
	• Some use of nonphysician clinicians
	• Evidence-based guidelines applied in some practices
	• Information technology broadly applied, but most applications are stand-alone
	• Formal recognition of patient preferences
4. High-performing health care organizations	• Highly coordinated care—across provider groups and settings of care—over time
	• Evidence-based practice the norm
	• Sophisticated information technology linking all systems and groups; automated decision support
	• Extensive clinical measurement and performance feedback to clinicians; continuous quality improvement
	• Extensive training and use of nonphysician clinicians
	• Patients actively involved in treatment decisions

Source: Adapted from IOM Board on Health Care Services, 2001.

not aligned to reward quality improvements at the systems level (Kimberly & Minvielle, 2003). Physicians and hospital administrators commonly complain that FFS incentives in the prevailing health care reimbursement systems reward quantity, not quality. As a result, when financial pressures on institutions are high, they may focus more on quantity and billings at the expense of quality. A widespread concern among management and financial staff at hospitals and physician groups has been the lack of evidence to support the business case for quality improvement efforts (Reiter et al., 2007). P4P programs can help to address that concern by linking reimbursement directly to quality measures and ensuring that the financial benefits from quality improvement efforts accrue to the organization that provided the investments required to implement them. Total quality management initiatives may be unsustainable without positive, systemwide financial incentives for improving quality.

Contingency Theory: A Multidisciplinary Perspective on P4P

As the preceding section indicates, developing a theoretical model of P4P requires a breadth of multidisciplinary perspectives: economic, organizational, psychological, and sociological. All of these perspectives include factors that can enhance or impede the intended impact of P4P programs. These perspectives must be accounted for in considering the range and complexity of policy, institutional, and technological factors at work in the health sector. As a result, P4P theories are likely to remain contingent, applicable under certain prescribed conditions but subject to reconsideration as factors from one or more of the disciplinary perspectives are modified. These theories will still be useful as long as policy makers understand that they apply to particular sets of institutional circumstances and that they can generalize to new circumstances only cautiously.

This type of theoretical situation is well known in management theory, in which "contingency theory" is one of the mainstream viewpoints (Shortell & Kaluzny, 2006). The central idea of contingency theory in management is that organizations and their subunits should develop structures, staff, cultures, and systems differently depending on the specific environments and technologies with which they are involved. Given that health care organizations operate in a very wide variety of environments and institutional relationships, and apply a broad range of different technologies, the contingency perspective has strong applicability (Shortell & Kaluzny, 2006). For example, quality improvement initiatives and P4P programs might well be organized differently depending on the local, state, and national policy environment each organization faces, the nature of the diseases and patients being treated, the types of physician and employee skills available, the internal organizational culture, the degree of teamwork among physicians and nonphysician health care professionals, and the extent of available health information technology.

However, this means that it will not be possible to develop a mathematical theory of P4P. Any mathematical theory that attempted to be comprehensive, accounting for all of these complexities of real-world policy environments, institutional arrangements, and health care organizations, would be analytically intractable (Escarce, 2004). Conversely, efforts to provide for analytical tractability could be successful only by a degree of simplification that would compromise the value of a mathematical theory in making testable predictions.

Nonetheless, the multidisciplinary model points to particular factors that policy makers can use to enhance the impact of P4P programs. Policy makers can consider these insights in the contingency theory perspective and apply them where the combination of policy, technological, and institutional circumstances indicate they are likely to be beneficial for P4P programs. The rest of this chapter describes three examples of these types of multidisciplinary perspectives: (1) reinforcing medical professionalism, (2) patient-centered teams and bundled payment, and (3) centers of excellence (CoEs).

Reinforcing Medical Professionalism

P4P can help physicians to regain some of the benefits of medical professionalism and the related intrinsic motivation in several ways. For example, P4P revenues can support medical practice innovations to contribute to physician satisfaction (Mechanic, 2008; Trisolini et al., 2008). Additional P4P funding may enable physicians to have more time to establish stronger partnerships with patients, promote competent practices based on best available evidence, improve chronic care management, and improve patient satisfaction (Mechanic, 2008). Similarly, cognitive services provided by primary care physicians suffer financially by being more tightly linked to time with patients, a factor often down-weighted in physician fee schedules in comparison with medical and surgical procedures. Many advocates of doctor-patient partnerships believe that primary visits lasting about 30 minutes are often needed, but this is a pattern of care that insurers are unlikely to reimburse adequately (Mechanic, 2008).

In this situation, health insurers can use P4P to supplement reimbursement to primary care physicians by focusing on primary care–oriented quality measures as the basis for P4P bonus payments. Longer patient encounters, often involving nonphysician clinicians, are more financially viable when extra P4P reimbursement will come from quality-of-care improvements achieved through those new patterns of care. Hence, P4P can open up other ways of practicing that may enable primary care physicians to escape the visit-centric emphasis of ambulatory care that is often their only way to gain adequate

FFS reimbursement (Trisolini et al., 2008). The economic incentives of P4P can reinforce both the sociological perspective on professionalism and the psychological perspective on intrinsic motivation that many physicians deem important. This will enable P4P programs to have improved opportunities for significant impacts on quality-of-care outcomes.

Patient-Centered Teams and Bundled Payment

Most P4P programs have opted to focus financial incentives for quality improvement not on individual physicians, but rather on higher levels of the health care system, such as multispecialty physician groups or hospitals. This approach recognizes the teamwork orientation of modern medical care organizations within the incentive system, providing incentives for collaboration among clinicians and recognizing better coordination of care. It is also consistent with Wynia's (2009) emphasis on focusing P4P on team or group rewards rather than individual physician rewards, to avoid or mitigate damage that financial incentives may do to intrinsic motivation. In addition, P4P programs could be targeted to lower organizational levels, such as a diabetes disease management program that requires teamwork among endocrinologists, primary care physicians, nurses, and diabetes educators.

P4P payment for episodes of care also make possible broader, cross-institutional teams. Episodes, which may last 30 days or more beyond a hospital discharge, allow bundling of P4P reimbursement across a range of providers, such as hospitals, physicians, and post–acute care providers. The opportunity to earn P4P revenue can enhance the integration of all of these different types of health care teams and reduce the risks of promoting competition and fragmentation of care if P4P focuses on the individual physician level.

Centers of Excellence

An alternative P4P approach, CoEs can also recognize and financially reward tightly integrated, high-performing, clinical care organizations. Physician-hospital or ambulatory primary care groups could receive a CoE imprimatur after a thorough examination of their quality-of-care performance. This approach has the advantage of more explicitly recognizing an organization's holistic performance, and P4P linked to CoE can provide incentives for organizational change toward higher stages of organizational development, described in Table 3-2. The CoE imprimatur could also enhance physicians' and other clinicians' reputations on the regional or national stage; this positive effect could complement the financial rewards that P4P programs provide and increase their impact.

In sum, theoretical perspectives from several different disciplines can aid in the design of P4P programs by identifying factors likely to enhance or inhibit the effects of P4P. A multidisciplinary or "composite" perspective from contingency theory will enable the design of P4P programs to better respond to the range of factors that may affect their success. This approach will enable P4P to move beyond the simpler theory underlying most early P4P programs, which focused on economic perspectives, and enable P4P to improve its impact on health care quality and cost outcomes.

References

Akerlof, G. (1970). The market for lemons: Quality uncertainty and the market mechanism. *Quarterly Journal of Economics, 84*, 488–500.

Arrow, K. J. (1963). Uncertainty and the welfare economics of medical care. *American Economic Review*, 53(5), 941–973.

Campbell, E. G., Regan, S., Gruen, R. L., Ferris, T. G., Rao, S. R., Cleary, P. D., et al. (2007). Professionalism in medicine: Results of a national survey of physicians. *Annals of Internal Medicine, 147*(11), 795–802.

Christianson, J. B., Knutson, D. J., & Mazze, R. S. (2006). Physician pay-for-performance. Implementation and research issues. *Journal of General Internal Medicine, 21*(Suppl 2), S9–S13.

Cockerham, W. (2007). *Medical sociology* (10th ed.). Upper Saddle River, NJ: Prentice Hall.

Eisenhardt, K. (1989). Agency theory: An assessment and review. *Academy of Management Review, 14*(1), 57–74.

Escarce, J. J. (2004). Assessing the influence of incentives on physician and medical groups: A comment. *Medical Care Research and Review, 61*(3 Suppl), 119S–123S.

Ferlie, E. B., & Shortell, S. M. (2001). Improving the quality of health care in the United Kingdom and the United States: A framework for change. *Milbank Quarterly, 79*(2), 281–315.

Golden, B., & Sloan, F. (2008). Physician pay for performance: Alternative perspectives. In F. Sloan & H. Kasper (Eds.), *Incentives and choice in health care* (pp. 289–318). Cambridge, MA: MIT Press.

Gray, B. H. (2004). Individual incentives to fix organizational problems? *Medical Care Research and Review, 61*(3 Suppl), 76S–79S.

Grol, R. P., Bosch, M. C., Hulscher, M. E., Eccles, M. P., & Wensing, M. (2007). Planning and studying improvement in patient care: The use of theoretical perspectives. *Milbank Quarterly, 85*(1), 93–138.

Herzberg, F. (1966). *Work and the nature of man.* Cleveland: World Publishing.

Institute of Medicine, Board on Health Care Services. (2001). *Crossing the quality chasm: A new health system for the 21st century.* Washington, DC: National Academies Press.

Jenkins, G. D. J., Mitra, A., Gupta, N., & Shaw, J. (1998). Are financial incentives related to performance? A meta-analytic review of empirical research. *Journal of Applied Psychology, 83*(5), 777–787.

Kimberly, J., & Minvielle, E. (2003). Quality as an organizational problem. In S. Mick & M. Wyttenbach (Eds.), *Advances in health care organization theory* (pp. 205–232). San Francisco: Jossey-Bass.

Kirk, L. M. (2007). Professionalism in medicine: Definitions and considerations for teaching. *Baylor University Medical Center Proceedings, 20*(1), 13–16.

Klein, K., & Sorra, J. (1996). The challenge of innovation implementation. *Academy of Management Review, 21*, 1055–1080.

Landon, B. E., Wilson, I. B., & Cleary, P. D. (1998). A conceptual model of the effects of health care organizations on the quality of medical care. *JAMA, 279*(17), 1377–1382.

Lukas, C. V., Holmes, S. K., Cohen, A. B., Restuccia, J., Cramer, I. E., Shwartz, M., et al. (2007). Transformational change in health care systems: An organizational model. *Health Care Management Review, 32*(4), 309–320.

Mechanic, D. (2008). Rethinking medical professionalism: The role of information technology and practice innovations. *Milbank Quarterly, 86*(2), 327–358.

Nicholson, S., Pauly, M. V., Wu, A. Y., Murray, J. F., Teutsch, S. M., & Berger, M. L. (2008). Getting real performance out of pay-for-performance. *Milbank Quarterly, 86*(3), 435–457.

Packwood, E. (2008). Using variable pay programs to support organizational goals. In L. Berger & D. Berger (Eds.), *The compensation handbook* (5th ed., pp. 215–226). New York: McGraw-Hill.

Reiter, K. L., Kilpatrick, K. B., Greene, S. B., Lohr, K. N., & Leatherman, S. (2007). How to develop a business case for quality. *International Journal for Quality in Health Care, 19*(1), 50–55.

Robbins, S., & Judge, T. (2009). *Organizational behavior* (13th ed.). Upper Saddle River, NJ: Pearson Prentice Hall.

Robinson, J. C. (2001). Theory and practice in the design of physician payment incentives. *Milbank Quarterly, 79*(2), 149–177, III.

Rynes, S. L., Gerhart, B., & Parks, L. (2005). Personnel psychology: Performance evaluation and pay for performance. *Annual Review of Psychology, 56*, 571–600.

Shortell, S. M., & Kaluzny, A. (2006). *Health care management: Organization design and behavior* (5th ed.). Clifton Park, NY: Thomson Delmar Learning.

Smalarz, A. (2006). Physician group cultural dimensions and quality performance indicators: Not all is equal. *Health Care Management Review, 31*(3), 179–187.

Town, R., Wholey, D. R., Kralewski, J., & Dowd, B. (2004). Assessing the influence of incentives on physicians and medical groups. *Medical Care Research and Review, 61*(3 Suppl), 80S–118S.

Trisolini, M., Aggarwal, J., Leung, M., Pope, G. C., & Kautter, J. (2008). *The Medicare Physician Group Practice Demonstration: Lessons learned on improving quality and efficiency in health care.* Waltham, MA: RTI International.

Wang, M. C., Hyun, J. K., Harrison, M., Shortell, S. M., & Fraser, I. (2006). Redesigning health systems for quality: Lessons from emerging practices. *Joint Commission Journal on Quality and Patient Safety, 32*(11), 599–611.

Wynia, M. (2009). The risks of rewards in health care: How pay-for-performance could threaten, or bolster, medical professionalism. *Journal of General Internal Medicine, 24*(7), 884-887.

CHAPTER 4

Quality Measures for Pay for Performance

Michael G. Trisolini

Concerns about quality of care have accelerated since the 1990s, as studies by Wennberg, Fisher, and others have documented large and unexplained variations in rates of health care utilization and clinical outcomes across geographic areas, calling into question the traditional approach of relying on the medical profession to deliver high-quality care uniformly (Davis & Guterman, 2007; Wennberg et al., 2002). Since about 2000, several landmark publications have highlighted widespread problems with patient safety and quality of care, most notably from the Institute of Medicine (IOM) and the RAND Corporation (IOM Board on Health Care Services, 2001; Kohn et al., 1999; McGlynn et al., 2003). These studies helped to galvanize a policy consensus, leading the federal government and private health insurance plans to increasingly focus policy, regulatory, and management interventions more directly on quality of care measurement, quality improvement programs, and financial incentives for quality improvement through pay for performance (P4P).

P4P programs have focused primarily on quality of care measures to assess provider performance. Although other performance evaluation approaches, such as efficiency measures, are possible for P4P, those in policy circles currently perceive the lack of incentives for improved quality in the prevailing fee-for-service (FFS) payment systems as a major problem in the US health care system. As a result, P4P programs have focused mainly on addressing this problem.

This chapter reviews issues regarding the application of quality measures in P4P programs. The first section of the chapter provides background, including conceptual frameworks for quality of care, and reviews organizations that develop and certify quality measures. The second section discusses different types of quality measures, including structure, process, and outcome measures (Donabedian, 1966). The third section reviews issues in selecting quality measures for P4P programs. The fourth section describes methods for analyzing quality measures for P4P. The fifth section discusses public reporting of quality measures and how that separate approach to quality improvement can be integrated with P4P programs.

Background

Two major conceptual frameworks have been developed for health care quality, one by Donabedian (1966) and the other by the IOM (IOM Board on Health Care Services, 2001). Researchers and policy makers can use both to guide development and implementation of quality measures for P4P. Other models are available (IOM Board on Health Care Services, 2006), but the Donabedian and IOM frameworks are the most widely used. This section describes both models, although we emphasize Donabedian's framework because developers of P4P programs use it more frequently than the IOM model.

Donabedian's model focuses on the concepts of structure, process, and outcome for defining quality of care. Despite being first published more than 40 years ago, this model remains a leading paradigm. The key elements can be described as follows:

Structure—the inputs into the health care production process. These include physicians, nurses, and other staff; medical equipment; facilities; information technology; administrative support systems; medical supplies; pharmaceuticals; and other resources. Problems may arise if inputs are not available when needed to treat a patient or when health professionals do not view the capabilities of inputs as optimal. For example, from a structure perspective, high-quality care may entail using clinical teams, including board-certified cardiologists, to treat patients with advanced heart failure rather than relying solely on primary care physicians.

Process—the procedures used to diagnose a patient, prescribe a course of testing or treatment, and ensure that the testing and treatment are carried out in accordance with clinical guidelines or norms of medical practice. Process problems are often classified as underuse, misuse, or overuse of tests or treatments (IOM Board on Health Care Services, 2001; Chassin & Galvin, 1998). For example, from a process perspective, high-quality care may be associated with laboratory testing of diabetic patients at least once a year for their levels of glycosylated hemoglobin (HbA1c). To be useful, process measures must have been demonstrated to be statistically and clinically associated with corresponding outcome measures. For example, appropriate colorectal cancer screening is a process measure known to reduce mortality attributable to colon cancer.

Outcome—the ultimate goals of reducing morbidity and mortality and improving quality of life (QOL) and patient satisfaction. Quality analysts can identify problems by comparing outcomes achieved for patients with the outcomes expected for similar patients with the same disease. For example, from an outcome perspective, high-quality care may be associated with a reduced frequency of relapses for patients with multiple sclerosis (MS) because such care reduces the morbidity that patients suffer.

Quality measures focused on structure are easier to measure, but they may have only limited impact on the final outcomes of interest. Process measures assess the actual medical treatment that physicians and other health professionals provide. However, they may require detailed data collection through costly medical record reviews to obtain the clinical data necessary to identify patients for the measures' denominators and clinical events for their numerators. Administrative data such as insurance claims may enable less costly measurement for some types of process measures, but most cannot be measured in this way. Combinations of administrative and chart review data collection ("hybrid" measures) have been encouraged by the National Committee for Quality Assurance (NCQA) as an efficient approach. Electronic medical records (EMRs) may someday reduce the data collection burden for process measures, but they are still not widely implemented. The technical specifications for process measures may also be costly to develop and keep updated because of changes in medical practices and technologies and development of new pharmaceuticals; such changes may cause shifts in the lists of inclusion and exclusion criteria for denominators and numerators used to calculate performance rates. Outcome measures may be ideal in theory, because they represent the ultimate goals of interest, but they are often difficult to measure, especially in a timely fashion. For example, variation in mortality outcomes may appear only many years after patients have received medical treatment. In addition, many factors can affect variation in mortality and other outcomes besides the quality of medical care, for example, age, comorbidities, diet, exercise, and risky behavior. Consequently, physicians and provider organizations, such as hospitals and physician groups, may consider it inappropriate to hold them accountable for quality of care measured by outcomes unless complex risk adjustments are applied. QOL and other patient-reported outcomes are costly to measure because they require primary data collection through direct responses from patients in formal surveys. Patient-reported outcomes data cannot be collected from secondary data, such as insurance claims, that do not include patient surveys. Finally, poor outcomes, which are often rare and therefore more difficult to measure, lead to sample size issues.

As a result, not one of these three categories—structure, process, or outcome—is always better than the others for quality measurement, and P4P programs have applied all of them in practice to assess quality. P4P programs use process measures more frequently than the other types of measures because they represent a middle ground, physicians and other clinicians are more familiar with them, and process measures make clear what must be improved in care processes in comparison with outcome measures. However, process measures also have shortcomings and are often complemented by structure measures, outcome measures, or both.

The IOM presented its conceptual model of factors affecting health care quality in its *Crossing the Quality Chasm* report (IOM Board on Health Care Services, 2001). This report focused on six goals for improving health care. As the report noted, health care should be

- **Safe**—avoiding injuring patients with the care that is intended to help them.
- **Effective**—providing services based on scientific knowledge to all who could benefit and refraining from providing services to those not likely to benefit (avoiding underuse, misuse, and overuse).
- **Patient-centered**—providing care that respects and is responsive to individual patient preferences, needs, and values and ensuring that patient values guide all clinical decisions.
- **Timely**—reducing waits and sometimes harmful delays for both those who receive and those who give care.
- **Efficient**—avoiding waste, in particular waste of equipment, supplies, ideas, and energy.
- **Equitable**—providing care that does not vary in quality because of personal characteristics such as gender, ethnicity, geographic location, and socioeconomic status (pp. 39-40).

The Donabedian and IOM models overlap in many areas. The IOM's goal of safe care relates to all three of Donabedian's concepts. For example, a structure intervention to implement computerized physician order entry

(CPOE) for drug prescriptions may prevent overdoses of chemotherapy drugs, or dangerous interactions between drugs a patient may be taking. A process intervention could also prevent overdoses by requiring multiple nurses to check dosages before they administer drugs. If these interventions are successful, then the beneficial outcomes are reduced rates of morbidity and mortality for patients taking these drugs.

The IOM's aims for effective, patient-centered, and timely care all relate to Donabedian's concept of process. The aim of patient-centered care also relates to outcomes that are measured using patient surveys of QOL, patient satisfaction, or experience of care.

However, the IOM model includes additional concepts of cost and access in its domains of efficient and equitable care. We prefer to maintain the conceptual distinctions between the overall health policy goals of increasing quality, reducing cost, and improving access, which are often used as a larger conceptual framework for analyzing health services. For example, quality is often associated with measuring the performance of clinicians or provider organizations. Access is often associated with measuring the performance of health care systems that may cover regions or an entire country. Cost is usually considered separately from quality, and discussions of cost tend to focus on the analysis of financial resources and budgets used for providing care.

Researchers and policy makers can use the IOM's aims for efficient and equitable care to develop performance measures for P4P programs that are separate from measures of quality. We view those concepts as useful for P4P programs, although we present efficiency measures in a separate chapter of this book (Chapter 5) to maintain the conceptual distinctions between quality, cost, and access goals. Researchers and policy makers can also develop access measures of performance that are separate from quality measures. The access measures may, for example, be included alongside quality measures in P4P programs that focus on vulnerable populations, such as Medicaid enrollees.

Types of Quality Measures

Following our focus on Donabedian's model, we categorize quality measures for P4P programs into structure, process, and outcome. This section describes examples of all three types and illustrates how P4P programs have applied them.

Structure Measures

Health professionals and policy makers sometimes view structure measures as less valuable than process or outcome measures because they are further removed from the ultimate goal of improving outcomes. Structure measures indicate only the potential for providing or improving quality of care; they do not directly measure the clinical processes of care or health outcomes that more closely represent true quality. Also, fewer structure measures are available for ambulatory care than for inpatient care (Birkmeyer et al., 2006), although that situation may change in coming years, with expanded emphasis on implementing systems that support health care delivery, such as EMRs and chronic disease registries.

Some individual health professionals may view structure measures as unfair if the individuals score low on them but have high quality in terms of outcomes (Birkmeyer et al., 2006). Moreover, linkages between structural measures and outcomes may be evident at health system or community levels, but they may not differentiate individual clinicians well.

In recent years, health care accreditation organizations have moved away from their traditional reliance on structure measures to focus more on process measures. For example, the Joint Commission has moved toward using measures of process and outcome (Hurtado et al., 2001). Several quality monitoring organizations have also begun to focus more on process and outcome indicators than they previously did. For example, the NCQA has developed and periodically updated a set of Healthcare Effectiveness Data and Information Set (HEDIS) indicators used to measure quality in private managed health care plans, Medicare, and Medicaid (National Committee for Quality Assurance [NCQA], 2006). Federal quality improvement efforts, including Medicare's Physician Quality Reporting Initiative and the Hospital Quality Initiative, have also focused on process and outcome measures (Centers for Medicare & Medicaid Services [CMS], 2008a, 2008b).

Nonetheless, some structure measures have been found effective in promoting quality. In addition, they are usually easier to measure than process or outcomes, so data collection is both less challenging and less expensive. They may also be efficient in the sense that one structure measure may relate to several different diseases or outcomes (Birkmeyer et al., 2006).

The Leapfrog Group is a proponent of several specific structure measures of quality (Birkmeyer & Dimick, 2004; Leapfrog Group, 2008). Its focus includes three structure measures for hospitals:

- **Computerized physician order entry**—Studies have shown that physicians can significantly reduce prescribing errors when they use CPOE to highlight incorrect dosages, drug interactions, or patients' allergies to prescribed drugs.

- **Intensive care unit (ICU) physician staffing**—ICUs staffed with critical care specialists (sometimes referred to as intensivists) can reduce the risk of patients' dying in the ICU.

- **Evidence-based hospital referral**—For patients needing certain types of complex medical procedures, referral based on scientifically evaluated factors, such as the number of times a hospital has performed a procedure each year, has been shown to reduce the risk of death.

All of these measures can potentially be applied in P4P programs, and the Leapfrog Group has provided assistance to health plans and payers using them in P4P. A recent study confirmed the Leapfrog Group's claims about the value of these structure measures of quality (Jha et al., 2008). It found that hospitals that implemented these three types of patient safety–oriented interventions also had improved process and outcome measures of quality—including lower 30-day mortality rates—for patients with acute myocardial infarction, congestive heart failure, and pneumonia.

Although the Leapfrog Group has received the most attention in policy circles for its focus on structure measures of quality, several payer groups have also used structure measures in their P4P programs. For example, some programs have provided incentives for health care professionals to invest in information technology, implement electronic health records (EHRs), or use EHRs. These programs include the Integrated Healthcare Association, Bridges to Excellence, and the Hawaii Medical Services Association (IOM Board on Health Care Services, 2007; Bridges to Excellence, 2008; Gilmore et al., 2007; McDermott et al., 2006).

Payers and health care plans are particularly interested in the development and use of EHRs or EMRs because they have the potential to improve coordination of care and reduce medical errors. For more than a decade, many commentators have noted this potential, but the high costs of these systems in relation to the benefits received at the physician practice level have hampered implementation. As a result, most small and medium-sized physician practices have been slow to adopt EHRs and EMRs. This means that P4P programs with incentives for EHR or EMR implementation could provide a useful catalyst for improving the business case for these systems at the physician practice level.

Organizational interventions are another type of structure measure. The United Kingdom's (UK's) P4P program applies a substantial number of these interventions to assess performance of family practitioners in the National Health Service (Department of Health, 2004; Doran et al., 2006a, 2006b). The UK P4P system is noteworthy because it has 146 quality measures, the largest number of any P4P program. Of these, 76 are classified as clinical quality indicators, and another 70 are classified as organizational and patient experience quality measures. Our review of these measures reveals that many of them are structure measures of quality. For example, the 76 clinical quality indicators are classified into 11 chronic disease domains. For each domain, the first indicator is a structure measure of whether the practice has a register of the patients with that disease (e.g., "DM 1. The practice can produce a register of all patients with diabetes mellitus"; Department of Health, 2004). Similar indicators are repeated for the other 10 disease domains (e.g., hypertension, chronic obstructive pulmonary disease, and cancer), thus the 76 clinical quality indicators are actually 11 structure measures and 65 clinical process measures.

The UK's 70 organizational and patient experience quality measures include 31 structure measures, according to our review (e.g., "Records 3: The practice has a system for transferring and acting on information about patients seen by other doctors out of hours"; Department of Health, 2004). The wording of this measure uses structure language about the presence of a system, rather than process language about the percentage of patients seen by other doctors who were transferred to new doctors and for whom clinical information was acted on. The other structure measures from among these organizational and patient experience indicators are worded in similar ways, often referring to the presence of a system rather than to how the system should be applied in medical practice.

Therefore, the UK's P4P program uses 42 structure measures, which is 29 percent of the overall total of 146 quality measures. This is a higher percentage of structure measures than most US P4P programs use, although the difference may reflect the much larger overall number of indicators in the UK program compared with US programs. The UK program uses 104 process measures—in absolute numbers, more process measures than are used in any US P4P programs, even if the percentage of process measures is lower in the UK program than in some US programs.

The current interest in "medical homes" in the United States can be viewed as another type of structure intervention for quality of care. Medical homes are sometimes proposed for additional per capita or per visit bonus payments to physicians because they are expected to improve coordination of care, case management, information technology, and continuity of primary care. As a result, medical homes could be a focus of implementation incentives in P4P programs as structure measures of quality, similar to P4P programs that provide incentives for implementing EMRs.

Process Measures

Process measures are procedures or treatments that are designed to improve health status or prevent future complications or comorbidities. In most cases, a process measure is a dichotomous indicator of whether the process was performed during the recent past (e.g., whether patients taking interferon drugs had liver function tests in the past 6 months). When characterizing health professional and provider organization performance, a process measure is expressed as the proportion of eligible patients who received the procedure. Process measures are often limited to certain subgroups of patients for whom a particular treatment process applies.

A benefit of process measures is that health professionals recognize them as reflecting routine clinical care. In many cases research studies have found them to be associated with outcomes, although this is not always well established and it is becoming increasingly less acceptable to use process measures that lack an evidence base. Process measures may also provide positive spillover effects, such as raising clinicians' awareness about quality measures and clinical guidelines (Birkmeyer et al., 2006).

Process measures are important to consider because they are usually more practical for data collection and monitoring than outcomes are for quality improvement programs (IOM Board on Health Care Services, 2006; Eddy, 1998; Jencks et al., 2000). Four characteristics of process measures make them more feasible than outcome indicators for routine quality monitoring. First, outcomes often occur with lower frequency than do associated process indicators. For example, breast cancer deaths occur at a rate of only about 1 per 1,000 women older than 50 years of age (an outcome indicator). In contrast, NCQA and Medicare apply process indicators specifying that all women ages 50 to 69 should be receiving biennial mammograms for breast cancer screening (Kautter et al., 2007; NCQA, 2006).

Second, outcomes often require long periods for evaluation of effects (Palmer, 1997). For example, to get outcomes measured as 5- to 10-year cancer survival rates, it will take at least 5 to 10 years and probably longer because of data reporting lags. Routine evaluation of process indicators can usually be done annually or even more frequently, depending on how many patients with a particular disease physicians treat in any given month or year.

Third, factors outside the control of health plans, health care organizations, or clinicians who treat patients with chronic diseases often affect outcomes. In contrast, process of care measures are, by definition, primarily under health professionals' control and usually do not require risk adjustment.

Fourth, significant improvements in processes are generally larger in relative terms than improvements in outcomes, which makes it easier to measure the former and easier to identify significant changes. This aspect enables P4P programs to base incentive payments on more statistically reliable data.

Process measures have another appealing aspect. One of the key steps in quality improvement is identifying the cause of problems and improving the associated care processes. Unlike outcome measures, process measures target which area of care needs to be improved, although the health care organization still needs to ascertain how to achieve the improvement needed.

Nonetheless, just because process measures are usually easier to specify, measure, and track from year to year does not mean that P4P programs should use them exclusively. An important consideration with process measures is whether they are clearly linked to improved outcomes or at least to a higher likelihood of improved outcomes. Researchers have developed a range of methods to assess the strength of scientific evidence that underlies clinical practice guidelines, quality measures, and quality improvement programs (Lohr, 2004). However, the extent of currently available evidence to support links between process indicators and outcomes varies widely (Birkmeyer et al., 2006). Process measures recommending routine laboratory testing may be good clinical practice, but the results of testing, and the degree of follow-up that health professionals provide, are more closely linked to outcomes than to whether testing was done.

Outcome Measures

Ultimately, people care most about outcomes, including morbidity, mortality, QOL, functioning, and patient satisfaction. Improved outcomes are the desired consequences of quality improvement efforts. For example, for treatment of MS, outcomes may be measured through physical and mental functional status indicators (which can be either physician reported or patient reported), disability, complication rates (e.g., urinary tract infections, pressure ulcers), frequency of relapses, standardized measures of health-related QOL, standardized measures of patient satisfaction, and other indicators.

In addition, implementing a system of outcome measurement may itself improve outcomes—a "Hawthorne effect"—beyond the interventions that may be related to particular outcomes (Birkmeyer et al., 2006). For example, surgical morbidity and mortality in VA hospitals fell dramatically after measurement began in 1991, to an extent too large to explain solely by organizational or process improvements.

To date, P4P programs have used outcome measures less frequently than process measures, even though outcome measures are preferable in theory because they represent the ultimate health care goals. As noted, researchers have raised concerns regarding the strength of the relationship between structure or process quality measures and the outcomes they target. In addition, focusing on outcome measures is expected to encourage innovation in health care services more than would focusing on process measures (Sorbero et al., 2006). Unless process measures are updated frequently, which could be costly, they may reinforce existing care patterns rather than encourage development of new treatment methods that improve outcomes even more.

Thus, moving P4P programs toward more direct use of outcome measures where possible may be beneficial. Physicians and other clinicians may want to maintain a mix of process and outcome measures in P4P programs, however, given that process measures provide more specific information about particular care processes that need to be improved.

One concern is that multiple factors outside of the health care system can affect outcome measures, a problem that is commonly cited. As a result, physicians and other clinicians may not consider it fair to be held accountable for outcome performance. For example, many different physicians and other health professionals may treat patients with cardiovascular disease, and patient factors regarding diet, exercise, and adherence to medications may play a large role in mortality rates. Risk adjustment for outcome measures can be expensive if it is done in detail using data from medical records, and it may be inadequate if done using administrative data that, though usually less expensive, contain less clinical detail (Birkmeyer et al., 2006). However, recent efforts to add present on admission (POA) codes in hospital medical records will enable better analysis of outcomes for hospitals (Jordan et al., 2007; Pine et al., 2007). POA codes help to determine whether complications and comorbidities were acquired by patients during a hospital stay, and thus can be attributed to the care provided at the hospital.

Another problem with measuring outcomes is that sample sizes may be small for surgical outcomes or rare diseases. This means that statistical analysis of performance improvement may be unreliable, so P4P programs cannot pay bonuses with confidence in these situations.

An issue with patient-reported outcomes, such as those reflecting QOL or health status, is that they require patient surveys, which may impose costs that provider groups find difficult to sustain. Lower cost options are not readily available because these types of outcomes require primary data collection from patients.

In general, outcomes can be categorized into two types: clinician-reported outcomes and patient-reported outcomes. Both can be applied in P4P programs.

Clinician-Reported Outcomes

Clinician-reported outcomes are those that physicians or other health care professionals measure and record. They can be further classified as "intermediate" outcomes (e.g., blood pressure levels or HbA1c levels that put patients at risk for severe complications or comorbidities, or stage of cancer at diagnosis) and final outcomes (e.g., decubitus ulcers causing morbidity, or mortality). Medical records are primary data sources for collecting clinician-reported outcomes, but P4P programs can also use laboratory databases and claims data for some types of outcome measures.

Intermediate outcomes. Intermediate outcomes measure clinical results, so they can be viewed as outcomes rather than process measures, but they are not final outcomes in the sense of being direct measures of morbidity or mortality. Blood pressure levels are important outcomes that provide information on patients' risks for heart disease and stroke. As a result, control of blood pressure is a goal that makes sense to reward through P4P programs. Similarly, HbA1c levels can indicate risks for diabetics to develop several severe complications, including retinopathy, nephropathy, and neuropathy.

A positive feature of intermediate outcomes is that they are closer than process measures to the final clinical outcomes of interest, so they provide a closer link to final outcomes. For example, HEDIS process measures include measuring blood pressure periodically for patients with heart disease or hypertension, and testing for HbA1c levels periodically for diabetics. However, just because the tests were conducted does not mean that the clinical indicators of interest were brought under control. Thus, focusing on the levels themselves—the intermediate outcomes—is preferable to targeting only the frequency of testing.

Another positive feature of intermediate outcomes is that they can be measured more frequently than final outcomes. As a result, P4P programs that focus on providing routine performance assessments and periodic (often annual) bonus payments to physicians and other health professionals can use them more easily. For example, among patients with diabetes, neuropathy can result in foot or leg infections that require amputations, but these events occur much less frequently than elevated levels of HbA1c. Amputation rates can be tracked as performance measures, but they may require much larger samples of diabetic patients than are available for most physician practices or even larger group practices. Because amputations occur less frequently than HbA1c tests, they may not provide the routine data needed for annual performance assessments for P4P bonus payment determinations.

A third positive feature of intermediate outcomes is that they may not require risk adjustment for appropriate performance assessment, in contrast to mortality and other final outcomes. Appropriate levels of blood pressure and HbA1c are standardized for most patients, and although patient factors enter into the levels achieved, physicians and other clinicians can usually be held accountable for average levels achieved over groups of patients. Physicians may not be able to control patients' diet and exercise patterns completely, but most accept responsibility for working with their chronic disease patients to control blood pressure and HbA1c, especially when patients are at risk for complications associated with elevated blood pressure or HbA1c. Risk adjustment may still be indicated for some types of intermediate outcomes, but it may be implemented more easily than for final outcomes, with fewer variables and data collection requirements.

As a result, a promising approach for P4P programs would be to work more aggressively to expand the range of intermediate outcomes that they use to assess provider performance. They can also increase the weighting provided to these measures relative to others. Intermediate outcomes represent a middle ground between the more controllable process indicators, which may not be closely linked to final outcomes, and the final outcomes of interest, which would be ideal performance measures—if they were easier to measure frequently and if it were easier for providers to link the final outcomes to their efforts.

In addition, Current Procedural Terminology (CPT-II) codes have now been developed for some intermediate outcomes, such as HbA1c levels for diabetics, so these outcomes can now be measured using administrative claims data instead of relying solely on more expensive chart review (American Medical Association, 2008). More work is needed to expand the list of CPT codes for intermediate outcomes, and to expand the extent to which health professionals use them for billing for clinical services, but the technical groundwork has been laid in the CPT coding system.

Final outcomes. Clinician-reported final outcomes can include a range of morbidity, functional status measures, and mortality measures. Morbidity measures include medical and surgical complications that can be used in P4P programs, although they apply primarily to hospitals or other institutional providers. Decubitus ulcers are an example of a preventable complication that can develop during hospital stays or among nursing home residents. Because they are preventable for most patients, they can serve as a useful outcome measure for P4P programs. They usually occur infrequently, however, so they may need to be measured as average rates over large groups of patients.

Other types of hospital-related complications, such as postsurgical infections, readmission rates within 1 to 3 months of discharge, and "never" events, such as surgery on the wrong body part, can also serve as final outcome measures. These outcome measures include patient safety quality indicators that the US Agency for Healthcare Research and Quality (AHRQ) and others have developed (AHRQ, 2003). An advantage of these complication-related outcome measures is that they do not require risk adjustment in most cases, because patient safety indicators such as avoiding postsurgical infections apply to most patients. In addition, these indicators are clearly under the control of hospitals and their medical staff because they occur during the patient's stay in the hospital, nursing home, or other medical facility. As a result, clinicians are more willing to accept responsibility for these types of final outcome measures. For example, when Medicare recently announced that it would not reimburse hospitals for admissions that resulted in "never" events, there was little resistance from the hospital or physician community.

Functional outcomes comprise measures of activities of daily living (ADLs), instrumental activities of daily living (IADLs), time to walk 25 feet, established scales such as the Functional Independence Measure (FIM), and others. Health professionals often use such outcomes in rehabilitation services assessments, to judge patients' progress in recovery from illness, or to assess levels of disability. These measures have promise for P4P because they can be measured frequently and can show significant changes resulting from effective treatment in many situations. Assessments of MS patients routinely use clinician-reported outcome measures of physical and cognitive function, including the Expanded Disability Status Scale, Multiple Sclerosis Functional Composite, neuropsychological tests, and others (Cohen & Rudick, 2007; Coulthard-Morris, 2000; Joy & Johnston, 2001; Rothwell et al., 1997).

However, functional outcomes suffer from at least two concerns. First, they may require risk adjustment like other types of outcomes, because factors unrelated to the quality of medical care can affect them. Second, many functional outcomes rely to some extent on the clinician's judgment for scoring each patient on the measures or scales. This can make the functional outcomes more vulnerable to gaming by health professionals and providers, especially when P4P programs use scores to calculate bonus payments.

Mortality is the ultimate final outcome, although mortality measures can be a sensitive topic for both patients and clinicians. In principle, P4P programs could use mortality rates or risk-adjusted mortality rates for performance assessment. Aligning health professionals' financial interests in keeping the patient alive as long as possible may improve mortality outcomes. However, patients and their families may understandably be concerned if the presence of a P4P program implied that physicians would not be doing all they could to keep patients alive in the absence of P4P financial incentives.

At the same time, researchers have conducted much statistical analysis in recent years to create risk-adjusted mortality rates for several diseases and populations. Quality improvement efforts and public reporting of mortality outcomes have used these rates. For example, risk-adjusted mortality rates have been reported publicly for several years on Medicare's Dialysis Facility Compare Web site (Trisolini & Isenberg, 2007), the State of New York has reported publicly on risk adjusted mortality rates for cardiac surgeons for many years (Jha & Epstein, 2006), and Medicare recently began reporting risk-adjusted mortality rates for some types of patients on its Hospital Compare Web site (CMS, 2009). These measures have been well tested, so they presumably could be extended for use in P4P programs.

Patient-Reported Outcomes

Patient-reported outcomes have the advantage of providing data on outcomes that can be collected only from patients; broadly speaking, such outcomes can include QOL, patient satisfaction, and patient experience of care. Patient satisfaction data are already used in P4P programs, including those sponsored by the Integrated Health Care Association, the Hawaii Medical Services Association, and the British National Health Service (Doran et al., 2006b; Gilmore et al., 2007; McDermott et al., 2006). Health professionals might be expected to object to P4P programs tying financial rewards to subjective indicators such as patient satisfaction, but the success of these three large P4P programs in implementing these patient-reported outcome measures indicates that clinician acceptance is possible.

Standardized patient satisfaction scales for quality measurement and public reporting have become widely accepted in recent years, which has helped to promote their use in P4P programs. AHRQ developed the Consumer Assessment of Healthcare Providers and Systems in the 1990s, and it now includes a family of standardized patient surveys that have broad acceptance for assessment of health plans, hospitals, physician groups, and other provider organizations (AHRQ, 2007, 2009).

Clinical trials of new drugs and evaluations of health service interventions have used QOL scales, such as the SF-36 or SF-12, to monitor outcomes of care, and these types of scales have potential for use in P4P programs. Medicare has also publicly reported QOL scales in recent years through the Health Outcomes Survey (NCQA, 2006). These scales can include broad, generic measures of functioning, such as the Physical Component Summary (PCS) and Mental Component Summary (MCS) scales for the SF-36 or SF-12. They can also include more specific measures of particular symptoms, such as the Modified Fatigue Impact Scale (MFIS; National Multiple Sclerosis Society, 1997a). With a wide range of both general health and symptom-specific QOL scales developed in recent years, P4P programs have many options if they wish to measure QOL performance.

Some QOL scales have been developed for particular diseases, such as the Kidney Disease Quality of Life Scale (Hays et al., 1994). The National Multiple Sclerosis Society developed a multipurpose patient survey instrument for measuring a range of outcomes, the Multiple Sclerosis Quality of Life Inventory (MSQLI; National Multiple Sclerosis Society, 1997a, 1997b). The MSQLI includes the 21-item MFIS and nine other scales that measure outcomes related to generic physical and mental health, pain, sexual satisfaction, bladder control, bowel control, visual impairment, perceived deficits (cognition), and social support. Several other disease-specific QOL measures for MS have also been developed in recent years (Burks & Johnson, 2000; Nortvedt & Riise, 2003).

Patient-reported outcomes can also include other types of functioning scales; some overlap with clinician-reported outcomes such as those assessing ADLs, IADLs, and mobility for rehabilitation programs. For example, for MS two disability scales focus mainly on walking ability: the Extended Disability Status Scale (EDSS), which neurologists assess, and the Patient-Determined Disease Steps (PDDS), which patients can assess. In theory, P4P programs could use either or both measures to assess performance, although in MS, the goal is usually slowing the decline in function rather than improving function.

It is interesting that P4P programs have used patient satisfaction scales to date but not QOL scales. This may stem from health professional and provider organization concerns that factors outside their control can affect QOL and thus would require risk adjustment. For example, one study that used QOL scales to assess the performance of Medicare providers used several demographic and comorbidity variables for risk adjustment (Trisolini et al., 2005). In contrast, patient satisfaction is more under the control of physicians and other health professionals and providers because it largely reflects the patient's experience of receiving care from the clinician. Moreover, private health insurance plans may include patient satisfaction in P4P programs because it helps them attract enrollees into their plans and thus affects their ability to compete against other health insurance companies.

Issues in Selecting Quality Measures for P4P Programs

Data Sources and Administrative Burdens

The three basic data sources for measuring quality of care indicators are medical records (paper-based or electronic), patient surveys, and administrative data (including enrollment records, insurance claims, and facility records). Each has advantages and disadvantages (Berlowitz et al., 1997).

Medical records. Medical records have the advantage of including much more detailed clinical information than do administrative data: for example, the specific clinical values provided by laboratory test results for HbA1c for diabetics, assessments of the patient's severity of illness, physical examinations, pharmaceutical prescriptions, neurological tests that physicians conducted, results of magnetic resonance imaging (MRI) tests and other radiology examinations, and clinicians' or providers' notes about treatments and the patient's status. They also provide more complete information than do claims data on diagnoses, complications, and comorbidities because claims rely on coding that information, and coding efforts may be incomplete for some types of diagnoses and complications.

As a result, medical record abstracts are important data sources for quality measures when process interventions or outcomes depend on identifying patients with a particular clinical or functional status that cannot be identified through claims data or patient surveys. For example, appropriate interventions and expected outcomes will vary between MS patients depending on whether they have relapsing-remitting or progressive forms of the disease (Noseworthy et al., 2000).

The main disadvantage of medical records is the high cost of collecting those data in many circumstances, particularly when the records are paper-based or when EMRs do not include the specific data necessary for quality measures. The manual medical record abstraction process necessary in such circumstances can be very labor intensive; usually a trained nurse must ensure accuracy, and medical record coders and administrators may also be involved. However, large sample sizes may become increasingly available in EMRs as implementation of EMRs spreads, at least for larger physician groups and integrated delivery systems. In theory, EMRs could reduce the cost of data collection substantially, by enabling access to data already stored in digital format, like claims data. However, at present EMRs are available only in a limited number of hospitals and physician groups, and smaller physician practices have had even lower implementation rates. Comprehensive availability of EMRs for all health professionals and provider organizations across the country remains a long-term goal that may take many years to achieve despite the initiative in the American Recovery and Reinvestment Act of 2009 to fund implementation of EMRs.

Another weakness of medical records, and even EMRs, is that a given patient's medical data can be fragmented across the multiple medical records maintained by the different physician practices, hospitals, and other providers treating the patient. Efforts to develop community-wide health information exchanges (HIEs), to enable more comprehensive access to a patient's data, are still in the pilot phase. EMR vendors are working to make their systems compatible with one another to better promote development of HIEs, but this effort, too, remains in the development phase.

Patient surveys. Surveys can provide unique types of data for measuring quality indicators. For example, some types of outcomes, such as patient satisfaction, can be measured only through patient surveys. Surveys can be used to collect data on physical functioning, mental functioning, and social support for a range of diseases. Disease-specific symptoms, such as fatigue, urinary dysfunction, bowel dysfunction, and sleep satisfaction, can also be captured in survey scales. Standardized QOL survey instruments often capture both generic and disease-specific outcomes data (e.g., National Multiple Sclerosis Society, 1997a, 1997b) by including a mix of scales. Researchers and policy makers can analyze those patient-reported quality measures independently or in conjunction with physician-reported measures of complementary outcomes that may be included in a patient's medical record.

Patient surveys have two main disadvantages, however. First, they can be costly, depending on how they are administered, whether by trained interviewers (in person or by phone) or not. Mail surveys may be a relatively low-cost option in many cases, but they often suffer from lower response rates and higher rates of missing data. Conversely, when trained interviewers conduct in-person interviews, the costs of administering the survey are higher, but the data may be more complete. Many studies have struck a middle ground, using telephone surveys, which can be conducted by interviewers or with computer assistance.

In recent years, online surveys have become more common, and they may enable less expensive survey data collection to become more widespread in the future. At present, the more limited availability of Internet access for low-income respondents and the more limited willingness of elderly or chronically ill patients to participate in online surveys pose problems. However, these concerns will likely diminish considerably in the future as online access and Web use become more routine for most Americans. Online surveys also have the advantage of enabling automated skip patterns and immediate prompts to respondents for out-of-range values and missing data.

The second disadvantage of patient surveys is reliance on patient recall. For infrequent events (such as use of some types of health services) or long recall periods, this drawback may result in inaccurate data. Where possible, combining patient surveys with administrative data can avoid this problem, such as by using surveys for QOL outcomes that require patient responses and administrative claims data for hospital days and other utilization or cost outcomes.

Researchers must also guard against unexpected variations in patient responses due to cultural, racial, ethnic, language, educational, or socioeconomic differences among respondents. Survey instruments often require translations into multiple languages, and researchers may conduct cognitive testing, reading level testing, and other types of pretesting with different patient groups prior to widespread implementation of surveys.

Administrative data. In quality measurement, analysts commonly apply two types of administrative data: enrollment records and insurance (billing) claims data. Quality measurement also sometimes uses clinical data systems—laboratory and pharmacy—although they more closely relate to EMRs while often containing some administrative data.

Administrative data have the advantage of being a low-cost data source: they are already stored in digital format for other purposes, so they are less difficult to access and analyze. Researchers often apply administrative data to identify denominator inclusions and exclusions for quality measures. For example, quality measures for treatment of diabetics are often limited to patients between the ages of 18 and 75 (NCQA, 2006), for whom age data and ICD-9 diagnosis code data used to identify the denominator population are often accessed through administrative data. A lack of detailed clinical information, however, such as the results of laboratory tests, is a common weakness of administrative data; in addition, diagnosis code data often need to be screened or validated to ensure accuracy.

Enrollment data are useful for the basic demographic information needed for both process and outcome indicators, such as age, gender, insurance coverage, and death dates. These data are usually included in databases with one record per patient; generally, they are easy to use for data analysis, but only rarely do they provide all of the information needed for quality measures.

Claims data are useful for some types of process measures, in situations in which the claims data are reasonably complete and provide sufficiently detailed clinical information. Two good examples are indicators for pharmaceutical utilization (e.g., whether MS immunomodulatory disease-modifying drugs have been in continuous use) and laboratory test utilization (e.g., whether patients taking interferons receive liver function tests and complete blood counts with platelet counts every 6 months). However, in a recent study on MS quality indicators (Trisolini et al., 2007), we found claims data to be limited in their applicability for MS quality measurement in many ways, for they did not have sufficiently detailed or consistent data on some types of important diagnoses (e.g., urinary tract infections), important treatments (e.g., intravenous corticosteroids), or episodes of illness (e.g., MS relapses). In addition, claims data do not contain any information on a patient's course of MS (i.e., relapsing-remitting, secondary progressive, primary progressive, or progressive relapsing); on patient-reported outcomes such as QOL, functional status, or satisfaction; or on physician-reported outcomes such as EDSS scores.

Claims data do have several advantages. First, they are reasonably complete for the data they collect, because they are used primarily for billing purposes; health professionals and providers thus have a direct financial incentive to ensure that all bills are submitted for reimbursement. Second, they usually include data on all of the clinicians and provider organizations treating a patient and thus avoid one of the weaknesses of medical records data: patient records that may be fragmented across the different health professionals and facilities providing treatment. Third, they enable analysis of quality measures using large sample sizes, including up to thousands of patients at a time. The large numbers of enrollees that many private health insurance plans cover, and even larger numbers that public payers such as Medicare and Medicaid cover, make this possible.

Risk adjustment of quality measures can also use claims data, because they include variables such as age, gender, diagnoses, and others that risk adjustment models often apply. For example, the risk adjustment model applied for hospital mortality measures in Medicare's Hospital Compare Web site (CMS, 2009) uses claims data. With the advent of Medicare's requirement for POA coding of comorbid conditions, the potential for more accurate coding in claims data has increased considerably.

In sum, all three data sources have advantages and disadvantages for quality measurement. Efforts to measure quality indicators for P4P programs should consider all three options before selecting the most suitable source—or sources—for each indicator. A comprehensive set of quality indicators can include contributions from all three sources. The choice for each P4P program may depend on a range of factors, including budget constraints, preferences for the types of quality measures to be collected, and the need for patient surveys, if the program desires data on QOL or patient satisfaction outcomes.

Number of Quality Measures

P4P programs have included widely varying numbers of quality measures. The United Kingdom's program includes 146 quality measures, far more than any of the P4P programs in the United States have used. In contrast, the Medicare Physician Group Practice Demonstration includes 32 quality measures, which were phased in over several years (Kautter et al., 2007). Private-sector P4P programs typically include fewer measures than those in the public sector.

How many quality measures to include in a P4P program depends on several considerations. Using a larger number of measures poses three risks: (1) increasing the administrative burden on both P4P program administration and on participating health professionals and provider organizations; (2) making the results more complex and cumbersome for health professionals and provider organization staff to interpret; and (3) requiring use of measures less closely linked to health outcomes or less well studied. Data verification and audit costs may increase greatly as the number of measures increases, although sampling providers or measures (or both) to be audited can reduce this burden. Physicians often express concern about the dozens of clinical guidelines and quality incentives they face, at the same time that they perceive themselves to be under pressure to see more patients and complete more paperwork. Under the circumstances, some quality measures may be ignored—especially in situations where P4P incentives for individual quality measures may affect only a small percentage of physician income or provider organization revenue. An advantage of including a larger number of quality measures is a more comprehensive evaluation of the care provided.

The pros and cons of including smaller numbers of quality measures in a P4P program are generally the converse of those for larger numbers of measures. The positives of fewer quality measures include less administrative burden, lower overall program costs, and easier interpretation of results. The negatives include the danger of focusing provider attention on a subset of the important clinical areas and the risk of financial incentives' being focused on just a few measures. The financial incentives could motivate clinician or provider behavior that focuses too much on the clinical conditions included in the P4P program. Studies have found that high performers in some clinical areas are not necessarily high performers in other clinical areas (Sorbero et al., 2006). Although professional ethics and peer review may blunt the impact, inappropriate financial incentives nonetheless remain a risk. Given that hundreds of thousands of physicians practice in the United States, it is likely that some percentage will succumb to financial temptation. Even if this portion represents only 1 percent of all physicians, it would mean that thousands of physicians could be involved in such dubious financial and clinical practices.

Another potential advantage of including fewer structure or process measures is that researchers can focus on measures more closely linked with outcomes. For example, the Leapfrog Group initially focused on just three structure measures that had clear links to outcomes. Similarly, process measures could focus more on immunization or pharmaceutical indicators that have more evidence for impacts on outcomes than on other measures that may be less closely associated with outcomes or that have less evidence to support the relationship.

Intermediate outcomes, such as blood pressure and HbA1c levels, could be substituted for process measures to provide closer links to final outcomes. For example, in its total of 146 quality measures, the UK P4P system included many structure measures, most of which have not been rigorously studied for impact on outcomes. The UK system will be a valuable test of a P4P program with a larger number of quality measures, but there is a need for a closer examination of the structure measures it used to reach its high number of quality measures.

Another issue is whether to weight all of the quality measures equally in calculating provider performance scores. Equal weighting makes it easier for health professionals, provider organizations, and policy makers to interpret the results but may not reflect the underlying value of the different measures or the underlying level of evidence supporting different measures. For example, HEDIS includes quality measures for treatment of diabetics that focus on both the frequency of HbA1c testing and the levels found in that testing. The quality measure focusing on the level of HbA1c could be weighted more heavily in calculating provider performance because it is more closely related to patient outcomes than is the frequency with which the HbA1c tests were conducted.

Types of Quality Measures to Include

The IOM's (2006) report on performance measurement criticized the focus of most current quality measures on specific types of health professionals, provider organizations, or settings of care, such as only on physicians, medical offices, or hospitals. That report recommended expanding quality measurement to include three other types of quality measures:

- **Composite measures**—documenting whether a patient has received all recommended services for a particular condition (and perhaps for multiple conditions). Composite measures of process and intermediate outcomes may show greater room for improvement than individual measures and may be more closely related to final outcomes than single measures are. In calculating composite measures, analysts can apply weighting schemes to give higher weight to quality measures identified as more closely related to either final outcomes or cost savings. All-or-nothing measures may require success on each of a set of measures to be considered at the same time.

- **Population-based measures**—aggregating results for a given region or with breakdowns by population subgroup on socioeconomic status, race, or ethnicity to test for the presence of disparities. These aggregations can be done on several levels, such as groups of clinicians and provider organizations, delivery systems, a community, or a geographic region.
- **Systems-level measures**—analyzing performance across diseases, conditions, clinical specialties, or departments. Researchers and policy makers can define systems to include a continuum of care across ambulatory, inpatient, and long-term care services within a given community.

These alternate approaches to quality measurement have the potential to broaden the focus of current P4P programs, moving beyond the current emphasis on individual clinician, clinician group, or hospital accountability. For example, using these alternate types of quality measures could promote more shared accountability for quality performance across multiple health professionals and provider organizations, a goal that the IOM (2006) report highlighted for development of a national system of performance measurement. This approach may include rewarding the complete set of clinicians and providers included in the care of a patient, or participating in a system of care in a community. Such a step does open up the risk of "free riders," however, in that some clinicians or provider organizations may not be fully motivated to improve quality, preferring to benefit from improvements in performance measures that result from the efforts of the other professionals or provider organizations being assessed with them. However, this broader approach is consistent with management literature that emphasizes the value of applying group incentives in addition to individual incentives (Packwood, 2008). P4P systems could also pursue such a strategy, for example, basing some bonus payments on physician group or provider organization incentives and some on incentives at the level of the community, region, or health care system. In that way, providers could earn bonus payments based on both their own work and their contributions as part of a broader community of professionals and provider organizations that are treating patients in a given region or system of care. This approach mitigates the risk of free riders by tying some incentives directly to provider performance but also preserves some incentives for broader regional or systemwide performance results.

Room for Improvement in Performance

Another consideration for selecting quality measures is the degree to which there is room for improvement in performance on the measure. Ideally, P4P systems would select measures that have large opportunities for improvement, both because this represents good public health practice and because it enables health professionals and provider organizations to demonstrate improvement in quality. Conversely, if there is little room for improvement in a quality measure, where providers have already achieved high performance scores, then payers have less motivation to reward improvements, and providers have fewer opportunities to demonstrate improvement. For example, in recent years the NCQA removed one of its hospital quality measures—beta blocker treatment after myocardial infarction—because hospitals had improved their performance to a high average level, leaving little room for additional improvement.

Cost Containment

P4P programs have focused mainly on quality improvement, but both public and private payers have major concerns about cost containment as well. Quality improvement advocates have claimed that improving quality may in some situations also reduce costs, but evidence for that dual benefit is limited. In theory, better care for diabetics can reduce complications such as retinopathy, nephropathy, and neuropathy, thereby reducing or avoiding the future costs of treating those complications. However, many other factors can affect the actual levels of costs incurred by diabetic patients, such as age, comorbidities, and low-income status. Moreover, for most payers, the time horizon required to reap cost savings for reduced complications of diabetics is too long, meaning that they lack strong incentives to implement programs that address such complications.

Several types of quality improvements are fairly closely linked to cost savings, however. First, patient safety measures that improve quality by reducing adverse drug events, hospital-acquired infections, or surgical errors will directly affect costs by reducing hospital admissions, lengths of stay, or readmissions. Disease management programs that target heart failure patients for more intensive ambulatory care, case management, and nurse-led home care can also reduce hospital admissions and result in cost savings (Anderson et al., 2005). Several chronic diseases known as ambulatory care sensitive conditions (ACSCs) provide opportunities for cost savings through reduced hospital admissions, and quality measures based on ACSCs have been published as Prevention Quality Indicators (PQIs; AHRQ, 2001). The essential idea of ACSCs is that through improved primary health care and preventive care, achieved by enhancing quality or access (or both), chronic disease patients will be less prone to complications or exacerbations of their illnesses that will result in hospitalizations. Given that hospital admissions are very high-cost events in health care, ACSCs have the benefit of linking quality improvement more directly to cost savings than many other types of quality measures, which may take many years to realize their cost impacts.

Although both public and private payers have goals to improve quality of care as an end in itself, both may sometimes opt to target quality measures for P4P programs that also have demonstrated cost savings. For private payers, such a strategy can help reduce the premiums they charge business customers, thus providing a competitive advantage. For public payers, limited governmental budget resources may lead to a dual focus on measures that can simultaneously promote both cost savings and quality improvement.

A related issue is how to fund the bonus payments to providers in P4P programs. Some programs require bonus payments to be funded by cost savings demonstrated by the participating health professionals or provider organization. This is the approach that Medicare's Physician Group Practice Demonstration took (Kautter et al., 2007). Other P4P programs, such as that of the Integrated Healthcare Association, have provided "new money" for P4P bonus payments.

Methods for Analyzing Quality Measures for P4P Programs
Risk Adjustment
Ensuring fair performance assessments when using outcome measures often necessitates risk adjustment or stratification of performance results by population subgroups. For example, many factors apart from the quality of medical treatment affect outcome measures such as patient mortality (most notably, the patient's age and the number and severity of diseases). As a result, when analysts or policy makers use mortality as a quality measure, comparing health professionals and provider organizations on raw mortality statistics can be misleading. At worst, those types of comparisons might encourage

clinicians and provider organizations to avoid treating older or sicker patients who most need their care, because such patients would adversely affect mortality performance measurements.

For example, the New York State cardiac surgery mortality report cards are based on data that are risk adjusted to better ensure fair performance assessment of surgeons (Jha & Epstein, 2006). Similarly, Medicare's Dialysis Facility Compare Web site provides public reporting of mortality data associated with kidney dialysis facilities only after risk adjustment using a broad range of variables (Trisolini & Isenberg, 2007). At the same time, existing statistical models used for risk adjustment do not fully explain the range of factors affecting mortality outcomes. That is why researchers still prefer randomization of patients in evaluating outcomes from new pharmaceuticals in clinical trials (Palmer, 1995). Randomization controls for unmeasured and unknown factors affecting outcomes, whereas statistical models used for risk adjustment can only apply factors that can be measured. As a result, payers and policy makers have not yet been comfortable with moving from public reporting of risk-adjusted mortality outcomes to including mortality outcomes in P4P programs. Public reporting can include caveats, but bonus payments in P4P programs must be based on specific quantitative results, which leaves less opportunity to include qualifying statements regarding interpretation of the results.

Researchers and policy makers sometimes propose risk adjustment for process measures of quality (although in practice they are less often risk-adjusted). One rationale is that patient adherence to prescribed tests and pharmaceutical treatments may be lower for patients in lower socioeconomic groups or different racial or ethnic minority groups than in other populations. As a result, some health professionals and provider organizations argue that process measures such as HbA1c testing for diabetics or blood pressure levels should be risk-adjusted to account for patient factors affecting adherence. For example, Zaslavsky and Epstein (2005) found that racial, income, and education variables affected some HEDIS quality measure scores for health plans significantly, although the rates for most plans changed by fewer than 5 percentage points. Similarly, Mehta et al. (2008) found that patient characteristics (including age, body mass index, race, and type of insurance) and hospital characteristics significantly, but modestly, affected hospital process measures for treatment of acute myocardial infarction.

Nonetheless, a countervailing concern is that one could interpret risk adjustment for these types of factors as endorsing lower-quality care for low-income or minority patients. One method proposed to mitigate this concern is to stratify quality results for public reporting by patient-level factors, including insurance status, low income, and minority status. For example, NCQA requires that HEDIS quality measures for health plans be presented separately for different types of health insurance, including commercial insurance, Medicaid, and Medicare (Zaslavsky & Epstein, 2005). This approach could be extended to include other sociodemographic variables where sample sizes permit. P4P programs, however, may still face challenges of variable incentives for health professionals and provider organizations if the perception remains that avoiding treatment of certain population subgroups could improve performance scores and increase bonus payments. This problem could be mitigated if payers could provide higher P4P bonus payments for quality performance in treating patients in population subgroups known to be associated in the aggregate with worse outcomes or lower adherence to prescribed treatments.

Another approach that some quality measurement efforts use is for clinicians to document the prescription or recommendation for testing and to use that as the measure of quality, thus removing the effect of patient adherence from quality measurement. In most cases P4P program analysts will need medical record data for this measurement, because administrative claims data do not yet routinely capture this type of information. That drawback may change, however, if the new codes for the CPT-II system become more widely adopted; they allow coding for "patient reasons" (including refusal or nonadherence) why a given patient may not have undergone a particular test (American Medical Association, 2008). This new type of CPT coding reduces physicians' incentive to avoid the more difficult patients who may adversely affect their measured quality performance. A potential risk is that physicians will overuse these codes for patient exclusions, and thereby game the performance assessment calculations to increase their bonus payments. Auditing patient records to verify the exclusions is one approach for mitigating this risk.

Identifying High-Quality Providers

P4P programs can take several different approaches to identifying high-quality health professionals and provider organizations that qualify for P4P bonus payments through meeting quality goals or targets. Three basic methods are

(1) threshold targets, (2) improvement-over-time targets, and (3) comparison with other providers.

Threshold targets. The most common method for identifying high-quality clinicians and provider organizations in P4P programs, the threshold approach mainly offers simplicity and ease of understanding for clinicians. For example, "For patients with diabetes, 75 percent will have an HbA1c test at least once per year." The target is clear from the outset so practitioners and provider organizations know what specific number to aim for. A disadvantage is that providers starting at lower levels of quality may perceive thresholds as unattainable if the thresholds are set very high. Another disadvantage is the lack of incentives for further quality improvement above the threshold.

Results from the P4P program in the United Kingdom provide some evidence to support this latter point: an evaluation study found that initial gains in quality in the first 2 years of that program were significant, but gains slowed markedly in the third and fourth years when there were few additional financial incentives for further improvement (Campbell et al., 2009). These results must be interpreted cautiously, however, because there was no comparison group available for this P4P program, given that all UK family practitioners were included in it. As a result, the evaluators had to rely on an interrupted time-series analysis in their study design. Nonetheless, the results are consistent with the concern about lack of further incentives once threshold targets are achieved by providers in a P4P program that relies on that type of target.

One way to mitigate concerns of initially low-performing providers is to establish a series of thresholds, with successive incentives for higher levels of performance in a "stair step" model. For example, threshold-based P4P bonus payments could start at 40 percent performance (where 100 percent is perfect performance, with all denominator patients receiving the indicated numerator interventions) and increase with every 5 percentage points achieved, up to 80 percent. In this way, providers may be able to achieve the first two or three levels of incentives even if they cannot achieve all nine possible levels. They can then aim to achieve higher levels of incentives in future years of the P4P program as they are able to further improve quality performance. In this way, the threshold approach can motivate providers at lower levels of initial performance because they can earn some performance payments in even the first year of a P4P program.

P4P programs can apply several methods for setting specific performance levels for the threshold targets. For example, programs can use (1) consensus goals that P4P payers and participating health professionals and provider organizations have set through joint discussion and agreement, (2) levels set by payers to promote a "reasonable" degree of quality improvement, (3) target levels benchmarked to levels that other high-quality clinicians and provider organizations already achieve, and (4) comparison with other quality measurement programs to find targets these programs may have set for similar populations or similar quality measures.

Improvement-over-time targets. Improvement-over-time targets establish a baseline from a provider's own prior performance level and then evaluate current period performance starting from that level. For example, "Providers should achieve at least a 5 percent increase in performance from the prior year." An advantage of this approach is that providers starting from low levels of initial performance can view these targets as attainable. However, this approach has two disadvantages. First, providers at high levels of prior performance may find additional improvement difficult to achieve. For example, if a provider is already at 90 percent performance or above on a particular quality measure, then a 5 percent improvement may be difficult. Second, payers may object to rewarding providers at low levels of performance even if they are achieving improvements from even lower performance in the prior year. For example, if a provider improves from 10 percent to 15 percent from one year to the next, that 50 percent improvement may still represent a much lower absolute level of performance than that of all other providers in the P4P program.

P4P programs can set improvement-over-time targets in several ways. They can use percentage improvements (e.g., 5 percent), percentage-point improvements (e.g., 5 percentage points), or reductions in performance gaps (e.g., 10 percent reduction in the gap between 100 percent performance and the prior year's performance level).

The Medicare Physician Group Practice Demonstration uses the reduction-in-performance-gaps approach (Kautter et al., 2007), which has the advantage of requiring larger percentage improvements at lower levels of initial performance and smaller improvements at higher levels of initial performance. For example, if the initial performance is 40 percent, then the gap from the perfect score of 100 percent is 60 percent, and the 10 percent improvement target represents a 6 percentage point improvement. As a result, the target would be 46 percent performance in the year being assessed. In contrast, if the initial performance is 80 percent, then the gap is 20 percent and the target is just 2 percentage points' improvement, or 82 percent. In this way the reduction-in-performance-gaps approach mitigates one disadvantage of improvement-over-time targets, by requiring more improvement from low performers and less improvement from high performers.

Another way to mitigate the disadvantages of both the threshold and improvement-over-time targets is to adopt a combined approach that includes both types of targets in one P4P program. The Medicare Physician Group Practice Demonstration adopted such an approach, which included both threshold targets and improvement-over-time targets (Kautter et al., 2007). Physician group practices participating in the demonstration can meet any of the targets to earn performance bonus payments. In this way, the program established positive incentives for physician group practices at both high and low initial levels of initial and ongoing performance.

Comparison with other providers. The third approach to identifying high-quality performance is to compare providers with one another. In this method, P4P programs consider only those who perform better than their peers to be high quality and deserving of P4P bonus payments (irrespective of their absolute levels of performance). For example, P4P programs could award incentive payments to the top 20 percent of providers. The Premier Hospital Quality Incentive Demonstration, which compared more than 200 hospitals using a range of different quality measures, used this approach (Lindenauer et al., 2007).

This approach contrasts with both the threshold and improvement-over-time approaches, in which P4P programs allow all providers the possibility of earning incentive payments. The comparison approach focuses on rewarding only the highest performers from among those participating in the P4P program.

The comparison approach has at least two disadvantages. First, even low absolute levels of performance may earn rewards, as long as any given provider's performance is higher than that of the others. Second, providers do not know in advance what their goal is, because it depends on their peers' performance levels. Some may consider themselves unlikely to perform in the highest 20 percent, and they may therefore lose their motivation to improve (at least by this incentive alone).

Another option with the comparison approach is to include penalties for low performers at the same time as providing rewards for high performers. This option may provide an additional (negative) incentive for those who do not think they have the potential to reach the top 20 percent.

Statistical Analysis of Quality Improvement

Statistical confidence in P4P results can be problematic when individual physicians or small physician practices are the units of accountability. In many of these situations, only small numbers of patients may be available for the denominator populations for some types of quality measures in any given practice; as a result, random statistical fluctuations may account for observed performance on quality measures. Minimum sample sizes per quality measure may need to be as high as 411 patients, a figure HEDIS used to indicate a sample size sufficient to provide confidence that the detectable difference in performance is 10 percentage points (NCQA, 2006).

Achieving sample sizes of 411 or more for diabetics, for example, may require a focus on large physician groups, hospitals, integrated delivery systems, combinations of smaller physician practices into networks or virtual groups, or a geographic area such as a city or county that contains a higher number of providers. It may be easier to achieve sufficient sample sizes for population-based quality measures that do not focus on patients with particular diseases such as diabetes. For example, quality measures for influenza vaccinations include all people ages 50 or older in the denominator population.

Analysts and policy makers sometimes consider smaller sample sizes acceptable if quality measurement can include the entire population of patients in a physician practice, rather than a sample, so that the observed number of patients can be considered the true number and not subject to random statistical fluctuation. However, a countervailing argument is that the observed patient population and quality performance levels may vary randomly over time, so, from that perspective, the population of patients a physician practice treats in any one year is still a sample of the patients treated over multiple years. From that perspective, application of statistical analysis and calculation of confidence intervals is still needed, and the intervals may be very wide when only small sample sizes are available.

Public Reporting of Quality Measures

Researchers and policy makers generally view public reporting of quality performance as a distinct approach to promoting quality improvement, separate from P4P programs. For example, the Medicare Web sites Hospital Compare, Nursing Home Compare, and Dialysis Facility Compare all provide online data that consumers and medical professionals can view to check on the quality-of-care performance of those types of provider facilities in all regions of the United States. Similarly, for many years the NCQA has provided comparative quality performance data on managed care organizations through its HEDIS program. These efforts and others aim solely to enable public reporting of quality-of-care performance data, unrelated to any direct financial incentives that P4P programs would include. Public reporting can provide indirect financial incentives, however, by potentially motivating patients to "vote with their feet" and thus increase utilization and revenue for higher quality providers. In most cases this is only a potential effect, however, and evidence of its impact on patient behavior is limited.

Despite the conceptual distinction between P4P and public reporting initiatives, several P4P efforts have integrated public reporting into their programs. Most notably, the IHA established public reporting of the quality performance results used in its P4P program as one of its program's guiding principles to promote public transparency of P4P incentives.

Many P4P programs adopt a different strategy, releasing quality performance data only to participating clinicians and provider organizations— and not to the public. This is consistent with the methods that many continuous quality improvement (CQI) programs use, giving feedback of data only to the providers that the program is assessing; the aims are to preserve confidentiality of performance results and to promote providers' willingness to participate in CQI initiatives. By avoiding public reporting, staff of these CQI programs argue that they are increasing provider participation, decreasing the risk of "defensive medicine," such as avoiding sicker or more difficult patients, and forestalling efforts by providers to game the data collection efforts.

In general, it seems appropriate to limit public reporting to situations in which practitioners or provider organizations have developed a good level of experience with a set of P4P quality indicators and the methods for identifying performance targets. Especially in the early stages of P4P programs, many clinicians may be concerned about the fairness of quality measurement methods and performance assessments. They may prefer that public reporting of results wait until the performance measurement system has been better tested, and better established through several years of measurement cycles, so that confidence in the accuracy and appropriateness of the quality data has become well established.

P4P programs might also consider several middle ground approaches that entail more limited public reporting. For example, public reporting could focus on aggregated results by region or for groups of providers rather than individual clinicians or provider organizations. In this way the public could view the P4P program's overall results, but those of individual clinicians and provider organizations would still remain confidential. In addition, P4P programs could present individual physician or physician group results while masking the names of the physicians or physician groups with code numbers to prevent performance results from being associated directly with them. Such middle ground public reporting efforts could facilitate some degree of public transparency while mitigating clinicians' and provider organizations' concerns.

References

Agency for Healthcare Research and Quality. (2001). *Prevention quality indicators, version 2.1.* Rockville, MD: Agency for Healthcare Research and Quality.

Agency for Healthcare Research and Quality. (2003). *Guide to patient safety indicators.* Available from http://www.qualityindicators.ahrq.gov/archives/psi/psi_guide_rev1.pdf

Agency for Healthcare Research and Quality. (2007). *CAHPS: Assessing health care quality from the patient's perspective.* Rockville, MD: Agency for Healthcare Research and Quality.

Agency for Healthcare Research and Quality. (2009). *Consumer assessment of healthcare providers and systems.* Available from https://www.cahps.ahrq.gov/default.asp

American Medical Association. (2008). *Current procedural terminology (CPT) 2009.* Chicago: American Medical Association.

Anderson, C., Deepak, B. V., Amoateng-Adjepong, Y., & Zarich, S. (2005). Benefits of comprehensive inpatient education and discharge planning combined with outpatient support in elderly patients with congestive heart failure. *Congestive Heart Failure, 11*(6), 315–321.

Berlowitz, D. R., Brandeis, G. H., & Moskowitz, M. A. (1997). Using administrative databases to evaluate long-term care. *Journal of the American Geriatric Society, 45*(5), 618–623.

Birkmeyer, J., & Dimick, J. (2004). *The Leapfrog Group's patient safety practices 2003: The potential benefits of universal adoption.* Washington, DC: The Leapfrog Group.

Birkmeyer, J., Kerr, E., & Dimick, J. (2006). Commissioned paper: Improving the quality of quality measurement (Appendix F). In Institute of Medicine (Ed.), *Performance measurement: Accelerating improvement.* Washington, DC: National Academies Press.

Bridges to Excellence. (2008). *BTE fifth anniversary report—Five years on: Bridges built, bridges to build.* Available from http://www.bridgestoexcellence.org

Burks, J., & Johnson, K. (Eds.). (2000). *Multiple sclerosis: Diagnosis, medical management, and rehabilitation.* New York: Demos Medical Publishing, Inc.

Campbell, S. M., Reeves, D., Kontopantelis, E., Sibbald, B., & Roland, M. (2009). Effects of pay for performance on the quality of primary care in England. *New England Journal of Medicine, 361*(4), 368–378.

Centers for Medicare & Medicaid Services. (2008a). *Hospital quality initiative: Overview.* Available from http://www.cms.hhs.gov/HospitalQualityInits/.

Centers for Medicare & Medicaid Services. (2008b). *Physician quality reporting initiative: 2007 reporting experience.* Available from http://www.cms.hhs.gov/pqri/.

Centers for Medicare & Medicaid Services. (2009). *Hospital Compare.* Retrieved August 14, 2009, from http://www.hospitalcompare.hhs.gov

Chassin, M. R., & Galvin, R. W. (1998). The urgent need to improve health care quality. Institute of Medicine National Roundtable on Health Care Quality. *JAMA, 280*(11), 1000–1005.

Cohen, J., & Rudick, R. (Eds.). (2007). *Multiple sclerosis therapeutics* (3rd ed.). London: Informa Healthcare.

Coulthard-Morris, L. (2000). Clincial and rehabilitation outcome measures. In J. Burks & K. Johnson (Eds.), *Multiple sclerosis: Diagnosis, medical management, and rehabilitation* (pp. 221–290). New York: Demos Medical Publishing, Inc.

Davis, K., & Guterman, S. (2007). Rewarding excellence and efficiency in Medicare payments. *Milbank Quarterly, 85*(3), 449–468.

Department of Health. (2004). *Annex A: Quality indicators—Summary of points.* Available from http://www.dh.gov.uk/assetRoot/o4/07/86/59/04078659.pdf

Donabedian, A. (1966). Evaluating the quality of medical care. *Milbank Memorial Fund Quarterly, 44*(3 Suppl), 166–206.

Doran, T., Fullwood, C., Gravelle, H., Reeves, D., Kontopantelis, E., Hiroeh, U., et al. (2006a). Online only appendix 1: Abbreviated clinical quality indicators. Supplement to Pay-for-performance programs in family practices in the United Kingdom. *New England Journal of Medicine, 355*(4), 375–384.

Doran, T., Fullwood, C., Gravelle, H., Reeves, D., Kontopantelis, E., Hiroeh, U., et al. (2006b). Pay-for-performance programs in family practices in the United Kingdom. *New England Journal of Medicine, 355*(4), 375–384.

Eddy, D. M. (1998). Performance measurement: Problems and solutions. *Health Affairs (Millwood), 17*(4), 7–25.

Gilmore, A. S., Zhao, Y., Kang, N., Ryskina, K. L., Legorreta, A. P., Taira, D. A., et al. (2007). Patient outcomes and evidence-based medicine in a preferred provider organization setting: A six-year evaluation of a physician pay-for-performance program. *Health Service Research, 42*(6 Pt 1), 2140–2159; discussion 2294–2323.

Hays, R. D., Kallich, J. D., Mapes, D. L., Coons, S. J., & Carter, W. B. (1994). Development of the kidney disease quality of life (KDQOL) instrument. *Quality of Life Research, 3*(5), 329–338.

Hurtado, M., Swift, E., & Corrigan, J. (Eds.). (2001). *Envisioning the national health care quality report.* Washington, DC: Institute of Medicine, National Academies Press.

Institute of Medicine, Board on Health Care Services. (2001). *Crossing the quality chasm: A new health system for the 21st century.* Washington, DC: The National Academies Press.

Institute of Medicine, Board on Health Care Services. (2006). *Performance measurement: Accelerating improvement* (Vol. 11517). Washington, DC: The National Academies Press.

Institute of Medicine, Board on Health Care Services. (2007). *Rewarding provider performance: Aligning incentives in Medicare.* Washington, DC: The National Academies Press.

Jencks, S. F., Cuerdon, T., Burwen, D. R., Fleming, B., Houck, P. M., Kussmaul, A. E., et al. (2000). Quality of medical care delivered to Medicare beneficiaries: A profile at state and national levels. *JAMA, 284*(13), 1670–1676.

Jha, A. K., & Epstein, A. M. (2006). The predictive accuracy of the New York State coronary artery bypass surgery report-card system. *Health Affairs (Millwood), 25*(3), 844–855.

Jha, A. K., Orav, E. J., Ridgway, A. B., Zheng, J., & Epstein, A. M. (2008). Does the Leapfrog program help identify high-quality hospitals? *Joint Commission Journal on Quality and Patient Safety, 34*(6), 318–325.

Jordan, H., Pine, M., Elixhauser, A., Hoaglin, D., Fry, D., Coleman, K., et al. (2007). Cost-effective enhancement of claims data to improve comparisons of patient safety. *Journal of Patient Safety 3*(2), 82–90.

Joy, J., & Johnston, R. (Eds.). (2001). *Multiple sclerosis: Current status and strategies for the future.* Washington, DC: Institute of Medicine, National Academies Press.

Kautter, J., Pope, G. C., Trisolini, M., & Grund, S. (2007). Medicare Physician Group Practice Demonstration design: Quality and efficiency pay-for-performance. *Health Care Financing Review, 29*(1), 15–29.

Kohn, L. T., Corrigan, J. M., & Donaldson, M. S. (Eds.). (1999). *To err is human: Building a safer health system.* Washington, DC: Institute of Medicine, National Academies Press.

Leapfrog Group. (2008). *Fact sheet.* Retrieved August 3, 2008, from http://www.leapfroggroup.org

Lindenauer, P. K., Remus, D., Roman, S., Rothberg, M. B., Benjamin, E. M., Ma, A., et al. (2007). Public reporting and pay for performance in hospital quality improvement. *New England Journal of Medicine, 356*(5), 486–496.

Lohr, K. N. (2004). Rating the strength of scientific evidence: Relevance for quality improvement programs. *International Journal on Quality in Health Care, 16*(1), 9–18.

McDermott, S., Williams, T., Lempert, L., & Yanagihara, D. (Eds.) (2006). *Advancing quality through collaboration: The California Pay for Performance Program, a report on the first five years and a strategic plan for the next five years.* Los Angeles: Integrated Healthcare Association. Available from http://www.iha.org/pdfs_documents/p4p_california/P4PWhitePaper 1_February2009.pdf

McGlynn, E. A., Asch, S. M., Adams, J., Keesey, J., Hicks, J., DeCristofaro, A., et al. (2003). The quality of health care delivered to adults in the United States. *New England Journal of Medicine, 348*(26), 2635–2645.

Mehta, R. H., Liang, L., Karve, A. M., Hernandez, A. F., Rumsfeld, J. S., Fonarow, G. C., et al. (2008). Association of patient case-mix adjustment, hospital process performance rankings, and eligibility for financial incentives. *JAMA, 300*(16), 1897–1903.

National Committee for Quality Assurance. (2006). *HEDIS 2007: Health plan employer data & information set.* Washington, DC: National Committee for Quality Assurance.

National Multiple Sclerosis Society. (1997a). *Multiple sclerosis quality of life inventory: A user's manual.* New York: National Multiple Sclerosis Society.

National Multiple Sclerosis Society. (1997b). *Multiple sclerosis quality of life inventory: Technical supplement.* New York: National Multiple Sclerosis Society.

Nortvedt, M. W., & Riise, T. (2003). The use of quality of life measures in multiple sclerosis research. *Multiple Sclerosis, 9*(1), 63–72.

Noseworthy, J. H., Lucchinetti, C., Rodriguez, M., & Weinshenker, B. G. (2000). Multiple sclerosis. *New England Journal of Medicine, 343*(13), 938–952.

Packwood, E. (2008). Using variable pay programs to support organizational goals. In L. Berger & D. Berger (Eds.), *Compensation handbook* (5th ed., pp. 215–226). New York: McGraw-Hill.

Palmer, R. H. (1997). Quality of care. *JAMA, 277*(23), 1896–1897.

Pine, M., Jordan, H., Elixhauser, A., Fry, D., Hoaglin, D., Jones, B., et al. (2007). Enhancement of claims data to improve risk-adjustment of hospital mortality. *JAMA, 297*(1), 71–76.

Rothwell, P. M., McDowell, Z., Wong, C. K., & Dorman, P. J. (1997). Doctors and patients don't agree: Cross sectional study of patients' and doctors' perceptions and assessments of disability in multiple sclerosis. *BMJ, 314*(7094), 1580–1583.

Sorbero, M., Damberg, C. L., Shaw, R., Teleki, S., Lovejoy, S., DeCristofaro, A., et al. (2006). *Assessment of pay-for-performance options for Medicare physician services: Final report* (RAND Health Working Paper WR-391-ASPE). Available from http://aspe.hhs.gov/health/reports/06/physician/report.pdf

Trisolini, M., Constantine, R., Green, J., Moore, A., Pope, G. C., & Miller, A. (2007). *Multiple Sclerosis Quality Indicators Project: Final report.* Prepared for the National Multiple Sclerosis Society. Waltham, MA: RTI International.

Trisolini, M., & Isenberg, K. L. (2007). Public reporting of patient survival (mortality) data on the Dialysis Facility Compare Web site. *Dialysis & Transplantation, 36*(9), 486–499.

Trisolini, M. G., Smith, K. W., McCall, N. T., Pope, G. C., & Klosterman, M. (2005). Evaluating the performance of Medicare fee-for-service providers using the health outcomes survey: A comparison of two methods. *Medical Care, 43*(7), 699–704.

Wennberg, J. E., Fisher, E. S., & Skinner, J. S. (2002). Geography and the debate over Medicare reform. *Health Affairs (Millwood), Supplement Web Exclusives*, W96–W114.

Zaslavsky, A. M., & Epstein, A. M. (2005). How patients' sociodemographic characteristics affect comparisons of competing health plans in California on HEDIS quality measures. *International Journal on Quality in Health Care, 17*(1), 67–74.

CHAPTER 5

Incorporating Efficiency Measures into Pay for Performance

John Kautter

The early pioneers of pay for performance (P4P), such as US Healthcare (now Aetna), launched P4P in the mid-1980s, and the movement grew dramatically in the 2000s. At the end of 2007, there were 148 P4P sponsors nationwide; commercial P4P sponsors were the most prevalent. P4P programs most often focus on clinical quality; however, as of 2006, 23 percent of P4P sponsors included efficiency or cost of care as one of their domains (Baker & Delbanco, 2007). This chapter examines the use of efficiency measures in P4P programs.

P4P was born during the nation's backlash against the cost-control emphasis of managed care. Hence, P4P programs tended to restrict their focus to quality, patient satisfaction, and, to some extent, adoption of information technology (Robinson, 2008; Robinson et al., 2009). Several seminal Institute of Medicine (IOM) reports on health care quality and safety also galvanized a call to action that led to the rise of P4P programs (e.g., IOM Board on Health Care Services, 2001; Kohn et al., 1999). However, health care cost growth in the United States has overshadowed the original concerns. Determinants of this cost growth include (1) population aging, (2) general economic growth, (3) expansions of insurance coverage, and most important, (4) expansion of technological capabilities of medicine (White, 2007). Technological advances are likely to yield new and desirable medical services in the future, fueling further spending growth and imposing difficult choices in spending on health care versus alternatives. Spending growth will depend largely on how the health care system responds to future technological change (Congressional Budget Office, 2008).

This chapter presents a broad overview of efficiency measures in P4P programs. After first providing the motivation for including efficiency in P4P, we review definitions of efficiency. We follow this with an examination of the measurement of efficiency and a discussion of the evaluation of efficiency measures and measurement challenges. Then we discuss risk adjustment and quality in the context of efficiency measurement. Finally, we offer conclusions.

Motivation for Including Efficiency in Pay for Performance

Evidence is strong that substantial inefficiencies exist in the US health care system (Safavi, 2006). First, per capita health care spending varies widely across the United States; substantial variations in cost per patient, however, are not correlated with overall health outcomes. For example, analysis of composite quality scores for medical centers and average spending per patient shows no correlation. Even among elite medical centers, costs vary substantially. Some regions are more likely than others to adopt low-cost, highly effective patterns of care, whereas some tend to adopt high-cost patterns of care and deliver treatments that provide little benefit (or even cause harm) (Orszag, 2008). Second, the per capita health care expenditure in the United States is 2 times greater than that of most other developed countries; it is nearly 1.5 times greater than the per capita spending of Switzerland, which is the second highest spending nation (Reinhardt et al., 2004). However, these expenditures in the United States result in quality outcomes that are indistinguishable from those in other nations (Hussey et al., 2004). In fact, a recent international survey finds that the United States lags behind other developed countries on important measures of access, quality, and use of health information technology (Schoen et al., 2009).

Researchers estimate that 30 percent of Medicare's costs could be saved without negatively affecting health outcomes if spending in high- and medium-cost areas were reduced to the level in low-cost areas; they further hypothesize that these estimates could be extrapolated to the health care sector as a whole (Fisher, 2005; Wennberg et al., 2002). Further, analysts should consider not only static estimates of one-time potential savings for the US health care system but also dynamic estimates of potential savings over time.

Unlike the health care industry, other industries have discovered efficiency improvements sufficient to lower the cost of services by 2.5 to 6.5 percent annually, thereby offsetting the cost-additive impact of new technologies. In contrast, annual efficiency gains achieved in the US health care system are much lower, leaving a 2.5 percentage point gap between health care spending growth and gross domestic product (GDP) growth (Milstein, 2008). If the gap between health care spending growth and GDP growth continued over this century, then more than 100 percent of the increase in GDP growth would be required for health care spending. However, if health care spending grew only one percentage point faster than GDP growth, health care spending over this century would be "affordable," although still about 50 percent of GDP growth (Chernew et al., 2009).

One can make a strong argument for including efficiency as a criterion for health care payment. One reason is simply the inefficiency in the health care system. Further, many costs have been attributed to inefficient practices within the control of providers and individual practitioners. This factor—combined with the relationship between health care users and providers regarding the cost of care—places a burden on payers to reward efficient behavior to stretch the available resources (Safavi, 2006). Consensus is growing that meaningful cost control will require changing the fee-for-service (FFS) system to reward both quality and efficiency.

Efficiency-based payments are, however, not new. For several decades, payers have compensated physicians based on relative value work units and have compensated hospitals based on patient diagnosis and complexity. Even under these systems, however, payers have not held costs in check adequately, efficiency is not what it should be, and further reform is necessary (Medicare Payment Advisory Commission [MedPAC], 2005a).

Defining Efficiency

To measure efficiency, and ultimately to apply efficiency measures to a P4P program, analysts must define efficiency. Several organizations have developed definitions of efficiency. For example, the IOM defines efficiency as avoiding waste, including waste of equipment, supplies, ideas, and energy (Berwick, 2002; IOM Board on Health Care Services, 2001). However, to date no broad consensus has emerged on how to define efficiency for the health care system.

In general, efficiency is concerned with the relationship between health care outputs and resource inputs. Outputs can be defined as health care services (e.g., episodes of care) or final health outcomes (e.g., quality-adjusted life years, or QALYs). Inputs can be defined as physical inputs (e.g., nursing days) or financial inputs (e.g., costs). In addition to the relationship between health care outputs and resource inputs, efficiency might also be concerned with the relationship between health care services and final health outcomes. We now present some of the general definitions of health care efficiency that have been used (or could be used) in establishing efficiency measures.

Cost Efficiency

Payers and purchasers of health care services (as well as many health economists) tend to define efficiency as *cost efficiency*, which is generally defined as either the maximization of health care services for a given cost or the minimization of cost for a given level of health care services. Such cost efficiency measures are independent of measures of health outcomes, but P4P programs should consider such outcomes along with available clinical effectiveness and patient experience measures when evaluating the performance of providers. In the context of this discussion, cost efficiency refers to the total cost for treatment of specific conditions relative to a cost standard. It reflects the combination of quantity and mix of health care services as well as the unit prices for these services, and generally it is risk adjusted (Thomas, 2006).

Economic Definitions

Health economists sometimes differentiate between three types of efficiency: technical efficiency, productive efficiency, and allocative efficiency (Palmer & Torgerson, 1999; Varian, 1992). *Technical efficiency* refers to the physical relation between physical inputs and outputs (in which outputs can be health care services or health outcomes). Technical efficiency is achieved when the level of output is maximized from a given set of physical inputs, but it cannot be used to compare alternative interventions, for example, in which one intervention produces the same output with less of one resource and more of another.

Productive efficiency refers to either the maximization of output for a given cost or the minimization of cost for a given level of output (note that when outputs are defined as health care services, then productive efficiency is equivalent to cost efficiency). Productive efficiency permits assessment of relative value for interventions with directly comparable outputs. It cannot, however, address the impact of reallocating resources at a broader level.

Allocative efficiency accounts for both productive efficiency and the efficiency of output distributed across the community. This type of efficiency occurs when resources are allocated to maximize the welfare of the community. Allocative efficiency implies productive efficiency, which in turn implies technical efficiency.

Cost-Effectiveness

Cost-effectiveness analysis is a method used to evaluate the costs and outcomes of interventions designed to improve health (Gold et al., 1996). For a given condition and population, treatment options 1 and 2 (e.g., new treatment versus old treatment) can be compared by calculating the incremental cost-effectiveness ratio (ICER), which is the difference in costs between options 1 and 2 divided by the difference in outcomes. The ICER is the "price" of the additional outcome purchased by using option 1 rather than option 2, generally in dollars per QALY. If the price is low enough, then option 1 is cost-effective (American College of Physicians, 2000). When option 1 has both lower costs and better outcomes than option 2, then option 1 is "dominant" relative to option 2. Thus, an efficient health care system necessarily would choose option 1 over option 2. However, when option 1 has both higher costs and better outcomes than option 2, then neither option is dominant relative to the other. In this case standard definitions of efficiency do not apply, and cost-effectiveness analysis could be used to develop efficiency measures.

At the present time, no agency in the United States formally establishes standards for cost-effectiveness analysis outcomes. However, most researchers consider interventions costing less than $50,000/QALY to be very cost-effective and those costing more than $100,000/QALY not to be cost-effective (Brown et al., 2008). Other countries and international organizations have formally established cost-effectiveness thresholds. For example, the United Kingdom's National Institute for Health and Clinical Excellence (NICE) recommends that a health care technology should have a cost-effectiveness threshold of £20,000 to £30,000 (approximately $31,000 to $46,000 in mid-2010 US dollars) per QALY gained (NICE, 2009; Culyer, 2009). The World Health Organization (WHO) recommends that countries use a cost-effectiveness threshold that is 1 to 3 times their per capita GDP (WHO, 2001).

Efficiency Measurement

Health care efficiency measurement has been a subject of intense research by academics, vendors, and various health care stakeholders such as payers, providers, and individual health professionals. The Southern California Evidence-Based Practice Center (McGlynn & Southern California Evidence-Based Practice Center, 2008; see also Hussey et al., 2009) has provided a useful typology for efficiency, which explicates the content and use of efficiency measures. Their typology for efficiency has three tiers:

- Perspective: Who is evaluating the efficiency of what entity and why?
- Outputs: What type of product is being evaluated?
- Inputs: What resources are used to produce outputs?

Unfortunately, much of the peer-reviewed research on efficiency measurement is fragmented; it tends to focus on the production of specific health care outputs and services without a general theoretical or methodological framework (Chung et al., 2008). Further, most measures that payers use have been developed by vendors and are proprietary. We now discuss the current state of efficiency measurement, focusing on hospital and physician efficiency measurement.

Hospital Efficiency Measurement

The majority of peer-reviewed literature on health care efficiency measurement relates to the production of hospital care. Academics often use sophisticated empirical techniques called "frontier modeling" to identify best-practice output-input (cost) relationships and to gauge how much efficiency levels of given hospitals deviate from these frontier values (Bauer, 1990). These empirical techniques include data envelopment analysis (DEA) and stochastic frontier regression (SFR). Although DEA and SFR models yield convergent evidence about hospital efficiency at the industry level, they produce divergent evidence about the individual characteristics of the most and least efficient hospitals (Chirikos & Sear, 2000).

Academic studies such as these generally measure hospital efficiency from the perspective of hospitals. In terms of P4P, however, payers and purchasers have perspectives different from those of hospitals. Therefore, to date, payers and purchasers have not shown much interest in the academic approach to hospital efficiency measurement. Fortunately, hospital efficiency indicators from the perspective of payers and purchasers have been developed (Thomas, 2006):

- Hospital stays: Several hospital efficiency indicators use hospital stays as the unit of analysis. These hospital efficiency indicators include average length of stay, early readmission rate, and hospital payments. These indicators generally adjust risk by adjusting hospitals' actual values upward or downward to account for the case mix (case type and severity) characteristics of the patients treated.

- Episodes of care: Evaluators use episodes of care to incorporate pre-hospital services (e.g., office visits, radiology examinations), post-hospital services (e.g., medications, physical therapy), and professional fees into efficiency calculations. Case-mix-adjusted episode payments can be calculated for a given condition group (e.g., stroke) or for multiple conditions.

- Cohort-based, longitudinal patient-level indicators: These indicators use the patient as the unit of analysis and note differences among cohorts of patients in outcomes occurring during an observation period.

According to MedPAC, "Ideally, we would want to limit our set of efficient hospitals to those that not only have high in-hospital quality and low unit costs but also have patients with low risk-adjusted overall (across all services) annual Medicare costs" (MedPAC, 2009, p. 65). However, MedPAC goes on to point out that the risk adjustment and standardization of these cost data still need refinement before they can be used for cross-sectional comparisons of efficiency. Thus, to measure hospital efficiency, MedPAC focuses on outcome measures (e.g., mortality, readmissions) and inpatient costs, but not overall costs. Inpatient costs per discharge are adjusted for factors beyond the hospital's control that reflect the financial structure of the hospital rather than efficiency. Specifically, costs are standardized by adjusting for case mix, area wage index, prevalence of outliers and transfer cases, and the effects of teaching activity and service to low-income Medicare patients on costs per discharge. MedPAC also adjusts for differences in interest expenses because those do not reflect operational efficiency. MedPAC developed efficiency rankings based on the dimensions of hospital outcomes and inpatient costs (MedPAC, 2009).

Physician Efficiency Measurement

Ratio-based efficiency measures have been used mostly to evaluate physician efficiency. For example, Pope and Kautter (2007) developed a population-based methodology for profiling the cost efficiency and quality of care of large physician organizations (POs) by comparing the efficiency index for a PO with an index for a peer group defined as all POs in the Boston metropolitan statistical area (Pope & Kautter, 2007; see also US Government Accountability Office, 2007). They assigned patients to POs based on the plurality of outpatient evaluation and management visits (Kautter et al., 2007) and standardized costs across the POs by adjusting for health status risk using the hierarchical conditions categories model (Pope et al., 2004), county, and teaching and disproportionate-share hospital payments. Using the patients assigned to each PO, Pope and Kautter defined an efficiency index for the organization as follows:

$$\text{Efficiency Index} = \frac{\text{Actual Per Capita Expenditures}}{\text{Predicted Per Capita Expenditures}}$$

When actual per capita expenditures equal predicted per capita expenditures, then the efficiency index equals 1.00; this means that the observed expenditures of patients assigned to the PO equal the expenditures expected for these patients. In this case, the PO is neither efficient nor inefficient relative to expectations. When the efficiency index is less than 1.00, actual expenditures are less than predicted, and the PO is more efficient than predicted. Conversely, if the index is greater than 1.00, the PO is less efficient than predicted. This is the standard statistic used in efficiency profiling exercises, and it is often referred to as "observed/expected" (Thomas et al., 2004).

Commercial vendors have developed most physician efficiency measures used by purchasers and payers; for that reason, most such measures are proprietary. The main application of these measures is to reduce costs through P4P, tiered product offerings, public reporting, and feedback for performance improvement. These vendor-based measures of efficiency generally fall into two main categories: population-based and episode-based (McGlynn & Southern California Evidence-Based Practice Center, 2008; see also Hussey et al., 2009).

Population-based measures classify a patient population according to the morbidity burden for a given period (e.g., 1 year). Efficiency is measured by comparing the costs/resources used to care for that risk-adjusted population for a given period, and a single entity such as a PO is responsible for the care of that defined population. Episode-based measures use diagnoses and procedure codes from claims or encounter data to construct discrete episodes of care. Efficiency is measured by comparing the physical/financial resources used to produce an episode of care; attribution rules based on the amount of care provided by each provider are applied to attribute episodes to particular providers, after additional risk adjustment is applied (McGlynn & Southern California Evidence-Based Practice Center, 2008; see also Hussey et al., 2009).

Population-based approaches to efficiency assessment include measuring the risk-adjusted rate at which a certain intervention is performed across physicians' patient populations (e.g., number of hospitalizations or diagnostic tests per 1,000 patients) or measuring the risk-adjusted total costs associated with primary care physicians' patient populations over a year (MedPAC, 2005a). Episode-based approaches are often considered more actionable and more applicable to specialists than population-based approaches are. However, population-based approaches can measure the overall performance for a population (Leapfrog Group & Bridges to Excellence, 2004) and may be more conducive to risk adjustment (Centers for Medicare & Medicaid Services [CMS], 2009a).

Although current strategies for addressing health care costs emphasize physician performance measurement and commonly use an efficiency index such as one of those described here, using an efficiency index for P4P at the level of individual health practitioners might hinder the goal of reducing overuse of services. An efficiency index might not always reflect costs generated by overuse: costs of increased but appropriate care and costs associated with correcting underuse also could result in a higher efficiency index. An alternative approach is to identify key cost drivers and then, instead of focusing on cost reduction per se, focusing on reducing unnecessary variation and eliminating overuse; this approach places cost reduction in the larger context of quality improvement (Greene et al., 2008).

Finally, cost-effectiveness analysis is an approach worth considering in measuring physician efficiency (Gold et al., 1996). Because the costs of treatments have finite limits, the largest incremental cost-effectiveness ratios, and hence the most inefficient uses of limited resources, occur when more expensive interventions provide little or no health benefit (American College of Physicians, 2008). Services with low cost per QALY (e.g., beta blockers for high-risk patients after heart attack) are cost-effective, meaning that these services deliver considerable value per unit cost. Services with a high cost per QALY (e.g., left ventricular assist device—as compared with optimal medical management—in patients with heart failure who are not candidates for a transplant) are not cost-effective (Cohen et al., 2008; Drexler, 2010). In this context, primary care physicians or groups that manage the overall care of attributed patients who receive a high rate of discretionary, low-value, high-cost services relative to their peers are relatively economically inefficient. For cases in which alternative treatments of varying known cost-effectiveness are available for the same condition, specialist physicians or groups that provide a higher rate of more cost-effective treatments are more economically efficient.

In addition, MedPAC (2005b) has suggested that Medicare could begin to use available cost-effectiveness analysis to prioritize P4P and disease management initiatives. As an example, for the screening of chronic kidney disease among the Medicare population, cost-effectiveness analyses could help inform policymakers about which populations (such as patients who have diabetes) would generate the most favorable ratios of health gain to spending.

Evaluation of Efficiency Measures and Measurement Challenges

For hospital or physician efficiency measurements to be widely accepted in the market, they should be feasible for health plans to implement, credible and reliable for consumers, and fair, equitable, and actionable for providers. Specifically, according to the Leapfrog Group & Bridges to Excellence (2004):

- Efficiency measures should be actionable by plans, providers, and clinicians, enabling them to identify opportunities for improvement and to compare their performance with that of others.

- Efficiency measures must be operationally focused and feasible for plans, benefit administrators, and health professionals to implement without creating undue burden on staff and resources.

- Methods used in calculating efficiency measures and the application of those methods should reflect the overall, true cost of care and the appropriate locus of control. The methods should allow for appropriate risk adjustment and for peer-to-peer comparisons.

- All efficiency measures should be sound, evidence-based, and valid, and they should produce timely results.

- Use of efficiency measures to evaluate providers should be reasonable and should avoid gaming by any party; publication of these measures should lead to overall improvements benefiting purchasers, plans, providers, health professionals, and consumers.

Ideally, efficiency measures would possess each of these attributes. Measurement challenges present a formidable barrier to achieving these attributes, however. Greenberg (2006) provides a good discussion of these measurement challenges and makes seven key points.

First, effective efficiency measurement may require data from multiple sources, which may not always be available or accessible. Second, pooling data across multiple payers can be a valuable approach to collecting information on provider performance. However, technical adjustments must be used to standardize the information. Third, attributing care to accountable health care providers is a key process step in evaluating performance. This is particularly true when physician incentives are tied to performance. Fourth, achieving a sufficient sample size is a challenge for many forms of measurement, especially evaluations of individual physicians' performance. The adequacy of sample sizes and adjustments for case mix have a great impact on validity of measurement. Fifth, performance of hospitals and physicians is usually not consistent across all efficiency measures. This factor makes it difficult to provide a simple ranking to guide consumer or purchaser choices and introduces challenges to reporting provider performance across multiple measures. Sixth, physicians and hospitals often want to understand the approaches and methods underlying performance measurement; thus, "showing the math" and offering tools for various users to understand the information is an important goal. Finally, measuring efficiency may have unintended consequences. For example, inadequate severity adjustment may cause providers with more complex patient populations to be designated "inefficient."

Risk Adjustment

Risk adjustment is potentially the biggest challenge to measuring efficiency. Risk adjustment is the statistical process used to identify and adjust for differences in patient characteristics (or risk factors) before comparing outcomes of care. The purpose of risk adjustment is to facilitate an equitable and accurate comparison of outcomes of care across health care organizations or providers (CMS, 2009b). Lack of adequate risk adjustment has been an important barrier to the widespread application of efficiency measurement in the Medicare program, including both hospital efficiency measurement (MedPAC, 2009) and physician efficiency measurement (CMS, 2009a).

Analyses of hospital cost as an efficiency indicator involve comparing patients' actual hospitalization costs with their expected costs, with expected cost estimates based on patients' diagnoses, severity, and demographics. In hospital efficiency calculations, the function of risk adjustment is to estimate an expected value for each hospital stay, outpatient visit, episode of care, or other unit of service being analyzed, so that efficiency estimates can properly account for differences among hospitals in the case mix, severity, and demographics of patients being treated (Thomas, 2006).

In general, analysts use one of two types of risk-adjustment methodologies: categorical risk adjusters and regression-based risk adjusters (Thomas, 2006). An example of a categorical risk adjuster is Medicare severity diagnosis-related groups (MS-DRGs), which are used for Medicare hospital inpatient FFS payment. MS-DRGs are a patient classification system that can relate the types of patients that a hospital treats (i.e., its case mix) to the costs incurred by the hospital (CMS, 2009c). An example of a regression-based risk adjuster is the proprietary Symmetry Episode Risk Groups, which predict current and future health care usage for individuals and groups by creating individual risk measures that incorporate episodes-of-care methodology, medical and pharmacy claims information, and demographic variables (Ingenix, 2006). The choice of the most appropriate risk-adjustment methodology depends on several factors, including predicted outcome, analytical time frame, relevant population, purpose, and performance (Thomas, 2006).

In measuring physician efficiency performance, a key statistical challenge is to minimize the influence of patient health status variation, and the health status of a panel of patients, on an individual physician's score. Separating the practice pattern of the physician from the health status variation of the patients is a key element of efficiency measurement. Several factors, if left uncontrolled, could influence the results of efficiency measurement. These include variation in (1) patient health status, (2) severity of illness (within the condition affecting the patient), (3) the case mix in each physician's panel of patients, and (4) the number of episodes (or patients) assigned to each physician and associated susceptibility to high outlier influences (Pacific Business Group on Health & Lumetra, 2005).

Thomas and colleagues (2004) examined the consistency among risk-adjusted efficiency measures for physicians, investigating whether different risk-adjustment methodologies produce differences in practice efficiency rankings for a set of primary care physicians. They calculated patient risk scores for six of the leading risk-adjustment methodologies and observed moderate to high levels of agreement among the six risk-adjusted measures of practice efficiency. They pointed out, however, that the consistency of measures does not prove that practice efficiency rankings are valid. For that reason, they advise that analysts should exercise caution when using practice efficiency information.

Efficiency and Quality

As the AQA Alliance (2009a, 2009b) has discussed, "efficiency of care" and "value of care" measures have not been evaluated in the same way as clinical quality measures have been. Cost of care measures can inform the development of true efficiency and value measures. Definitions related to performance measures are as follows:

- Cost of care is a measure of total health care spending.

- Efficiency of care is a measure of cost of care associated with a specified level of quality of care.

- Value of care is a measure of a specified stakeholder's (e.g., payer's) preference-weighted assessment of a particular combination of quality and cost of care performance.

Although most of the literature on hospital efficiency does not account for quality outcomes, some does. For example, a study sponsored by The Leapfrog Group (Binder & Rudolph, 2009; Robinson & Center for Health Systems Research and Analysis, University of Wisconsin–Madison, 2008) rated efficiency on four procedures or conditions: coronary artery bypass graft, percutaneous coronary intervention, acute myocardial infarction, and pneumonia. To assess resource utilization, the study measured severity-adjusted average length of stay (ALOS), inflated by readmission rate. For outcomes, it considered risk-adjusted mortality rates. For the resource utilization measure, it calculated the observed ALOS in the facility relative to the expected ALOS in the facility, in which the expected ALOS was based on a linear regression model calibrated on all-payer National Hospital Discharge Survey data.

Medicare publicly reports hospital outcome measures on its Hospital Compare Web site (http://www.hospitalcompare.hhs.gov); this information includes 30-day readmission measures for acute myocardial infarction, heart failure, and pneumonia (CMS, 2009d). Given that hospital readmissions can be considered both a quality of care measure and a cost-efficiency measure, one could argue that these measures bridge the gap between quality of care and efficiency.

Physician efficiency measures ideally should be combined with measures of quality of outputs. Unfortunately, to date, most physician-oriented, episode-based measures of efficiency do not control for patient outcomes (Safavi, 2006). There are practical reasons for this. In many areas of health care, no good quality indicators exist; in others, outcome information is not readily available because of ongoing reliance on paper medical records (Milstein & Lee, 2007). However, payers and other stakeholders have begun testing models for rewarding both quality and efficiency (Davis & Guterman, 2007). For example, responding to soaring health care costs and double-digit increases in health insurance premiums, the Integrated Healthcare Association, an association of health plans, hospital systems, and medical groups in California that manages the state's P4P program, has expanded the program to include efficiency. For the first time, the new measures add information on cost and resource use alongside existing P4P quality measures (Robinson et al., 2009;

Romano, 2007). Another example is the Medicare Physician Group Practice Demonstration, which is Medicare's first physician P4P initiative. The demonstration established P4P incentives for quality improvement and cost efficiency at the level of the large physician group practice. The P4P incentives include "shared savings" in which the physician group practices that control Medicare costs, while simultaneously improving quality, share in the cost savings (Kautter et al., 2007). Results of the demonstration to date indicate that the P4P incentives that the demonstration provides have resulted in modest cost savings (CMS, 2009e).

Conclusions

The single most important factor influencing the US federal government's long-term fiscal balance is the rate of growth in health care costs. Rising health care costs per patient are more important to long-term fiscal challenges than demographic changes are. Many other factors that play a key role in determining future fiscal conditions, such as Social Security, pale in comparison to containing the cost growth for federal health insurance programs. Without changes in federal law, health care spending will rise to 25 percent of GDP by the year 2025 (Congressional Budget Office, 2007). Containment of health care costs will, therefore, be an especially important societal goal to achieve in the coming years. Incorporation of efficiency measures into P4P programs has shown promise as a strategy to control health care costs (Cutler et al., 2009). This chapter has provided a broad overview of efficiency in P4P.

On March 23, 2010, President Barack Obama signed into law the Affordable Care Act, the most comprehensive health reform legislation in half a century. The legislation recognizes the urgent need to address health care costs and will initiate a variety of P4P and other payment reform initiatives. These include allowing providers that are organized as accountable care organizations and that voluntarily meet quality thresholds to share in the cost savings they achieve for the Medicare program. Also, an Innovation Center within the CMS will test, evaluate, and expand different payment structures to reduce program expenditures while maintaining or improving quality of care (Kaiser Family Foundation, 2010).

Finally, a broad consensus holds that spending on new medical technologies and drugs is the primary driver of health spending growth in the United States (Smith et al., 2009). This implies that, even if the health care system were perfectly efficient based on standard definitions of efficiency (e.g., productive efficiency), the growth of health care spending may still be unsustainable in the long run. Because of this, cost-effectiveness analysis should be seriously considered as one of the tools to "bend the cost curve." Using cost-effectiveness as one of the criteria for covering new medical technologies has been controversial in the United States (Neumann et al., 2005). As we discuss in this chapter, however, P4P programs could use cost-effectiveness analysis to develop efficiency measures, which would give incentives for providing the most cost-effective health care services.

References

American College of Physicians. (2000). Primer on cost-effectiveness analysis. *Effective Clinical Practice, 5*, 253–255.

American College of Physicians. (2008). Information on cost-effectiveness: An essential product of a national comparative effectiveness program. *Annals of Internal Medicine, 148*(12), 956–961.

AQA Alliance. (2009a, June). *AQA principles of efficiency measures.* Retrieved April 19, 2010, from http://www.aqaalliance.org/performancewg.htm

AQA Alliance. (2009b, June). *AQA parameters for selecting measures for physician and other clinician performance.* Retrieved April 19, 2010, from http://www.aqaalliance.org/performancewg.htm

Baker, G., & Delbanco, S. (2007). *Pay for performance: National perspective, 2006 longitudinal survey results with 2007 market update.* Retrieved November 9, 2009, from http://www.medvantage.com

Bauer, P. W. (1990). Recent developments in the econometric estimation of frontiers. *Journal of Econometrics, 46*(1–2), 39–56.

Berwick, D. M. (2002). A user's manual for the IOM's 'Quality Chasm' report. *Health Affairs (Millwood), 21*(3), 80–90.

Binder, L. F., & Rudolph, B. (2009). Commentary: A systematic review of health care efficiency measures. *Health Services Research, 44*(3), 806–811.

Brown, M. M., Brown, G. C., Brown, H. C., Irwin, B., & Brown, K. S. (2008). The comparative effectiveness and cost-effectiveness of vitreoretinal interventions. *Current Opinion in Ophthalmology, 19*, 202–207.

Centers for Medicare & Medicaid Services. (2009a). *Medicare resource use measurement plan.* Retrieved October 1, 2009, from http://www.cms.hhs.gov/QualityInitiativesGenInfo/downloads/ResourceUse_Roadmap_OEA_1-15_508.pdf

Centers for Medicare & Medicaid Services. (2009b). *CMS measures management system blueprint* (Version 6.2). Baltimore, MD: Centers for Medicare & Medicaid Services.

Centers for Medicare & Medicaid Services. (2009c). *Medicare claims processing manual.* Retrieved November 9, 2009, from http://www.cms.hhs.gov/Manuals/IOM/list.asp

Centers for Medicare & Medicaid Services. (2009d). *Hospital Compare.* Retrieved August 14, 2009, from http://www.hospitalcompare.hhs.gov

Centers for Medicare & Medicaid Services. (2009e). *Physician Group Practice Demonstration evaluation report—Report to Congress.* Retrieved April 19, 2010, from http://www.cms.gov/DemoProjectsEvalRpts/downloads/PGP_RTC_Sept.pdf

Chernew, M. E., Hirth, R. A., & Cutler, D. M. (2009). Increased spending on health care: Long-term implications for the nation. *Health Affairs (Millwood), 28*(5), 1253–1255.

Chirikos, T. N., & Sear, A. M. (2000). Measuring hospital efficiency: A comparison of two approaches. *Health Services Research, 34*(6), 1389–1408.

Chung, J., Kaleba, E., & Wozniak, G. (2008). *A framework for measuring healthcare efficiency and value, clinical performance evaluation.* Retrieved November 9, 2009, from http://www.ama-assn.org/ama1/pub/upload/mm/370/framewk_meas_efficiency.pdf

Cohen J. T., Neumann P. J., & Weinstein M. C. (2008). Does preventive care save money? Health economics and the presidential candidates. *New England Journal of Medicine, 358,* 661–663.

Congressional Budget Office. (2007). *The long-term outlook for health care spending.* Retrieved November 9, 2009, from http://www.cbo.gov/ftpdocs/87xx/doc8758/11-13-LT-Health.pdf

Congressional Budget Office. (2008). *Technological change and the growth of health care spending.* Retrieved November 9, 2009, from http://www.cbo.gov/ftpdocs/89xx/doc8947/01-31-TechHealth.pdf

Culyer, A. (2009, June 22). How nice is NICE? A conversation with Anthony Culyer. *Hastings Center Health Care Cost Monitor.* Retrieved January 20, 2010, from http://healthcarecostmonitor.thehastingscenter.org /admin/how-nice-is-nice-a-conversation-with-anthony-culyer/.

Cutler, D. M., Davis, K., & Stremikis, K. (2009, December). Why health reform will bend the cost curve. *The Commonwealth Fund Issue Brief, 72.* Retrieved April 20, 2010, from http://www.commonwealthfund.org/Content /Publications/Issue-Briefs/2009/Dec/Why-Health-Reform-Will-Bend-the-Cost-Curve.aspx

Davis, K., & Guterman, S. (2007). Rewarding excellence and efficiency in Medicare payments. *Milbank Quarterly, 85*(3), 449–468.

Drexler, M. (2010). Can cost-effective health care = better health care? An interview with Harvard School of Public Health's Milton Weinstein. *Harvard Public Health Review.* Retrieved April 22, 2010, from http://www .hsph.harvard.edu/news/hphr/winter-2010/winter10assessment.html

Fisher, E. (2005, January). More care is not better care. *Expert Voices (National Institute for Health Care Management), 7.* Retrieved January 20, 2010, from http://www.nihcm.org/~nihcmor/pdf/ExpertV7.pdf

Gold, M. R., Siegel, J. E., Russell, L. B., & Weinstein, M. C. (Eds.). (1996). *Cost-effectiveness in health and medicine.* New York: Oxford University Press.

Greenberg, L. (2006). *Efficiency in health care: What does it mean? How is it measured? How can it be used for value-based purchasing? Highlights of a May 2006 national conference co-sponsored by AHRQ and the Employer Health Care Alliance Cooperative.* Report prepared by Academy Health under AHRQ contract #290-04-0001. Available from http://www.academyhealth.org/files/publications/EfficiencyReport.pdf

Greene, R. A., Beckman, H. B., & Mahoney, T. (2008). Beyond the efficiency index: Finding a better way to reduce overuse and increase efficiency in physician care. *Health Affairs (Millwood), 27*(4), w250–w259.

Hussey, P. S., Anderson, G. F., Osborn, R., Feek, C., McLaughlin, V., Millar, J., et al. (2004). How does the quality of care compare in five countries? *Health Affairs (Millwood), 23*(3), 89–99.

Hussey, P. S., de Vries, H., Romley, J., Wang, M. C., Chen, S. S., Shekelle, P. G., et al. (2009). A systematic review of health care efficiency measures. *Health Services Research, 44*(3), 784–805.

Ingenix. (2006). *Symmetry episode risk groups* (Product Sheet). Retrieved October 1, 2009, from http://www.ingenix.com/content/attachments/SymmetryEpisodeRiskGroupsproductsheet.pdf

Institute of Medicine, Board on Health Care Services. (2001). *Crossing the quality chasm: A new health system for the 21st century.* Washington, DC: The National Academies Press.

Kaiser Family Foundation. (2010, April 8). *Summary of new health reform law. Focus on health reform.* Retrieved April 19, 2010, from http://www.kff.org/healthreform/upload/8061.pdf

Kautter, J., Pope, G. C., Trisolini, M., & Grund, S. (2007). Medicare Physician Group Practice Demonstration design: Quality and efficiency pay-for-performance. *Health Care Financing Review, 29*(1), 15–29.

Kohn, L. T., Corrigan, J. M., & Donaldson, M. S. (Eds.), Institute of Medicine (1999). *To err is human: Building a safer health system.* Washington, DC: National Academies Press.

Leapfrog Group & Bridges to Excellence. (2004). *Measuring provider efficiency version 1.0.* Retrieved November 9, 2009, from http://www.bridgestoexcellence.org/Documents/Measuring_Provider_Efficiency_Version1_12-31-20041.pdf

McGlynn, E. A., & Southern California Evidence-Based Practice Center. (2008, April). *Identifying, categorizing, and evaluating health care efficiency measures* (Agency for Healthcare Research and Quality Publication No. 08-0030). Prepared by the Southern California Evidence-Based Practice Center—RAND Corporation, under Contract No. 282-00-0005-21. Retrieved November 9, 2009, from http://www.ahrq.gov/qual/efficiency/efficiency.pdf

Medicare Payment Advisory Commission. (2005a, March). *Report to Congress: Medicare payment policy.* Retrieved November 9, 2009, from http://www.medpac.gov/publications/congressional_reports/Mar05_EntireReport.pdf

Medicare Payment Advisory Commission. (2005b, June). Using clinical and cost-effectiveness information in Medicare. In *Report to the Congress: Issues in a modernized Medicare program.* Washington, DC: Medicare Payment Advisory Commission.

Medicare Payment Advisory Commission. (2009, March). *Report to the Congress: Medicare payment policy.* Retrieved November 9, 2009, from http://www.medpac.gov/documents/Mar09_EntireReport.pdf

Milstein, A. (2008, June 2). Toxic waste in the US health system. *Health Affairs (Millwood)*, (blog). Retrieved November 9, 2009, from http://healthaffairs.org/blog/

Milstein, A., & Lee, T. H. (2007). Comparing physicians on efficiency. *New England Journal of Medicine, 357*(26), 2649–2652.

National Institute for Health and Clinical Excellence. (2009, January). *The guidelines manual 2009.* London: National Institute for Health and Clinical Excellence. Available from http://www.nice.org.uk/aboutnice/howwework/developingniceclinicalguidelines/clinicalguidelinedevelopmentmethods/GuidelinesManual2009.jsp

Neumann, P. J., Rosen, A. B., & Weinstein, M. C. (2005). Medicare and cost-effectiveness analysis. *New England Journal of Medicine, 353*(14), 1516–1522.

Orszag, P. R. (2008). *Opportunities to increase efficiency in health care, statement of the Congressional Budget Office at the Health Reform Summit of the Committee on Finance, United States Senate.* Retrieved November 9, 2009, from http://www.finance.senate.gov/healthsummit2008/Statements/Peter%20Orszag.pdf

Pacific Business Group on Health & Lumetra. (2005, September). *Advancing physician performance measurement: Using administrative data to assess physician quality and efficiency.* Retrieved November 9, 2009, from http://www.pbgh.org/programs/documents/PBGHP3Report_09-01-05final.pdf

Palmer, S., & Torgerson, D. J. (1999). Economic notes: Definitions of efficiency. *BMJ, 318*(7191), 1136.

Pope, G. C., & Kautter, J. (2007). Profiling efficiency and quality of physician organizations in Medicare. *Health Care Financing Review, 29*(1), 31–43.

Pope, G. C., Kautter, J., Ellis, R. P., Ash, A. S., Ayanian, J. Z., Lezzoni, L. I., et al. (2004). Risk adjustment of Medicare capitation payments using the CMS-HCC model. *Health Care Financing Review, 25*(4), 119–141.

Reinhardt, U. E., Hussey, P. S., & Anderson, G. F. (2004). US health care spending in an international context. *Health Affairs (Millwood), 23*(3), 10–25.

Robinson, J., & Center for Health Systems Research and Analysis, University of Wisconsin–Madison. (2008, March). *Development of severity-adjustment models for hospital resource utilization data. A white paper analysis for the Leapfrog Group.* Retrieved November 9, 2009, from http://www.leapfroggroup.org/media/file/RiskAdjustmentWhitePaper.pdf

Robinson, J. C., Williams, T., & Yanagihara, D. (2009). Measurement of and reward for efficiency in California's pay-for-performance program. *Health Affairs (Millwood), 28*(5), 1438–1447.

Robinson, J. C. (2008, May 29). Pay for performance: From quality to value. *Health Affairs (Millwood),* (blog). Retrieved November 9, 2009, from http://healthaffairs.org/blog/.

Romano, M. (2007). Efficiency counts. IHA adds measures to pay-for-performance formula. *Modern Healthcare, 37*(8), 10.

Safavi, K. (2006). Paying for efficiency. *Journal of Healthcare Management, 51*(2), 77–80.

Schoen, C., Osborn, R., Doty, M. M., Squires, D., Peugh, J., & Applebaum, S. (2009, November 5). A survey of primary care physicians in eleven countries, 2009: Perspectives on care, costs, and experiences. *Health Affairs (Millwood), Supplemental Web Exclusives,* W1171–W1183.

Smith, S., Newhouse, J. P., & Freeland, M. S. (2009). Income, insurance, and technology: Why does health spending outpace economic growth? *Health Affairs, 28*(5): 1276-1284.

Thomas, J. W. (2006, January). *Hospital cost efficiency measurement: Methodological approaches.* Retrieved November 9, 2009, from http://www.pbgh.org/programs/documents/PBGHHospEfficiencyMeas_01-2006_22p.pdf

Thomas, J. W., Grazier, K. L., & Ward, K. (2004). Economic profiling of primary care physicians: Consistency among risk-adjusted measures. *Health Services Research, 39*(4 Pt 1), 985–1003.

US Government Accountability Office. (2007, April). *Medicare: Focus on physician practice patterns can lead to greater program efficiency* (GAO-07-037). Washington, DC: US Government Printing Office.

Varian, H. R. (1992). *Microeconomic analysis* (3rd ed.). New York: W. W. Norton and Company.

Wennberg, J. E., Fisher, E. S., & Skinner, J. S. (2002). Geography and the debate over Medicare reform. *Health Affairs (Millwood), Supplemental Web Exclusives,* W96–W114.

White, C. (2007). Health care spending growth: How different is the United States from the rest of the OECD? *Health Affairs (Millwood), 26*(1), 154–161.

World Health Organization. (2001). *Macroeconomics and health: Investing in health for economic development* (Report of the WHO Commission on Macroeconomics). Available at http://whqlibdoc.who.int/publications/2001/924154550x.pdf

CHAPTER 6

Who Gets the Payment Under Pay for Performance?

Leslie M. Greenwald

Pay for performance (P4P) models involve several complex design elements. One of the most difficult—but important—of these design elements is determining whom P4P should reward. P4P models, in theory, work because they closely link positive incentives (the reward, or payment) with measurable performance achievements. If the performance and reward are not clearly related and/or if a program makes additional payments to providers or clinicians who are not directly responsible for performance, incentives to change behavior and improve care may not be effective, and the program may be misspending the scarce resources devoted to performance payments.

Many organizations that have experimented with P4P models have struggled with the issue of whom to pay. In their article describing practical issues related to P4P systems, Young and Conrad (2007) describe the problem of whom to pay in terms of "units of accountability" and consider this topic one of four key design issues for P4P programs. (Chapter 2 of this book also provides an overview of common P4P models.)

Many models focus payments on physicians and other clinicians, whereas others pay institutions (such as hospitals) for improved performance (Young et al., 2005). In the United States, performance-based payments vary widely, even among models that focus on physicians. They range from payments directed at individual physicians to those made to large group practices (which may include nonphysicians) (Felt-Lisk et al., 2007; Landon et al., 1998). International models also vary; P4P models in the United Kingdom direct quality-improvement payments to physician practices, not to individual physicians (Smith & York, 2004). Although some existing models suggest whom to pay, literature summarizing and evaluating current P4P systems often notes that this area warrants more research (Folsom et al., 2008; Young & Conrad, 2007).

This chapter discusses topics related to whom to pay in P4P. Although some literature on specific P4P models exists, publications on broader issues related to the design of implementable P4P initiatives are limited. Our experience in the design, implementation, and evaluation of many Medicare P4P projects enables us to observe and formulate solution options for key implementation issues—such as whom to pay and what to pay for—under different P4P models. First, we discuss why deciding whom to pay can be such a complex issue in P4P models, and we note factors that can influence this decision. Second, we outline the options for specific health care provider entities who might receive payments under P4P. In discussing the options for whom to pay, we consider the related topic, *what to pay for*. This chapter concludes with a discussion of the respective pros and cons of the options for making payments to different health care providers.

What Makes the Issue of Whom to Pay So Complex in Designing Pay for Performance?

Determining whom to pay is a central design issue. Lack of a single "right" answer or even consensus around best practices highlights the difficulty in choosing among the options. Ultimately, practical options for whom to pay include clinical providers of health care (individual physicians, physician groups, hospitals or integrated delivery systems), insurers (managed care organizations), and other care managers (such as case management organizations). Which option is most practical and appropriate often depends on the primary goals and incentives in the P4P model. Further complicating the issue of whom to pay is the related issue of what to pay for. Identifying the most appropriate entity to reward with performance payments is difficult until one considers what we are paying for.

Rewarding Clinical Providers vs. Other Organizations

Sponsors of P4P models have to make an initial decision on whether to focus rewards directly on clinical providers and health care practitioners (often physicians but sometimes also hospitals) or on other contracted organizations such as managed care plans or case managers (e.g., primary care case managers). Ideally, as several evaluations of existing models suggest, P4P programs must clearly tie performance payments to measurable improvement; otherwise, the incentive to change behavior for quality improvement is less effective (Folsom et al., 2008; Young et al., 2005). Many P4P programs have interpreted this assumption as indicating direct measurement and incentive payments to physicians, physician groups, or hospitals. (Chapters 3 and 8 of this book provide a more detailed examination of the economic theory behind health care provider incentives and payments under P4P.)

The decision to focus payment on clinical providers involves the complex issue of whether to focus on individual health care professionals or provider groups. Focusing on individual clinicians has the potential to offer the strongest direct incentives for behavioral change; however, because of the limited number of cases, this method presents the greatest challenges to valid measurement calculation, and assigning clinical attribution is extremely complex. Offering incentives to provider groups weakens behavioral incentives for improvement, but it involves more cases, thus improving the validity of measurement calculation, and acknowledges the team effort of clinical management.

Sponsors of P4P programs must also sometimes provide coverage through insurance intermediaries, such as managed care plans; this requirement makes direct reimbursement of providers and clinicians more complex. The majority of Medicaid performance-based systems, in which managed care plans and primary care case managers are the dominant entities that engage with states in P4P, face the problem of incentive weakening caused by the separation of P4P payments from direct providers of care (Kuhmerker & Hartman, 2007). (This problem occurs because most states require managed care enrollment for the majority of their Medicaid population.)

Paying indirect providers, such as managed care organizations, offers some advantages. These organizations offer larger and more diverse groups of patients, which in turn improves quality measurement validity. Thus, these organizations, more often than other smaller insurers or self-insured groups, are able to collect and submit performance-based data. For example, managed care plans already collect and submit Health Plan Employer Data and Information Set (HEDIS) performance data, which several state Medicaid P4P models already use (Kuhmerker & Hartman, 2007). In addition, modifying managed care organizational contracts, which potentially cover networks of clinicians and large blocks of patients, is far less complex than making modified payment arrangements with many individual health professionals.

Directing performance payments to clinical providers and practitioners rather than to insurers and other organizations has both advantages and disadvantages. Performance payments made to providers have the benefit of improved incentives and greater ability to actually change patient outcomes. Yet incentives that under some circumstances prompt improvements in care can become perverted if health care providers use their proprietary knowledge to select patients for care in ways that maximize their performance bonuses rather than optimize care. Alternatively, performance payments made to insurers can include broader and more diverse groups of patients, leading to better performance measurement accuracy and therefore wider buy-in for participation. However, managed care organizations can be too far removed from clinical care to actually change physician and hospital behavior. Although further removed from actual patient care, insurers can engage in biased selection to avoid high-risk patients that might have a negative impact on measured performance.

Unfortunately, there is no single best approach. Adopting a particular approach may depend on whether a P4P sponsor has the technical and/or legal ability to modify payment arrangements with providers to incorporate payment-based performance standards. Adoption may also depend on the technical requirements of the performance measures that the program sponsor chooses.

Attribution of Responsibility

Attributing responsibility for outcomes is another critical factor in determining whom should be paid for performance (discussed in greater detail in Chapter 7). Presumably, performance measures create the strongest incentives for improvement when providers and clinicians clearly define and accept responsibility for both success and failure. Furthermore, literature on the structure of P4P models suggests that parties held responsible for performance outcomes must have the means within their control to meet the targets (Durham & Bartol, 2000).

In clinical settings, determining who is responsible for specific performance outcomes is far from easy. For example, P4P programs may be able to assign responsibility for certain process quality measures, such as rates of immunizations or other disease screenings, but they may be unable to ascertain who is responsible for a hospital readmission when a patient has multiple comorbidities. Quality measures that focus on clinical outcomes rather than specific, observable processes of care create other problems of attribution.

Although programs commonly hold physicians responsible for performance targets, program sponsors note that this approach neglects the interdependency of physicians with other clinical professionals and support staff. Moreover, some staff in these categories, unlike physicians, have no direct incentive to change behavior to meet specific targets (Young et al., 2005). Evaluators and policy makers often criticize current P4P models for their inability to clearly identify (and reward) the part of the health care system that affects a target outcome (Evans et al., 2001). This criticism assumes, of course, that clinical outcomes are affected primarily by a single segment of the total health care system, which may not be the case.

Limited patient resources may also affect clinical providers' or individual practitioners' ability to meet performance targets. Such limitations might involve treating patients who do not have insurance or are otherwise unable to pay for certain elements of care, affecting clinical outcomes caused by factors beyond the control of the health care provider. When insurers (such as managed care organizations) are held responsible for performance, however, impacts on outcomes caused by lack of coverage or access may be more appropriate. In such instances, managed care organizations or other insurers may legitimately be responsible for performance because determining coverage and benefits is within their control. Clinical providers and individual practitioners may be unable to gain access to either clinical information (such as electronic medical records or comprehensive health information systems) or other resources necessary to manage care for general quality outcome measures. Once again, insurers or other organizations may control these resources.

Young and colleagues (2005) suggest that, because of these attribution problems, future programs may link performance incentives to health care delivery teams. However, they note that the concept of setting clinical responsibility at a diverse team level—possibly including clinical providers or practitioners, as well as nonclinical and even insurer partners—may upset traditional notions of clinical responsibility and professional independence. Assigning responsibility for performance to nonclinical providers, such as managed care plans, in some ways inverts the concept. Managed care organizations do not directly provide care. They do, however, provide the basic resources, benefit structures, and other management rules that govern what care is provided to whom.

This discussion on the difficulty of assigning appropriate responsibility—and rewards—for clinical outcomes again highlights the lack of a consensus on the best approach. Difficulty assigning clinical responsibility for outcomes may, in reality, reflect the current fragmented system of care in the United States more than problems inherent in P4P systems alone. Additionally, in attributing responsibility, we need to consider the role of patients themselves. Patient adherence or nonadherence to clinical recommendations may have a large effect on outcomes. Are providers and clinicians responsible for changing the behavior of nonadherent patients or only for recommending behavioral changes? Should provider and clinician performance be risk adjusted for patient characteristics or clinicians? Should some of the financial incentives of P4P be directed to patients rather than providers? We have yet to figure out how best to account for patient responsibility.

Defining the Unit of Care for Payment: What to Pay For

Closely related to determining whom to pay is defining the unit of care for which P4P programs measure and reward performance. To determine *whom* to pay, programs must also consider *what* to pay for—for example, care settings, services, diseases, and events to include. They must also consider the unit of care for which performance is measured. The unit of care could include individual services, episodes, or all services over a unit of time such as a year (capitation).

Current P4P models define the unit of care in various ways. Programs commonly pay for a specific scope of care. For example, performance measures can be created that account for either all care or only inpatient, outpatient, or other care settings. Including a broader set of clinical care settings may be more likely to improve overall care than focusing on a narrower set because the performance measures will not artificially place care-setting boundaries on providers. Including more care settings may, however, also compound the problem of attributing responsibility for performance measures. If performance is based on a narrow setting (e.g., only inpatient care), then there is a greater focus on fewer potentially responsible providers, and appropriate assignment of clinical responsibility on which performance is based is more feasible. This narrow focus also allows programs to more closely align incentives to change behavior through performance payments.

The scope of care subject to performance incentives can also include specific diseases, such as chronic illnesses. This condition-specific approach is typically used when care management organizations are a focus of P4P models because they most often develop a specific protocol for improving care for specific disease categories.

Finally, P4P models can also focus on specific events, such as "never events"—preventable medical errors that result in serious consequences for the patient. In these instances, variants of P4P models might either withhold usual payments when certain events occur (as with the never events proposed in the CMS Inpatient Prospective Payment System rule effective for hospital discharges on or after October 1, 2009). Examples of these events include foreign objects retained after surgery and catheter-associated urinary tract infections. P4P models may also base performance awards on low rates of medical errors or other similar negative clinical outcomes.

P4P sponsors face a key question in defining the scope of services to be included in the unit of care for which they wish to measure performance. The most narrow, and traditional, unit is the individual service, which corresponds to the unit of payment in traditional fee-for-service (FFS) medicine. The problem with this narrow definition is that it provides a very limited basis on which to judge performance. Quality of care typically requires a more global perspective on patient management than the individual service. Similarly, efficiency in providing individual services is important but does not capture the number or intensity of services provided in the course of a patient's treatment.

At the other end of the spectrum from the individual service is the capitated unit of service. Here, a single entity, such as a health plan, is contractually responsible for all medical services provided to an enrollee over a fixed enrollment period, typically a year. Capitation clearly attributes all responsibility for care to a single organization. This can be a strength, but global capitation may be too aggregated to measure performance and focus incentives on individual provider organizations. Thus, interest is increasing in the episode of care as a unit of accountability.

An episode of care can be defined in several ways, including the following:

- **Annual episodes.** All care, or care related to the condition, provided to a patient with a chronic condition during a year or some other prespecified calendar time interval. For example, management of a congestive heart failure patient over an annual period might constitute a practical or easy-to-measure episode of care.

- **Fixed-length episodes.** All care, or care related to a condition, provided within a fixed time window preceding and/or following an index event, such as initial treatment, surgical procedure, or a hospitalization for a medical condition. For example, a payer might bundle all services from 3 days prior to until 30 days after a hospital stay for coronary artery bypass surgery into one episode.

- **Variable-length episodes.** Care related to a medical condition (e.g., ankle fracture), from the initial treatment for the condition (e.g., diagnosis) until the course of treatment is completed (e.g., recovery and follow-up).

Which episode definition is appropriate depends on the performance measures that a P4P program chooses. For example, programs typically define specific screening process measures—such as rates of immunization, screening tests, or periodic medical tests or check-ups—for an episode, corresponding to the appropriate clinical indication. For immunizations or certain screening tests, such as mammography, this may be annually; for other screening tests, such as colonoscopy, this may be every 5 years, depending on current clinical guidelines. Guidelines might recommend that patients have diabetic eye and foot examinations every year or every 2 years.

As another example, programs may appropriately assess other performance measures, such as lowering rates of rehospitalization, using a fixed-length post-discharge episode (e.g., 7, 14, or 30 days after discharge). Measured performance may be highly sensitive to the length of time after initial discharge that programs include in the defined episode. Using a narrow window, such as a 7-day post-discharge rehospitalization window, is more likely to attribute rehospitalization appropriately to the effects of the initial hospitalization, but it may not capture all of the sequelae of the initial care. A 30-day post-discharge definition will include more of the initial discharge's subsequent effects, but it may also capture events unrelated to the initial hospital stay. Further complicating this approach is the ongoing debate over appropriate lengths of stay and post-discharge settings, which vary considerably based on the specific clinical diagnosis. Variable-length episodes most precisely measure care related to a particular medical condition, but they are the most complex to define. Attributing medical services to particular conditions and deciding when episodes begin and end may be difficult, especially for patients with multiple, coexisting medical problems (e.g., as is the case for many elderly patients).

Pros and Cons of Episode Length

Some general observations highlight the pros and cons of longer versus shorter episodes defined for the purposes of P4P. Longer episodes allow better evaluation of real clinical changes. Including patient outcomes over months or potentially even years is more likely to document lasting changes in clinical performance, particularly among complex patients. For example, changes in care for chronic diseases are unlikely to manifest significantly improved clinical outcomes over a period of only months. A recent evaluation of the Massachusetts Health Quality Partners program suggests that some performance improvements may take 2 years to show results (Folsom et al., 2008). Longer episodes also allow programs and providers or clinicians to identify more clinical complications that may not necessarily be the focus of the performance intervention but may affect outcomes. This makes assessing performance more complex but may yield more accurate measurements of lasting quality improvement. Finally, using longer episodes of care across multiple provider settings allows programs to evaluate potential cost shifting. This may be particularly important for performance measures that focus on reducing cost. Programs must consider reductions in costs caused by decreased hospital lengths of stay and/or fewer rehospitalizations, given corresponding use of alternative services such as post-acute care.

Applying longer episodes of care to defining what to pay for has downsides for P4P, too. The largest disadvantage is the potential dilution of the impact of incentive rewards for performance improvements. Evaluations of the Rewarding Results Demonstration sites suggest that quality improvement incentives should be made rapidly and frequently, potentially even multiple times per year (Folsom et al., 2008). This approach allows providers and individual practitioners to gain immediate feedback on their progress toward improved quality of care and strengthens their motivation to continue. However, although this approach is desirable for strengthening the impetus to improve care, it is potentially inconsistent with including longer episodes of care as a basis for determining what to pay for.

Using longer, bundled episodes also substantially increases the financial risk that provider organizations face. The variance of episode costs rises with length of episode and number of services bundled into the episode. For example, placing hospitals at risk for readmissions—especially over a longer period—puts them at a great deal of financial risk. Some of this risk is incurred by their own performance in avoiding readmissions, but over longer periods, most of this risk can be attributed to insurance and is unrelated to the hospitals' performance. To ameliorate this insurance risk, providers should receive bundled episode payments to treat a large number of cases to average out the risk; payments should incorporate risk adjustment; and outlier, reinsurance, or other exceptions policies for catastrophic cases should be in place.

The alternative to defining longer episodes of care is defining shorter ones. Shorter episodes of care more consistently link incentive payments to changes in provider behavior. Moreover, providers can more easily connect positive payment incentive rewards with more recent behavior changes. Also, shorter-term outcomes are easier to attribute to specific instances of provider care, such as specific hospitalizations.

However, using shorter episodes of care may reward providers for only short-term improvements rather than more substantive and lasting clinical outcomes. For example, a program may reward providers for savings observed during a 6-month episode; for the same patient, however, these savings could evaporate by the end of 12 months.

Who Gets the Money?

When moving from a disaggregated, FFS payment system to a more bundled, episode-based system, determining which organization receives payment for the episode may be contentious. For example, if post-acute care is bundled with hospitalizations, the post-acute providers (e.g., home health agencies or skilled nursing facilities) may feel threatened if a single lump-sum episode payment is made to the acute care hospital, which then has responsibility for paying the post-acute providers. Currently, these health care provider organizations are paid separately under distinct payment rules. P4P models that would place payment of currently independent providers under the control of another organization may lead to concerns for fiscal and/or clinical independence. In some cases, these various provider types will all be members of a unified integrated delivery system that could accept payment on behalf of all the providers. In many cases, however, such providers represent unrelated organizations.

P4P programs will reward accountability and efficiency if payment is made to a single entity that is responsible for an episode and can coordinate care decisions and reap the rewards of improved quality and efficiency. This single entity may decide to reconfigure current care patterns—which is the point of episode payment—threatening the livelihood of existing provider organizations by reducing the services provided. Various bundling policies, transition policies, and gainsharing could be designed to try to soften the impact of bundled episode payment, but any dislocation in current arrangements would be likely to prove controversial. If a single bundled episode payment is made, the organization that receives it should be capable of and willing to manage the entire episode of care. For example, if an acute care hospital is to be paid for episodes that include post-acute care, the hospital must be willing to contract for and manage post-acute services such as home health and skilled nursing services.

Options for Whom to Pay

Determining whom to pay is a difficult but critical decision for P4P programs. For most P4P models, we have observed six possible options:

1. individual physicians

2. physician group practices

3. hospitals or hospital systems

4. physician-hospital organizations (PHOs) and integrated delivery systems (IDSs)

5. health maintenance organizations (HMOs) and other managed care organizations (coordinators of care that are also directly responsible and financially at risk for care delivery)

6. disease management organizations (organizations that coordinate care but do not have direct responsibility for care management).

The choice of option depends in part on the structure and quality-improvement focus of the specific P4P program to be implemented. These options are not mutually exclusive. For example, both physicians and hospitals can be simultaneously involved in coordinated or uncoordinated P4P programs. Still, each option can be evaluated for distinct advantages and disadvantages, generally determined by the following criteria:

- ability to convey strong incentives for performance improvement,
- locus of care specificity,
- ability to muster sufficient sample size for valid performance measurement,
- ability to connect control over outcomes, and
- ability to match with varied scopes of care (short or long episodes of care).

We discuss each option for whom to pay in terms of these criteria.

Paying Individual Physicians

Monitoring and rewarding individual physicians offers strong incentives for specific individuals to improve their performance. By rewarding, not rewarding, or penalizing each individual physician, payers clearly link performance and payment, which is ideal for P4P models. Of course, this link is strongest when rewards are based on a clearly accepted method of performance for a specific physician. This method of choosing whom to pay can be appropriate for several types of care because the physician is a primary driver of care in almost all clinical settings. The central role of the physician in directing care can, in theory, make it feasible to match paying individual physicians for performance with a range of units of payment. In theory, payers could match individual physicians with payments based in many care settings, particularly those with an outpatient focus. Also, payers theoretically could match individual physicians to short and long episodes of care. However, the broader the setting of care (e.g., if settings include inpatient services) and the longer the episode, the lower the probability that an individual physician will be able to exert sufficient clinical control to be held accountable for all but very focused performance outcomes.

Paying individual physicians for performance faces numerous difficulties. One particularly persistent problem is gathering sufficient sample size for valid performance measurement. Depending on the focus of performance standards and the expected incidence of patients, individual physicians will not have a sufficient number of patients; patient outliers and large variation in performance metrics driven by small numbers thus become problematic. For example, performance measures based on general process measures, such as rates of immunizations, may be feasible for performance payments directed at physicians with a reasonable practice size (e.g., 2,000 patients). However, more clinically focused measurements, pertaining only to a potentially small subset of a clinical practice, can yield numbers too small to detect the difference between actual performance and expected variance among a small group of patients.

This "sample size" problem is compounded by difficulties that a single physician may encounter in controlling all clinical activity that might affect performance measurement, even if that physician is a primary care specialist. For example, if a patient sees multiple physicians or has a hospitalization (or both), these health care system encounters may affect the performance of a single P4P participating physician, yet that participating physician could not control all factors that may have affected his or her performance (for better or worse). FFS insurance structures just exacerbate this problem because physicians have only limited control over the health care services that patients consume; patients are free to see specialists or other physicians and to receive a wide range of health care services that do not require prior authorization or review by a single physician. Exceptions to this general rule are demonstration or other pilot projects requiring primary care management. This problem is less frequent in managed care settings in which primary care physicians may act as gatekeepers. Still, the compliance of other providers and the patient with review of services by a primary care or other coordinating physician can vary widely.

Paying Groups of Physicians

Moving away from paying individual physicians solves some problems but also causes new ones.

Groups of physicians who are rated and rewarded collectively are more likely to generate sufficient sample size for valid performance measurement than are single physicians. A larger group of physicians, particularly a multispecialty group working together in a coordinated way, is also more likely to control patient outcomes over a wider scope of care settings. For example, if the group includes both primary care and internal medicine, medical subspecialties, and surgeons, care is far more likely to be coordinated and controlled over a broader scope of services, making an understandable connection between outcomes and reward. Working examples of physician groups that have been organized successfully for this purpose include the Medicare Physician Group Practice Demonstration sites (this demonstration is described in detail in Chapter 9). Physician groups may also have management- or peer-based mechanisms to provide performance feedback to individual physicians that can facilitate initiatives to improve organizational care.

However, even groups of physicians will still be unable to have complete control over care in FFS environments when patients are not required to gain prior approval or are not subject to care coordination. Although patients can be encouraged to seek care within a group practice and many may find it convenient and beneficial to do so, fragmented care is still possible. Payers could theoretically match groups of physicians, like individual physicians, with payments based on many settings of care, particularly settings that primarily serve outpatients. Similarly, payers can, in theory, link physicians in groups to both short and long episodes of care. Also, as with P4P programs targeting individual physicians, the broader the setting of care and the longer the episode of care, the lower the probability that groups of physicians will be able to exert sufficient clinical control to be held solely accountable.

Paying Hospitals

With respect to P4P payments, hospitals have distinctly different advantages and disadvantages from physicians. As an advantage, hospitals have potentially large numbers of patients around whom to fashion a P4P model. If hospitals are rewarded for their good performance of inpatient services, they can also be highly motivated to meet performance goals. Hospitals can also have powerful incentives to improve care, control costs, and meet other performance standards and can set facility practices, procedures, and resource allocations that reflect these incentives.

Balancing these advantages as a basis for P4P payment, hospitals must work through physicians, who likely have the most powerful role in clinical decision making, to implement most clinical outcome and process improvements. If hospitals are not also rewarded downstream in some way, physicians may actually have outright disincentives to comply with performance standards that reduce lengths of stay, numbers of procedures, or other care that could negatively affect their income. For this reason, models such as the Medicare Centers of Excellence and the Medicare Gainsharing Demonstrations have incorporated physician rewards into hospital-focused P4P models. These demonstrations are described in detail in Chapter 9.

Moreover, hospitals are, by their nature, limited to P4P models that focus on inpatient care. Thus, they may also lack the ability to control performance for longer episodes of care that cut across outpatient, inpatient, and post-acute care settings. P4P initiatives often attempt to eliminate unnecessary hospitalizations, and creating effective incentives for hospitals to cut their own revenue is likely to be difficult. Finally, hospitals rarely admit patients on their own. With the exception of admissions generated from emergency rooms, this function is largely performed by physicians.

Paying Physician-Hospital Organizations and Integrated Delivery Systems

Among the six practical options presented, PHOs and IDSs may inherently have the fewest limitations for whom to pay. PHOs and IDSs are likely to control sufficient numbers of patients to minimize sample size problems and to produce reliable performance results. By coordinating care among physicians, hospitals, and other providers, PHOs and potentially broader IDSs theoretically control care and outcomes in ways that payers can measure across various scopes of care, including both outpatient and inpatient settings. Consequently, PHOs and IDSs may also have a greater span of clinical control appropriate for both short and longer episodes of care, particularly if such organizations also coordinate post-acute care providers.

The weakness of PHOs and IDSs as units of reward in P4P models is their somewhat diffuse ability to convey strong incentives for behavioral change. Although the relative clinical breadth of providers in these multi-provider organizations theoretically expands the span of control over farther-ranging clinical outcomes, it also weakens any direct incentive that individual providers have to change their specific behaviors. In performance rewards made to larger organizations, inherently fewer direct links exist between the rewards and individual members of the organization. Internal reward allocation strategies to strengthen these ties may be difficult to develop and implement. Also, in an FFS environment, PHOs and IDSs rarely receive any direct payments for care. Their lack of control over any major flow of funds reduces their influence. These organizations could, however, serve as possible recipients for bundled physician-hospital or episode payments.

Paying Health Maintenance Organizations and Other Managed Care Organizations

Managed care organizations, such as the still-dominant HMO model, are financially responsible for the purchasing and financing aspects of care and, as such, have (at a minimum) a role in the oversight and management of clinical care. A purchaser (such as an employer or a state Medicaid agency) might reward a managed care organization (such as an HMO) if, for example, the managed care organization met specific performance measures such as patient satisfaction, HEDIS, or other quality of care standards. The twin abilities of the managed care organization to control financial and other resources and to set standards for clinical care are this model's strengths.

In theory, managed care organizations have the potential to control care in a wide range of clinical settings and across both short and long episodes of care. Managed care organizations also typically control large numbers of patients. However, depending on the relationships between managed care organizations and providers and the relative power of the managed care organization to enforce clinical guidelines and policies, P4P programs may have only a weak or moderate ability to connect clinical control over outcomes with incentives for behavioral change. This factor will vary significantly among managed care models.

HMOs, which have closed provider networks, have the greatest clinical control over providers. Preferred provider organizations, which have looser networks and allow patients to use out-of-network providers with reasonable cost sharing, are less able to control provider behavior and patient outcomes. At the furthest end of the control scale, the private FFS model of managed care organizations, operating without provider networks, has virtually no control over physician behavior. Thus, managed care organizations that have competitive, closed networks or that operate on a staff model are more likely to exert strong clinical control over provider behavior and thus to have the technical and data availability to make strong connections between performance and reward. More loosely organized managed care organizations that have limited or open provider networks neither exert the same clinical control nor have the informational resources to create incentives for providers to modify their behavior according to performance standards.

Paying Disease or Care Management Organizations

Paying disease or care management organizations for performance has the same limitations as paying institutions (such as hospitals) without including the financial and resource control of IDSs. Disease or care management organizations are used in models of P4P most typically when these organizations are rewarded for meeting clinical outcome and/or cost savings performance measures for a specific set of patients.

These organizations may serve large numbers of patients across multiple practices. This factor introduces problems of clinical control and coordination, depending on how well the model incorporates direct care providers, such as physician offices. Frequently, coordination of disease/care managers with physicians and other providers is difficult because these managers typically operate outside physician offices.

In addition, care management organizations often operate on an at-risk basis for their fees. This practice gives them a strong incentive to lower health care expenditures. It also places them in conflict with physicians, who generally do not share these savings. This in turn weakens the ability of care management P4P models to create links between system rewards (which may not be shared with providers at all) and the performance of physicians and other clinicians, who make most care decisions. When P4P programs operate in managed care settings and program participation and cooperation are mandatory, care management organizations can exert more control over the clinical care that network providers give.

Care management models often focus on specific subsets of diseases, such as diabetes, cancer, or congestive heart failure; thus, most typically apply only to very limited scopes of care. Because they generally lack a direct relationship with direct care providers, care management organizations have only a limited ability to provide care, through either short or longer episodes. On the positive side, disease management organizations enjoy economies of scale because they can replicate their model and operations nationally and even worldwide. Given the high fixed costs of developing care management models, hiring and training the necessary specialized personnel, and developing the technology and data management platforms, there may be a role for disease or care management organizations in supporting and collaborating with health insurers and with physician practices—especially smaller ones—for which these large investments are impractical.

Summary and Discussion

The previous section highlights the lack of a single best approach for whom to pay under P4P models. The decision is often driven by other, more fundamental design issues; of particular importance are the presence of an insurance intermediary and the nature of the target performance measures and populations. In short, no universally accepted best practices exist, and those aiming to implement P4P programs must contend, instead, with a range of different strengths and weaknesses.

Table 6-1 summarizes the relative strengths and weaknesses of each of the six practical provider P4P models, addressing the criteria discussed earlier in this chapter.

Directing P4P payments to individual physicians is likely to provide the strongest incentives to change clinical behavior. Although other practical and technical problems arise in directing performance payments to individual physicians, the direct tie between individual performance and individual payment is clearly the strongest. On the other end of the scale, directing performance payments to disease management organizations, which have loose ties to and little control over key providers—particularly physicians—are likely to result in the weakest incentives for behavioral change.

Regarding the appropriateness of each payment unit for various scopes of care, the provider payment options differ widely. The organizations that include and control the widest range of providers—PHOs/IDSs and managed care organizations—are most likely to have multiple scopes of care.

Table 6-1. The strengths and weaknesses of six options for whom to pay

Whom To Pay	Ability to Convey Strong Incentives	Locus of Care Specificity
Individual physicians	Strong	Most appropriate for outpatient care; may be appropriate for inpatient care
Groups of physicians	Moderate	Most appropriate for outpatient care; may be appropriate for inpatient care
Hospitals	Moderate	Appropriate for inpatient care
Physician-hospital organizations/ integrated delivery systems	Weak to moderate, depending on provider specific allocation methods	Range in applicability to various scopes of care
Managed care organizations	Moderate to weak, depending on managed care model	Range in applicability to various scopes of care
Disease or care management organizations	Weak	Range in applicability to various scopes of care

The six provider payment options also vary considerably in their ability to obtain sufficient patient sample size for accurate performance measurement. Unfortunately, individual physicians (who have the strongest incentives to respond to performance payments made directly to them) are unlikely to have a large enough patient base to deliver the sample sizes necessary for accurate and reliable performance measurement. Organizations that control multiple providers—PHOs/IDSs and managed care organizations (which are also appropriate for a range of scopes of care)—have the strongest ability to obtain sufficient patient sample sizes to make payments based on accurate performance measures.

For some of the criteria, few of the provider payment options have clear relative strengths. For example, none of the models has a strong ability to connect payments to control of patient outcomes. The models that perform best on this criterion are, once again, those with the broader range of provider participants (PHOs/IDSs and managed care organizations). Disease management organizations, having the least direct control over providers,

Sufficient Sample Size for Accurate Performance Measurement	Ability to Connect Payments to Control Over Outcomes	Ability to Match with Appropriate Payment Unit
Weak	Weak	Most appropriate for short episodes of care
Moderate to strong	Weak to moderate	Most appropriate for short episodes of care
Strong	Weak to moderate	Most appropriate for short episodes of care
Strong	Moderate	Appropriate for short and long episodes of care
Strong	Moderate to high, depending on provider relationship	Appropriate for short and long episodes of care
Moderate to strong	Weak	May face difficulty matching payments with episodes as opposed to specific services

consequently have only a weak ability to connect performance payments to outcomes. Finally, these multiprovider organizations also have the greatest ability to match payments with both long and short episodes of care.

The results presented in Table 6-1 do not suggest that any single approach is best. The considerations presented suggest that options that include multiple providers may work best for a greater number of applications. These alternative options balance key factors such as the strong ability to convey incentives for performance improvement, offer at least a moderate ability to connect rewards with control over outcomes, and ensure the availability of sufficient patient sizes for valid performance measurement. Such options include paying groups of physicians, hospitals, PHOs/IDSs, and managed care organizations. Various Medicare demonstrations are either currently operating or developing all these payment options.

References

Durham, C. C., & Bartol, K. M. (2000). Pay for performance. In E. A. Locke (Ed.), *Blackwell handbook of principles of organizational behavior* (pp. 150–165). Oxford: Blackwell.

Evans, D. B., Edejer, T. T., Lauer, J., Frenk, J., & Murray, C. J. (2001). Measuring quality: From the system to the provider. *International Journal for Quality in Health Care, 13*(6), 439-446.

Felt-Lisk, S., Gimm, G., & Peterson, S. (2007). Making pay-for-performance work in Medicaid. *Health Affairs (Millwood), 26*(4), w516–w527.

Folsom, A., Demchak, C., & Arnold, S. B. (2008). *Rewarding results pay-for-performance: Lessons for Medicare.* Washington, DC: AcademyHealth.

Kuhmerker, K., & Hartman, T. (2007). *Pay-for-performance in state Medicaid programs: A survey of state Medicaid directors and programs.* Available from http://www.commonwealthfund.org/usr_doc/Kuhmerker_P4PstateMedicaidprogs_1018.pdf?section=4039

Landon, B. E., Wilson, I. B., & Cleary, P. D. (1998). A conceptual model of the effects of health care organizations on the quality of medical care. *JAMA, 279*(17), 1377–1382.

Smith, P. C., & York, N. (2004). Quality incentives: The case of UK general practitioners. *Health Affairs (Millwood), 23*(3), 112–118.

Young, G. J., & Conrad, D. A. (2007). Practical issues in the design and implementation of pay-for-quality programs. *Journal of Healthcare Management, 52*(1), 10–18; discussion 18–19.

Young, G. J., White, B., Burgess, J. F., Jr., Berlowitz, D., Meterko, M., Guldin, M. R., et al. (2005). Conceptual issues in the design and implementation of pay-for-quality programs. *American Journal of Medical Quality, 20*(3), 144–150.

Attributing Patients to Physicians for Pay for Performance

Gregory C. Pope

Pay for performance (P4P) programs often focus on physicians because of physicians' control over the provision of medical services to patients. Physician P4P requires the attribution of patients to physicians who will be held responsible for their care. This chapter addresses the issue of attributing (or assigning) patients to physicians for the purpose of P4P when there is no enrollment or designation process in which patients choose a physician or a health care organization to manage their care. A major focus of the chapter is the traditional Medicare fee-for-service (FFS) program. The chapter also reviews private insurance examples and literature, and many of the issues and concepts are similar in the Medicare and private insurance contexts.

Importance of Attribution

A major criticism of much of current US health care, and the Medicare FFS program in particular, is the lack of accountability for patients' care (Hackbarth, 2009). P4P programs are intended to remedy this deficiency by assigning responsibility, measuring performance, and rewarding results. The necessary first step in P4P is to assign physicians responsibility for a defined group of patients or episodes of patient care. Assignment is controversial; physicians contend that if their performance is to be evaluated fairly, they should have control over the patients or episodes of care assigned to them (Beckman et al., 2007). In addition to face validity, assignment affects the number and proportion of patients assigned to a physician. Assignment thus affects the statistical validity of performance assessment and the proportion of physicians' care used to judge their performance. The effect of alternative assignment rules on measured physician performance is an empirical question addressed later in this chapter.

Patient assignment is an important operational issue in P4P programs. It defines the population for which performance will be measured. For example, for larger health care organizations, should programs base assignment on primary care services or physicians only, or also on specialists and their services? If P4P programs use only primary care for assignment, they may be excluding from assignment sicker patients who are using mostly or exclusively specialist services. But organizations may not consider themselves responsible for—or as having control over—patients whom their primary care physicians (PCPs) are not managing.

Patient assignment may also have important incentive implications. In utilization-based attribution, physicians may attempt to avoid caring for high-cost or nonadherent patients so that these patients are not included in their cost or quality profiles. Attribution can be made narrowly to individual physicians, or more broadly to groups of physicians and other clinicians. Narrow assignment may lead to more specific accountability, but broader assignment may foster care team collaboration (Beckman et al., 2007).

Attribution of patients and care episodes may have high stakes for physicians. Many health plans perform economic (efficiency) and quality profiling of network physicians, using methods that entail attributing episodes of patient care to individual physicians (Adams et al., 2010; Mehrotra et al., 2010; Thomas, 2006; Thomas & Ward, 2006). P4P programs may use the results of such profiles for determining differential reimbursement, placing clinicians and providers in cost and quality tiers that have differential enrollee cost sharing, or providing enrollees with information to aid their choice of clinician or provider.

The need to attribute patients to physician units is not uncommon in the US health care system, which is dominated by insurance models with fragmented lines of clinical responsibility. Preferred provider organizations (PPOs) are a dominant type of health insurance in the commercial, employer-based sector. PPOs typically do not require enrollees to designate a personal PCP who approves services (a gatekeeper). Patients may self-refer to in-network or out-of-network providers with differential cost sharing. The dominant portion of the Medicare program also does not require beneficiaries to designate a responsible physician. About 80 percent of beneficiaries were enrolled in the traditional FFS program as of 2008 (Medicare Payment Advisory Commission [MedPAC], 2009a). Medicare FFS beneficiaries have near-complete freedom of provider choice with no physician "in charge."

In this chapter we first discuss the challenges to patient attribution, with selected examples of real-world attribution strategies. Then we consider basic concepts and alternatives, followed by elements of patient attribution in an FFS situation. The chapter concludes with comments about whether there is a "best" attribution rule, and the role of patient attribution in Medicare reform.

Challenges in Patient Attribution

The patient attribution process poses a complex problem in that FFS has no enrollment or designation process—as do managed care, medical home,[1] or programmatic contexts—whereby patients select a physician to manage their care. Instead, an insurer or other entity has access to data on utilization (the patient's medical services) and other information that medical provider claims (bills) and administrative files typically contain; such entities retrospectively assign patients to physicians who may reasonably be held responsible for their care.

Unlike the practice of attributing a population or panel of patients to a PCP who performs a gatekeeping function, but does not necessarily directly provide any services to a patient, P4P programs must assign patients on an activity or utilization basis to the physician(s) providing their care (that is, assignment must be based on medical services received) (Ingenix, 2007). With retrospective virtual attribution of patients, P4P programs may assign physicians responsibility for services that they neither delivered nor managed (Crosson et al., 2009); the physicians may not even know whether they are responsible for a patient or episode at the time of service.

Patients often see several physicians for potentially overlapping care. For example, Medicare FFS beneficiaries annually see a median of two PCPs and five specialists working in four different practices (Pham et al., 2007). When Pham and colleagues assigned beneficiaries to the physician who provided a plurality of their annual evaluation and management visits, they found that a median of 35 percent of beneficiaries' visits each year were with their assigned physicians. The assigned physician changed from one year to the next for 33 percent of beneficiaries. Considering all visits to any physician, a PCP's assigned patients accounted for a median of 39 percent of his or her Medicare patients and 62 percent of visits. These findings of such extensive dispersion of care were present using alternative assignment rules. Pham and colleagues' findings raise doubts about whether claims data can retrospectively identify a single physician who is primarily responsible for a Medicare beneficiary's care, and whether the physician delivers a meaningful proportion of the care for assigned patients.

Besides the issue of numerous physicians providing care, typically FFS systems cannot assign clear overall clinical or financial responsibility

---

[1] In the medical home model, patients designate a physician practice to be their "medical home," and a personal physician to coordinate and oversee their medical care.

for a patient's care over time. Current visit- and procedure-based FFS reimbursement systems pay physicians to treat particular medical problems, not to manage a patient's overall care. Physician specialists focus on narrow medical issues that are referred to them.

Attributing physician responsibility for care is further complicated in that FFS patients have the freedom to choose their providers; they face few limitations or preapproval requirements on seeing multiple doctors and seeking multiple opinions or treatment options. If one physician does not give a patient what he or she wants, the patient may consider moving on to another physician who will satisfy those demands. Patient adherence to physician recommendations is of course imperfect; this feature of health care can have direct implications for physician quality scores that are based on the percentage of patients who actually receive the services recommended and comply with physician instructions.

The FFS medical care delivery system is further fragmented between physicians, providers, and other sources of care. Compounding the problem is a lack of cooperation and integration across provider organizations, including their information technology and administrative systems. Pinpointing responsibility in such a system is challenging. Many patient and system factors that are beyond the control of individual physicians and group practices can affect the care and outcomes of patients attributed to a physician.

Real-World Examples of Patient Attribution to Physicians

Despite the obstacles in the typical FFS health care delivery system described above, successful examples emerge of patient attribution for the purposes of physician P4P.

Medicare Physician Group Practice Demonstration

The Medicare Physician Group Practice Demonstration, for example, creates incentives for participating physician group practices (PGPs) to improve the quality and efficiency of their care for assigned Medicare FFS beneficiaries (Kautter et al., 2007). In the demonstration, patients are assigned to participating physician groups if the plurality of their "allowed charges" for office and other outpatient evaluation and management services is with the group. That is, patients who receive more outpatient evaluation and management services (measured by allowed charges) from a given group than from any other physician organization are assigned to that group.

In general, the participating physician groups found the demonstration's patient assignment methodology to be a reasonable approach: it assigned a set of beneficiaries for whom the groups could be held accountable for annual cost and quality performance (Kautter et al., 2007). Participating groups provided services constituting an average of 85 percent of their assigned beneficiaries' evaluation and management allowed charges, and 5.4 evaluation and management visits per beneficiary annually. An average of about 70 percent of the groups' assigned beneficiary population carried over from one year to the next. These data indicate that the physician groups participating in the Medicare Physician Group Practice Demonstration had significant opportunities to manage and coordinate the care provided to their assigned beneficiaries.

Massachusetts Health Quality Partners

The Massachusetts Health Quality Partners (MHQP) attributed care for enrollees in Massachusetts commercial and public insurance plans (including Medicare) to PCPs (MHQP, 2008) for the purpose of quality measurement and public reporting. MHQP attributed care for patients in managed care insurance to the PCP whom the health plan assigned to the patients. It attributed care for patients in PPO and Medicare FFS products to the PCP who had the highest volume of evaluation and management office visits with that patient in the 18 months before the end date of the measurement period. For PPO/FFS patients with no visits to a PCP in the specified measurement period, MHQP attributed care to a visited specialist relevant to the quality indicator (e.g., a cardiologist for cardiac measures).

MHQP validated its attribution method by interviewing PCP physicians to whom patients were assigned and found the method to be "very accurate." Physicians agreed that they had seen 93 percent of the patients who had been attributed to them, and for 96 percent of these patients they perceived themselves to be at least partially responsible for managing the patient's preventive and chronic care or to be their PCP. In most cases in which physicians did not see the patient or did not see themselves as responsible for that patient, another physician in their group was following the patient (MHQP, 2008).

California Physician Performance Initiative

The California Physician Performance Initiative (CPPI) is a multi-stakeholder initiative to measure and report on the performance of physicians throughout California (CPPI, 2010). It conducts its work under the auspices of the California Cooperative Healthcare Reporting Initiative, a statewide collaborative of physician organizations, health plans, purchasers, and consumers. The CPPI has developed a system to measure and report the quality of patient care provided by individual physicians in California. It assigns CPPI patients to a single PCP with whom the patient had the most ambulatory/outpatient evaluation and management visits during the measurement year and 1 year prior. CPPI changed this rule from an earlier one that permitted multiple PCPs to be assigned to a single patient. The rule uses a 2-year assignment period to create a greater sense of ownership and responsibility on the part of PCPs for their assigned patients. If the visit count is equal for two or more PCPs, the rule assigns the patient to the PCP with the most recent visit.

For indicators relevant to specialists, in contrast to those relevant for PCPs, CPPI assigns patients to any specialist with whom they had an evaluation and management visit during the attribution period. For example, it could assign patients with diabetes who qualify for the low-density lipoprotein screening measure to an endocrinologist and to a cardiologist if they had a visit with both during the attribution period. For patients with no evaluation and management visits in the measurement year, or without one with a physician of the specialty relevant for a measure, CPPI does not assign a physician for that measure.

CPPI also assigns patients to *practice sites*, defined as physicians of the same specialty who share a practice address. CPPI (2009, p. 10) states, "While narrow accountabilities may be more consistent with physician perceptions of the care they provide to patients, broader accountabilities (i.e., multi-provider) emphasize joint responsibility for ensuring the proper management of the patient during each and every care encounter." CPPI uses office-based, hospital outpatient, and independent clinic visits to assign patients.

Basic Concepts and Alternatives in Patient Attribution

Prospective vs. Retrospective

The first key aspect to consider in trying to assign patients to physicians is whether attribution is prospective or retrospective. In *prospective* assignment, physicians know in advance the patients they are responsible for. This is true of managed care gatekeeper situations in which enrollees must choose their PCP. It is also true of medical homes, where the patient typically must designate a personal physician and medical home. Also, it would be true when patients must agree to and enroll in patient care interventions, such as the provider care management programs in Medicare's Care Management for High-Cost Beneficiaries Demonstration.

To the extent that physicians have discretion to designate assigned patients, prospective attribution creates incentives for risk selection, that is, designating healthier, lower cost, or more adherent patients. However, programs can prospectively assign patients to physician groups or other care management entities without allowing discretion, using historical utilization or diagnosis data. For example, Medicare's Health Support Pilot Program assigned all beneficiaries in a geographic area who met certain diagnostic and other criteria to a third-party disease management company or to a control group (Centers for Medicare & Medicaid Services [CMS], 2006b).

Finally, programs use random prospective assignment algorithms in some cases—for example, to allocate patients who have not chosen a responsible health professional or provider. Some state Medicaid programs and the Medicare Part D prescription drug program use "auto assignment" to assign some enrollees to health plans or clinicians (CMS, 2006a; Llanos & Rothstein, 2007).

Generally, this chapter addresses situations of *retrospective* patient attribution, which is based on utilization data, and for which the beneficiaries themselves do not designate a preferred physician. The following discussion explores the key elements of retrospective assignment.

Assignment to Single or Multiple Physicians

Another important element of patient attribution is whether the rule assigns a patient exclusively to a single physician (unique assignment) or jointly to multiple physicians. For example, algorithms that assign patients to the (single) physician who provides the "most" care to the patient are exclusive, or unique, attribution algorithms. An algorithm that attributes a patient to any physician who "touches" the patient (that is, directly provides any care to that patient) is an example of a nonexclusive, or joint, assignment algorithm. Joint assignment acknowledges that no one physician has complete control over all aspects of a patient's care. A PCP, for example, does not have complete control over specialist care, even if the PCP made the referral. Joint attribution algorithms may have weighting schemes that assign patients differentially to different providers—one example is assignment in proportion to the amount of care provided. In joint attribution algorithms, a single patient's experience may be counted in multiple providers' quality or cost profiles or performance.

The Medicare Payment Advisory Commission (MedPAC) (2009b) compared single-physician and multiple-physician attribution rules for Medicare episodes of care. MedPAC found that measures of resource-use efficiency for individual physicians that were calculated using single (compared with multiple) attribution of episodes were highly correlated (correlation coefficient of 0.95 or greater). Also, both single and multiple attribution rules yielded efficiency ratios that were stable across years for individual physicians. Multiple attribution increased the proportion of physicians who could be attributed the minimum number of episodes (20) specified as necessary for efficiency profiling. MedPAC concluded that statistical analysis did not lead to a clear-cut preference for single or multiple attribution. In choosing one or the other, P4P programs probably need to base the decision on qualitative criteria such as whether a payer wanted profiled physicians to focus on their referrals (which would favor single attribution) or to collaborate in patient care (which would favor multiple attribution).

Attribution to Individual Physicians vs. Physician Organizations

Also important is whether P4P programs attribute care to individual physicians or to physician groups, such as group practices or hospital medical staffs. Although attributing care to individual physicians is most specific, such an approach can result in insufficient patient sample sizes to measure performance reliably and can also pose difficult analytic problems when multiple physicians are providing care. By contrast, attributing care to physician *groups* may be more feasible and reliable, but doing so diffuses responsibility and incentives for care and does not measure intragroup variation. One advantage of attributing care to groups is that some, such as traditional integrated physician group practices, often have internal organizational mechanisms for providing performance feedback and incentives to member physicians.

Fisher and colleagues (2007) argue that payers potentially could assign patients to the "extended hospital medical staff," which would take responsibility for patient care. "The extended hospital medical staff is essentially a hospital-associated multispecialty group practice that is empirically defined by physicians' direct or indirect referral patterns to a hospital" (Fisher et al., 2007, p. w45). The authors propose to assign physicians to the hospitals at which they do most of their inpatient work and would assign those who do not do inpatient work to the hospitals that admit the plurality of their patients. In the case of Medicare enrollees, for example, Fisher et al. would assign beneficiaries to the physician who provided most of their ambulatory care, hence also assigning them to that physician's primary hospital and its extended medical staff.

The extended hospital medical staff concept provides a means of aggregating small and solo physician practices and institutional providers into larger multispecialty entities to which Medicare could assign all Medicare beneficiaries and physicians. However, extended hospital medical staff are not organized in any legal or practical way to receive payment and manage patient care in an integrated fashion. Payers would need to expend substantial effort to legally define the extended medical staff organization and prepare it to accept responsibility for managing patient care.

Exclusion of Patients from Attribution

A final issue is whether physicians can exclude patients from being attributed to them. Typically, a P4P program establishes a set of criteria for determining whether they will exclude or include patients in assignment in a given context. These criteria can encompass basic insurance eligibility, length of enrollment with a given insurer or program, and eligibility for the quality or cost indicator under consideration. For example, only patients diagnosed with diabetes are eligible for evaluation on indicators of the quality of care provided to diabetics.

More controversial than this type of assignment decision is whether physicians should be allowed to exclude *selected* patients for reasons such as nonadherence to physician recommendations or ineligibility for performance indicators for reasons that only the physician can ascertain. This policy creates obvious incentive issues for physicians to object or otherwise act to exclude patients from certain measures when the payer is going to judge their performance on assigned patients. Nevertheless, to a certain extent, issues such as patient nonadherence may legitimately be beyond physicians' control.

Patient vs. Episode Assignment

Another aspect of attribution is the range of health care services included—for example, a patient's entire care over a period such as a year, or episodes of care related to particular medical conditions, acute events, or procedures. Attributing a patient's entire care over a specified period reflects an integrated, holistic perspective on the patient's care; this approach best suits PCPs who are managing a patient's overall care, especially chronic condition care. Episodic care would describe care by specialists or acute care physicians, such as hospital-associated physicians managing an acute care episode centered on a hospitalization.

Patient Notification and Lock-In

Payers do not necessarily notify patients (beneficiaries, enrollees) about their assignment to providers who are participating in P4P programs. In fact, as long as patients' insurance benefits, cost sharing, and freedom of provider choice are unaffected, P4P programs for physicians typically do not inform patients. However, patients may have interest in knowing that their physicians are participating in or subject to incentive programs that could affect their choice of treatment for patients, particularly programs that include incentives to reduce costs.

Patient notification has several levels. A first level could be a posting in a physician's office or a letter stating that the physician is participating in an incentive program to improve quality and efficiency. A second level of notification could entail identifying the specific physician to whom the incentive program assigned the patient, perhaps including quality and cost-profiling results for that physician. A third use of assignment would be some form of incentives (e.g., differential copayments) or even lock-in for the patient to obtain services from the physician or physician organization to which he or she was assigned. Patient lock-in would require substantial patient education and the patient's commitment to receive all or most of his or her services from a designated provider organization. Devers and Berenson (2009) have discussed these options—no lock-in or even notification, soft lock-in (incentives), or hard lock-in (requirements to use selected providers)—in the context of accountable care organizations for Medicare.

Geographic Unit of Assignment

Instead of utilization-based attribution methods, a different, more aggregated approach is to assign all physicians in an area responsibility for all residents of an area. An advantage of this approach is that it assigns physicians responsibility for nonusers as well as users of medical care. It also promotes a sense of collective responsibility of an entire physician community for an entire community of residents. Assigning patients to individual physicians—an often difficult problem—is not an issue. Geographic assignment promotes a focus on community-wide public health, prevention, and outcome measures, which can be measured precisely because of the large sample size and can be compared across areas.

A geographic unit of accountability has several disadvantages, however. The major downside is its inability to distinguish between individual physicians' or practices' performance. A more disaggregated approach to individual physician or practice performance measurement could be a complement to an additional geographic unit of accountability; they are not mutually exclusive. An aggregate, geographic unit could also require considerable coordination and integration among different payers and provider organizations, which may be difficult to achieve in practice. Another disadvantage of area alliances is that comprehensive, standardized data submission and analysis for performance measurement often presents a major hurdle.

Elements of Patient Attribution in a Fee-for-Service Situation

Without patient enrollment or the designation of a responsible physician through managed care or medical homes, utilization of services is the only observable element in medical claims data that links patients to specific physicians or practices. The basic idea is that the physician(s) providing services can be considered responsible for their patients. The obvious drawback of this approach is that individuals who have not had any health care services will not be assigned to *any* physician. That is, for nonusers, P4P system administrators simply have nothing to go on in attributing responsibility. This lack of information can pose a serious issue in young, healthy populations that have many nonusers. Even in the high-utilizing Medicare FFS population, about 6 percent of beneficiaries do not use any Medicare-covered services in a year, and roughly 15 percent do not have any office or other outpatient evaluation and management visits, leaving little or no basis for assignment to a physician. Over multiyear periods, more enrollees can be assigned given that some of the nonusers in year 1 will have service utilization in year 2 that can be the basis of assignment.

The three major elements of a patient assignment algorithm based on utilization are as follows:

- the type of services that a program uses for attribution,
- the rules for determining responsible physicians from services it uses for attribution, and
- the type(s) of providers eligible for assignment or used in assignment algorithms.

Type of Services Used for Attribution

P4P programs generally use physician or professional services to attribute patients to physicians.[2] Of course, physicians provide many different types of services. To deal with this complexity, a program is likely, logically, to use the type of service most closely related to the care it is attributing. For example, for an episode of care centered on a major surgical procedure, a P4P program logically attributes the patient to the surgeon who performs the procedure. Under *one-touch rules*, an attribution algorithm may assign a patient receiving *any* services from a physician—over a period such as a year or in connection with an episode of care—to that physician.

In contrast, attribution rules, appropriately, assign management of chronic diseases to the physician(s) providing evaluation and management services. Typically, as in the Medicare Physician Group Practice Demonstration, P4P programs further restrict these services to those a physician provides in the office or other outpatient settings;[3] in other contexts, programs may further restrict the services to "established patient" visits. Hospital inpatient evaluation and management services are likely to focus narrowly on the reason for the hospital admission; thus they are less appropriate for assigning overall responsibility for a patient's care management than are ambulatory services.

Programs can also attribute acute care episodes based on evaluation and management services. For example, the "first contact" rule attributes the acute episode to the physician billing for the episode's first evaluation and management visit.

Rules for Determining Responsible Physicians

The simplest approach to determining responsible physicians from utilization data is to assign the episode or patient to the physician providing the largest quantity, cost, or share of the type of services used for attribution (*plurality rule*). For example, in the Medicare Physician Group Practice Demonstration, the demonstration algorithm assigns patients to a participating group practice if the practice provided the plurality of office or other outpatient evaluation and management services for that patient. Pham et al. (2007) and McCall et al. (2000) also use the plurality rule as their baseline assignment method.

---

[2] Conceivably, programs could attribute care to physicians based on nonphysician services. For example, a hospital admission could be used to attribute care to the (physician) medical staff of the hospital (Fisher et al., 2007). Another example might be immunizations—given by a nurse associated with a doctor or practice in situations in which the patient never sees the doctor in question.

[3] However, Pham et al. (2007) include inpatient claims in their baseline assignment method.

Quantity or cost. Either quantity or cost can be the basis for defining the largest amount of services provided—number of visits or total payments for visits or professional services (Mehrotra et al., 2010). Using visit cost is likely to assign patients to physicians providing higher-cost visits, that is, specialists rather than PCPs (Sorbero et al., 2006). Total rather than visit cost is an alternative basis for assignment: P4P programs can attribute patients or episodes of care to the physician providing the highest total cost of services during a specified period (e.g., a year) or during the episode; in this approach, visits and procedures might both count.

A very different approach is to assign a patient to any physician who provides any services (one or more) of a specified type to him or her (the *one-touch rule*). This is a nonexclusive assignment algorithm in which multiple physicians may be responsible for a single patient. Sharing responsibility could increase the chances of patients receiving recommended care because more than one physician may be responsible. Also, it encourages collaboration among physicians involved in a patient's care. However, shared responsibility creates "free rider" incentive problems. Physicians may receive credit for a patient receiving recommended care even if they were not the physician who ensured that the patient receive the care. Conversely, when physicians ensure that recommended care is provided, they may be improving performance scores of other physicians in their peer comparison group in addition to their own score.

Minimum quantity or share rules. Patient assignment algorithms may incorporate *minimum quantity or share rules*, for example, assigning a patient to a physician who provided the most visits to the patient, assuming that the physician provided at least two visits within the assignment period (typically a year). The rationale for a minimum quantity of services is that a physician may need some minimum level of contact with a patient to be able to manage the patient's care. Note that the minimum two-visit rule would exclude all patients with only one visit overall from assignment to anybody; obviously, this is also true for patients with no visits. Patients with no visit or one visit are likely to be healthier than average, thus a two-visit minimum for assignment skews the assigned population toward the sick.

As another illustration, a *minimum-share rule* might specify that a physician or practice be responsible for patients to whom they provided the most care, if the share of the patient's care was at least 50 percent (*majority rule*). The rationale for specifying a minimum share is that holding a physician responsible for the care of a patient to whom he or she is providing a minority of total care might be considered inappropriate. Note, however, that this rule would exclude all patients for whom no physician provided a majority of care from assignment. These cases could encompass the sickest patients, who are seeing multiple physicians for multiple illnesses and thus are most in need of care coordination. Also, over time the stability of assignment of the responsible provider is lower with a majority than with a plurality rule (Pham et al., 2007).

The attribution rule could also specify lower minimum shares—for example, 30 percent or 20 percent—with the rationale that a physician has to provide some minimal share of care to be held accountable. As one example, in its initial analyses, MedPAC assigned beneficiary episodes to physicians responsible for at least 35 percent of episode evaluation and management dollars (MedPAC, 2009b). Pham et al. (2007) simulate a rule that assigns beneficiaries to any physician billing for at least 25 percent of the patient's evaluation and management visits. This *multiple physician assignment rule* assigned physicians a higher proportion of their Medicare patients and visits but assigned fewer beneficiaries to any physician at all, compared with the baseline plurality assignment rule.

Using private insurer claims data, Thomas and Ward (2006) simulated minimum-share rules of 20, 30, and 50 percent in assigning episodes to four specialties (cardiology, family practice, general surgery, and neurology). With the 20 percent rule, for example, they assigned a physician responsibility for an entire episode if he or she accounted for at least 20 percent of total professional and prescribing costs for the episode. They would assign the episode to multiple physicians if two or more physicians each accounted for at least 20 percent of total costs. If *no* physician accounted for at least 20 percent of total costs, the episode would not be assigned to any physician. As expected, Thomas and Ward found that as the minimum share rose from 20 to 30 to 50 percent, more episodes were unassigned to any physician, but fewer were assigned to multiple physicians.

These authors also found that cost efficiency scores for physicians were not highly sensitive to the attribution method because most episodes were assigned to only one physician regardless of the episode-attribution methodology. For example, when Thomas and Ward assigned episodes using the 30 percent minimum-share rule, they saw the following results: more than 93 percent of cardiology episodes, 95 percent of family practice episodes, 99 percent of general surgery, and 95 percent of neurology episodes were assigned to a single physician (Thomas, 2006). Thomas and Ward (2006) did not find a single episode-attribution rule to be uniformly superior for all specialties. But they found it "reasonable" to attribute episodes to physicians who accounted for at least 30 percent of episode professional and prescribing costs.

Simulations of alternative rules. Mehrotra et al. (2010) simulated 12 attribution rules for the purpose of assigning physicians to one of four cost tiers: low cost, average cost, high cost, or low sample size (fewer than 30 assigned episodes). The 12 rules were combinations of four dimensions: unit of analysis (patient versus episode of care); signal for responsibility (professional costs versus evaluation and management visits); number of physicians assigned responsibility (single versus multiple); and minimum threshold for responsibility (majority versus plurality of visits or costs). The authors analyzed data on 1.1 million 2004–2005 commercial insurance enrollees between the ages of 18 and 65 in four Massachusetts health plans.

Mehrotra et al. (2010) found that the percentage of episodes that they could assign to a physician varied substantially across the attribution rules, from 20 to 69 percent. The mean percentage of costs billed by a physician that they included in that physician's own cost profile also ranged widely, from 13 to 60 percent. Further, using the 11 alternative rules, between 17 and 61 percent of physicians would fall into a different cost category than when using the "default" rule (episode-based, costs, plurality). The authors conclude that the attribution rule can substantially affect the cost category to which a physician is assigned.

As part of investigating estimates of physician cost efficiency using episode groupers, Adams et al. (2010) simulated the reliability of three alternative rules for attributing episodes to physicians. They conducted the analysis using commercial insurance claims for residents of Massachusetts between the ages of 18 and 65. The baseline rule was to attribute episodes to the physician who billed the highest proportion of professional costs in that episode as long as the proportion was greater than 30 percent. If no physician met these criteria, Adams and colleagues dropped the episode from analysis. Across all specialties, 52 percent of episodes could not be assigned to a physician.

The researchers also examined two alternative rules. One rule was to attribute episodes to the physician who accounted for the highest fraction (minimum of 30 percent) of face-to-face encounters (based on the number of evaluation and management visits) within the episode. In this alternative, 50 percent of episodes could be assigned. The other alternative was a *patient-based rule* that assigned all episodes for a patient to the physician who accounted for the highest fraction of professional costs for that patient (minimum of 30 percent) over the 2-year study period. With this rule, 39 percent of episodes could be attributed. The authors found that the reliability of physician cost profiles was lower using the two alternative rules than with their baseline rule (Adams et al., 2010).

McCall et al. (2000) explored alternative Medicare FFS beneficiary assignment criteria for process quality performance assessment of large PGPs, using claims data on and interviews with four "study partner" group practices. The researchers based assignment on evaluation and management visits, and the baseline attribution rule was to assign beneficiaries receiving a plurality of their visits from a PGP. McCall and colleagues tested a variety of alternative assignment algorithms, which included the following: varying the minimum number of evaluation and management visits and share of evaluation and management care provided by the PGP; weighting by the share of care at the PGP rather than using a plurality assignment rule; counting only visits provided by primary care or medical specialties rather than all specialties; and excluding from assignment patients residing outside the practice's service area.

McCall et al. (2000) found that measured PGP quality indicator performance was not very sensitive to the assignment algorithm. Moreover, almost three-quarters of PGP physicians felt definite ownership of patients assigned to them using the baseline attribution algorithm (plurality of all-specialty evaluation and management services). A patient survey showed that 88 percent of beneficiaries recognized that most of their doctor visits had occurred at their assigned PGP. PGP physicians felt more responsible for patients with more visits and a higher share of care at the PGP.[4] But the number of beneficiaries assigned to the practices dropped considerably with high thresholds for the number of visits or share of care required for assignment. In consultation with their technical experts panel, McCall et al. (2000) recommended the use of "minimalist assignment criteria" to enhance the statistical validity of performance measurement by increasing the number of assigned patients.

---

[4] Interestingly, physicians' perceived responsibility for recommending care varied considerably by process quality indicator. Almost three-quarters of physicians felt responsible for recommending mammography to qualifying assigned patients, but less than one-half for recommending retinal eye exams, less than one-third for prescribing beta blocker after a heart attack, and less than one-fifth for follow-up after a mental health hospitalization. These variations were related to the perceived clinical consensus around performance measures and physician responsibility for the type of care involved (PGP physicians felt more responsible for outpatient than inpatient care, and less responsible for mental health care).

Type of Physicians Used for Assignment

The third and final element of an attribution rule is the type of physician eligible for use in the assignment rule. For example, in a group practice context such as the Medicare Physician Group Practice Demonstration, patients assigned to any physician who is a member of the group could be assigned to that group. Alternatively, the rule could use only PCPs as the basis for assignment (patients assigned to the group's PCPs are assigned to the group). Limiting the assignment to PCPs reduces the number of patients who can be assigned because PCPs do not provide some beneficiaries with any services; these are often the sickest patients, whom specialists primarily treat. Pham et al. (2007) found that 94 percent of Medicare beneficiaries could be assigned to a physician of any specialty, but only 79 percent to a PCP, based on evaluation and management visits. In the context of an inpatient episode, payers could consider radiologists, anesthesiologists, and pathologists ineligible for assignment. The rationale for excluding certain specialties was that they are unlikely to be responsible for overall management of a patient's hospital stay.

By contrast, certain specialists—for example, internal medicine subspecialists such as cardiologists and endocrinologists—may in effect manage the care of many individuals, especially those with chronic illnesses (e.g., heart disease, diabetes) that the specialty treats. For this reason, eliminating all specialists and specialty care from assignment is probably unwarranted, even when the aim is to attribute responsibility for either overall care or, at least, primary care. This is especially true in a population encompassing many individuals with multiple chronic illnesses, such as the Medicare population. A compromise would include in an assignment both PCPs and selected specialists who may be responsible for managing patient care.

In attributing responsibility for episodes of care, the rule may logically hold that specialists are responsible when they are the primary caregiver for many episodes of care. In the inpatient hospital setting, payers could conceivably hold several physicians singly or jointly responsible; these physicians might include the admitting physician, the attending, hospitalists or intensivists managing care in the hospital, PCPs providing inpatient visits, specialists doing inpatient consultations, and, in surgical cases, the primary and assisting surgeons. For post-acute or post-discharge care, physicians who are candidates to be held responsible include the principal surgeon in surgical cases, primary care doctors, and doctors participating in post-acute care, such as physicians ordering home health care or managing skilled nursing care.

Concluding Comments

Is There a Best Attribution Rule?

Most studies reviewed here have not found physician cost and quality performance assessment to be very sensitive to the rules used to attribute patients to physicians. But some (notably Mehrotra et al., 2010) have found important effects of attribution on results, indicating that the effects of attribution may be sensitive to the particular context in which it is studied and the range of attribution rules that are considered. Assignment rules continue to be controversial and can influence face validity, statistical validity, and the proportion of care and patients included in performance assessment. More restrictive patient-assignment criteria, such as requiring more visits or a higher share of care for assignment, tend to increase face validity to physicians and their acceptance of responsibility for assigned patients. But more restrictive assignment criteria also reduce the number of patients assigned, which may threaten statistical validity, base physician performance assessment on a smaller share of total care provided, and leave more patients unassigned to any provider.

Although P4P programs commonly use certain attribution rules (e.g., a plurality rule, sometimes with a minimum number of visits or share of care), no clear consensus has yet emerged on the "best" attribution rule(s). As Mehrotra et al. (2010) point out, there may be no uniformly best rule. The preferred rule may depend on the purpose, context, and stakeholder perspective. The same rule may not be best from the perspectives of purchasers, providers, and consumers. The conclusion is that choice of attribution rule should be evaluated on a case-by-case basis to satisfy the purpose at hand.

Patient Attribution in Medicare Reform

Patient attribution in an unmanaged, fragmented FFS environment poses many challenges. Exactly how successful patient attribution can be in this setting remains unclear. Also still uncertain is whether institutional changes such as requiring patients to choose managed care gatekeepers or medical homes are necessary to attribute responsibility for care. Reflecting this uncertainty, current Medicare reform efforts, such as those incorporated in the Patient Protection and Affordable Care Act of 2010 and in previous legislation, include a spectrum of relationships between beneficiaries and physicians and other providers. Traditional FFS Medicare, with no explicit responsibility or assignment of patients to physicians, continues. Patients will likely be assigned to the newly established (2012) accountable care organizations on a retrospective basis using service utilization, without any beneficiary involvement or acknowledgment. In Medicare's medical home demonstrations, participating patients will likely have to prospectively designate a personal physician and sign an agreement specifying their obligations. But they will not be subject to a lock-in requiring them to obtain care only from their designated physician. The Medicare Advantage program will continue, in which patients actively enroll in private health plans and, depending on the plan's benefit design, may be required to obtain care exclusively from the plan's network of contracted providers (or at least enjoy discounted cost sharing if they do so). The private sector is also trying a wide variety of approaches. Experience and careful evaluation studies should reveal which of these models is most successful and under what circumstances.

References

Adams, J. L., Mehrotra, A., Thomas, J. W., & McGlynn, E. A. (2010). Physician cost profiling—Accuracy and risk of misclassification. *New England Journal of Medicine 362*(11),1014–1021.

Beckman, H. B., Mahoney, T., & Greene, R. A. (2007). *Current approaches to improving the value of care: A physician's perspective.* Report to the Commonwealth Fund. Retrieved April 22, 2010, from http://www.commonwealthfund.org/.

California Physician Performance Initiative. (2010). San Francisco: California Cooperative Healthcare Reporting Initiative. Available from http://www.cchri.org/programs/programs_CPPI.html

California Physician Performance Initiative. (2009). *Methodology for physician performance scoring. Cycle 4.* July 31, 2009. San Francisco: California Cooperative Healthcare Reporting Initiative.

Centers for Medicare & Medicaid Services. (2006a). *Auto-enrollment and facilitated enrollment of low income populations.* Retrieved September 6, 2009, from http://www.cms.hhs.gov/States/Downloads/AutoversusFacilitatedEnrollment.pdf

Centers for Medicare & Medicaid Services. (2006b). *Medicare Health Support: Overview.* Retrieved September 6, 2009, from http://www.cms.hhs.gov/CCIP/downloads/Overview_ketchum_71006.pdf

Crosson, F. J., Guterman, S., Taylor, N., Young, R., & Tollen, L. (2009). *How can Medicare lead delivery system reform?* Issue Brief, 71(1335). New York: The Commonwealth Fund.

Devers, K., & Berenson, R. (2009). *Can accountable care organizations improve the value of health care by solving the cost and quality quandries?* Retrieved April 23, 2010, from http://www.rwjf.org/files/research/acobrieffinal.pdf

Fisher, E. S., Staiger, D. O., Bynum, J. P., & Gottlieb, D. J. (2007). Creating accountable care organizations: The extended hospital medical staff. *Health Affairs (Millwood), 26*(1), w44–57.

Hackbarth, G. (2009). *Reforming America's health care delivery system.* Statement to the Senate Finance Committee, April 21, 2009. Retrieved April 22, 2010, from http://www.medpac.gov/documents/Hackbarth%20Statement%20SFC%20Roundtable%204%2021%20FINAL%20with%20header%20and%20footer.pdf

Ingenix. (2007). *Symmetry episode treatment groups: Issues and best practices in physician episode attribution.* Available from http://www.ingenix.com/content/attachments/Symmetry_EpisodeAttribution_WP_FINAL_112007.pdf

Kautter, J., Pope, G. C., Trisolini, M., & Grund, S. (2007). Medicare Physician Group Practice Demonstration design: Quality and efficiency pay-for-performance. *Health Care Financing Review, 29*(1), 15–29.

Llanos, K., & Rothstein, J. (2007). *Physician pay-for-performance in Medicaid: A guide for states.* Report prepared for the Commonwealth Fund and the Robert Wood Johnson Foundation. Hamilton, NJ: Center for Health Care Strategies, Inc.

Massachusetts Health Quality Partners. (2008). *Validation of MHQP's method for patient attribution.* Watertown, MA: MHQP.

McCall, N. T., Pope, G. C., Griggs, M., et al. (2000). *Research and analytic support for implementing performance measurement in Medicare Fee for Service.* Report prepared for the Health Care Financing Administration. Waltham, MA; Health Economics Research, Inc.

Medicare Payment Advisory Commission. (2009a, March). *Report to the Congress: Medicare payment policy.* Available from http://www.medpac.gov/documents/Mar09_EntireReport.pdf

Medicare Payment Advisory Commission. (2009b, June). *Report to the Congress: Improving incentives in the Medicare program.* Available from http://www.medpac.gov/documents/Jun09_EntireReport.pdf

Mehrotra, A., Adams, J. L., Thomas, J. W., & McGlynn, E. A. (2010). The effect of different attribution rules on individual physician cost profiles. *Annals of Internal Medicine 152*(10), 649–654.

Pham, H. H., Schrag, D., O'Malley, A. S., Wu, B., & Bach, P. B. (2007). Care patterns in Medicare and their implications for pay for performance. *New England Journal of Medicine, 356*(11), 1130–1139.

Sorbero, M., Damberg, C. L., Shaw, R., Teleki, S., Lovejoy, S., DeCristofaro, A., et al. (2006). A*ssessment of pay-for-performance options for Medicare physician services: Final report* (RAND Health Working Paper WR-391-ASPE). Available from http://aspe.hhs.gov/health/reports/06/physician/report.pdf

Thomas, J. W. (2006). *Economic profiling of physicians: What is it? How is it done? What are the issues?* (A guide developed for the American Medical Association.) Available from http://www.mag.org/pdfs/tiering_amaguide.pdf

Thomas, J. W., & Ward, K. (2006). Economic profiling of physician specialists: Use of outlier treatment and episode attribution rules. *Inquiry, 43*(3), 271–282.

CHAPTER 8

Financial Gains and Risks in Pay for Performance Bonus Algorithms

Jerry Cromwell

The burgeoning research on the wide geographic variation in surgery rates (Weinstein et al., 2004; Weinstein et al., 2006; Wennberg et al., 2006), the prevalence of medical errors, and the generally unacceptable quality of care in a variety of settings (Institute of Medicine [IOM] Board on Health Care Services, 2001; Chassin & Galvin, 1998) has motivated both public and private health insurers to incorporate financial incentives for improving quality into their payment arrangements with care organizations. Insurers are using both reward and risk—carrot and stick—approaches (Bokhour et al., 2006; LLanos et al., 2007; Epstein, 2006; Rosenthal & Dudley, 2007; Trude & Christianson, 2006; Williams, 2006). Payers may simply provide an add-on or allow higher updates to a provider's fees, or they may pay an extra amount for a desired service (e.g., a $10 payment for a mammogram): a reward (carrot) strategy. Alternatively, payers may reduce payments or constrain fee updates for unacceptable quality performance: the risk (stick) strategy.

A hybrid of the two approaches uses self-financing quality bonuses. Under a self-financing scheme, as with Michigan Medicaid's Health Plan Bonus/Withhold system (LLanos et al., 2007, p. 15), payers pay for quality improvements out of demonstrated savings generated by health care providers or managed care organizations.

Pay for performance (P4P) arrangements use financial incentives to encourage changes in patient care processes that, in turn, should lead to improved health outcomes. Evidence-based patient care studies have produced a list of care processes that lead to better outcomes (Agency for Healthcare Research and Quality, 2006; IOM Board on Health Care Services, 2006; National Committee for Quality Assurance, 2006; National Quality Forum, 2006; see Chapter 4 of this book for examples). Studies have paid much less attention, however, to the payout algorithms themselves. Yet the structure of the incentive arrangement may be as important as—or more so than—the quality indicators in encouraging quality improvements.

This chapter first presents several possible P4P payment models and investigates their key parameters. As part of this exercise, we highlight the effects on bonus levels of increasing the number of indicators, of how they are weighted, and of how targets are set. We then simulate actual quality performance against a preset target and test the sensitivity of a plan's expected bonus and degree of financial risk to different bonus algorithms and key parameters. Finally, we conclude by suggesting a few steps for payers to follow in designing P4P incentive programs that maximize the likelihood of positive responses on the part of managed care and provider organizations.

General Pay for Performance Payment Model

Many private and state Medicaid P4P programs use a simple payment scheme that pays a certain amount for providing a quality-enhancing service (e.g., mammograms, a primary care visit). Service-specific P4P payment, however, is an inadequate incentive to encourage higher quality in managing the chronically ill with multiple health problems. The following subsection summarizes the general theory of P4P payment arrangements. We emphasize the distinction between arrangements paying on relative levels vs. rates of improvement in performance because they have a material effect on rewards and penalties. After that, we describe six common approaches, or algorithms, that either the Medicare Health Support Pilot Program uses or other programs could use, to illustrate the differential impacts of "levels" vs. "rates of improvement" payment strategies.

P4P Payment Using Levels or Rates of Improvement in Performance

One likely bonus (or penalty) payment model uses several P4P indicators and is based on an organization's actual performance relative to each target rate. The target rate for the *i*-th quality indicator in the *p*-th arrangement, t_{ip}, is based on an improvement over the initial baseline rate, $\lambda_{base,ip}$:

$$t_{ip} = \lambda_{base,ip}(1 + E[\alpha_{ip}]) \tag{8.1}$$

where $E[\alpha_{ip}]$ = the payer's implicit expected rate of improvement over baseline for the *i*-th indicator in the *p*-th payment arrangement. For example, the payer might set a target of 75 percent based on a baseline rate of 60 percent with an expected improvement of 25 percent (or 15 percentage points). The payer might set the rate of improvement unilaterally or negotiate it with the organization.

We further assume that the organization (e.g., a primary care practice or commercial disease management company) has formed its own *expected lev*el of performance, $E[\lambda_{ip}]$, based on a likely rate of quality improvement, $E[\rho_{ip}]$, due to its intervention:

$$E[\lambda_{ip}] = \lambda_{base,ip}(1 + E[\rho_{ip}]). \tag{8.2}$$

The organization's expectation of financial success or failure depends on its clinicians' and managers' opinions of the likelihood of their intervention's effectiveness in improving quality.[1] Managers, for example, may be expecting a one-third improvement over the baseline 60 percent, with an expected rate $E[\lambda] = 80$ percent $> t = 75$ percent. Their expected level of performance also is conditional on the level of investment that they intend to make in trying to meet the target plus a random component that would occur in any single year as a result of other factors (e.g., shift in patient case mix, influenza epidemic). We assume that the organization's investment is at some "reasonable" level—possibly to ensure that the organization has at least a 50 percent chance of achieving or exceeding the target rate. We also assume that managers' own expectations of success dominate any random temporal risk, although random risk could dominate at high baseline quality levels with little opportunity for improvement.

Six Common Payment Algorithms

P4P quality payouts (or penalties) to an organization depend on the way the payer sets the target relative to managers' expectations. We consider six bonus algorithms:[2] (1) all or nothing; (2) a continuous unconstrained percentage between zero and 100 percent; (3) a continuous percentage constrained by a lower limit (LL) and upper limit (UL), or corridor; (4) a composite percentage score allowing above-target gains in some indicators to offset failures in other

---

[1] The rest of this chapter uses the term managers to mean either the managers of a commercial disease management organization (e.g., Health Dialog, Aetna, Healthways) or managers of a clinical practice involved in a P4P payment arrangement.

[2] Disease management organizations in the Medicare Health Support Pilot Program used the first five algorithms described in Table 8-1 (McCall et al., 2008; see also Cromwell et al., 2007). This program is described in more detail in Chapter 9.

indicators; (5) statistical differences above the baseline rate; or (6) rate of improvement over the target rate (see Table 8-1).

In the *all-or-nothing* algorithm, an organization would expect a full bonus for a given indicator, $E[B_i] = 1$, if the managers' own expected performance level, $E[\lambda_{ip}]$, equaled or exceeded the target rate; otherwise, the expected bonus for the indicator would be zero. With the *continuous unconstrained* arrangement, the expected bonus percentage is simply the ratio of managers' expected rate of success to the payer's target rate, up to a maximum of 100 percent. For example, if managers' expected success rate was 74 percent, with a target of 75 percent, then the organization would expect to receive 98.7 percent (74 percent / 75 percent) of the full bonus. A *constrained* version of the continuous algorithm produces no expected bonus if managers' expected success-to-target ratio is below a preset LL. For example, if the payer set an LL of 70 percent, the organization would expect no bonus for an indicator if its expected success rate was below 70 percent. Above the LL but below the UL, the payer pays a bonus based on how close the organization comes to meeting its target rate. The bonus within the LL–UL range may also be a fraction, $\theta < 1.0$, of success in achieving the target. For example, if t = 75 percent and actual success is 72 percent, then if the payer set $\theta = 0.80$, the payer would pay 76.8 percent [0.80 × (72 percent / 75 percent)] of the maximum bonus. Usually the upper limit is the target rate, but the payer might even pay an additional bonus where $\lambda > t$.

Table 8-1. Six common P4P payment algorithms

Payment Arrangement	Expected Bonus Percentage: $E[B_i]$	Expected Success-to-Target
1. All or Nothing	0	$E[\lambda_{ip}] / t_{ip} < 1.0$
	1	$E[\lambda_{ip}] / t_{ip} \geq 1.0$
2. Continuous Unconstrained	$E[\lambda_{ip}] / t_{ip}$	$0 \leq E[\lambda_{ip}] / t_{ip} \leq 1.0$
3. Continuous Constrained	0	$E[\lambda_{ip}] / t_{ip} < LL$
	$\theta \times E[\lambda_{ip}] / t_{ip}$	$LL \leq E[\lambda_{ip}] / t_{ip} \leq UL: 0 < \theta < 1.0$
	UL	$E[\lambda_{ip}] / t_{ip} > UL$
4. Composite	$\Sigma_i \omega_i E[\lambda_{ip}] / t_{ip}$	$0 \leq E[\lambda_{ip}] / t_{ip} \leq \infty$
5. Statistical	0	$E[\lambda_{ip}] < 1.96 SE_{\lambda base}$
	1	$E[\lambda_{ip}] \geq 1.96 SE_{\lambda base}$
6. Rate of Improvement	0	$E[\rho_{ip}] / E[\alpha_{iip}] < 1.0$
	1	$E[\rho_{ip}] / E[\alpha_{iip}] \geq 1.0$

The first three bonus arrangements evaluate each indicator's performance and payout percentage, $E[B_i]$, separately. A group's expected total percentage bonus, $E[TB]$,[3] in meeting a set of prespecified quality and satisfaction targets can be expressed as the maximum percentage of outlays (or management fees) eligible for bonuses, MPCT, multiplied by a weighted average of the bonus percentages that an organization might expect to achieve on each indicator (deleting the p-subscript for simplicity):

$$E[TB] = MPCT \times \{\Sigma_i^N \omega_i E[B_i]\} : \{\Sigma_i^N \omega_i E[B_i]\} \leq 1.0, \qquad (8.3)$$

where ω_i = the weight that a payer assigns to the i-th indicator, and $E[B_i]$ = the percentage of the bonus a group expects to achieve for each indicator. For example, if a payer proposed to increase fees to the group by a total of MPCT = 5 percent across 10 equally weighted indicators ($\omega_i = 1/N$), then success on each indicator would raise outlays by $0.05(0.1)(1.0) = 0.5$ percent $= 0.005$. If the group's expected success on each indicator was 80 percent, then the overall average expected bonus percentage would be $E[TB] = MPCT \times (N \times \omega \times 0.8) = 0.05 \times (10 \times 0.1 \times 0.8) = 4$ percent add-on to fees. Some experts have argued that expected bonus fractions of 4 percent are too low. Fractions may need to be at least 10 percent to motivate behavioral change in physicians (LLanos et al., 2007, p. 22).

The fourth, *composite*, bonus algorithm does not evaluate each indicator separately. Rather, it first calculates totally unconstrained actual-to-target performance ratios for each indicator and then applies the weights. It then determines the final bonus only after averaging the unconstrained actual-to-target ratios across all indicators.[4] Because individual indicator $E[B_i]$ ratios could be greater than 1.0, overachievement on some indicators can offset underachievement on others. It is likely that the payer would constrain the total bonus percentage to not exceed 1.0.

A fifth arrangement would require only that actual performance be *statistically higher* than the baseline rate (e.g., $1.96 \times SE$ at the 95 percent confidence level, where SE = standard error of mean λ_{base}). This approach adjusts for random variation only and implicitly assumes (near-) zero intervention effectiveness. It is reasonable for payers to expect a sizable, positive intervention effect on most quality indicators over and above random annual variation.

---

[3] Expected percentage gains can be converted to absolute dollars by multiplying by $E[TB]$ total fees paid out.
[4] This approach essentially replaces $\{\Sigma_i^N \omega_i E[B_i]\}$ in equation 8.3 with $\times \{\Sigma_i^N E[B_i]/N\} : E[B_i] \geq 0$.

The sixth arrangement pays on *relative rates of improvement*, ρ / α, instead of *relative levels*, λ / t. This subtle difference can introduce substantially greater financial risk, as shown in the next section, Setting Targets.

The next three sections focus on characteristics of P4P payment arrangements that entail more or less financial risk for organizations affected by the arrangements. We begin with the theory behind setting targets using *levels* or *rates of improvement* in performance, followed by similar discussions of the financial risks implicit in the number of targets and how quality indicators might be weighted to reflect their link to health outcomes.

Setting Targets (t)

Except for the all-or-nothing arrangement, the other payment arrangements in Table 8-1 are flawed in that they give "partial credit" for simply reproducing the baseline rate. We can see this in the following conversion formula between performance levels and rates of improvement:

$$\lambda / t = \lambda_{base}(1 + \rho) / \lambda_{base}(1 + \alpha) = (1 + \rho) / (1 + \alpha) \qquad (8.4)$$

$$= [1 / (1 + \alpha)] + [\alpha / (1 + \alpha)] \times (\rho / \alpha)$$

where α and ρ represent the payer- and organization-determined expected rates of improvement, respectively. Relative performance *levels*, λ / t, in the three payment arrangements depend not only on relative rates of expected-to-required *improvement*, ρ / α, but on the preset α target improvement rate as well. For example, if a group made no improvement in the target indicator, then ρ = 0. Yet according to the ratio of expected-to-target performance, the success-to-target ratio λ / t = 1 / (1 + α). If the payer sought a rate of improvement of 25 percent, then even with no improvement in performance the organization could enjoy as much as 80 percent of its bonus (= 1 / 1.25) by simply achieving the baseline level under an unconstrained payment levels arrangement.

Another way of setting targets assumes that an ideal performance level exists, λ_{ideal}, that can apply to all regions and groups. The ideal level could be (1) clinically based on "perfect practice," (2) based on local "best practice" among high-performing groups, or (3) nationally based when historical quality levels are averaged across all provider groups. A flexible payment approach would base an indicator's target on the difference between the baseline and ideal rates, λ_{base} and λ_{ideal}:

$$t_i = \lambda_{base} + \psi[\lambda_{ideal} - \lambda_{base}] = (1 - \psi)\lambda_{base} + \psi\lambda_{ideal}, \quad (8.5)$$

where $\psi \leq 1.0$ is the required fraction of the difference between the ideal and base rates of performance that must be closed in any performance period. The ψ parameter functions as an "ideal standard" weight, making target t_i a weighted average of the base and ideal performance levels. When $\psi = 1.0$, equation 8.5 reduces to $t = \lambda_{ideal}$. Any 50:50 actuarially fair α rate of improvement used by payers has a ψ target analog weight for the ideal quality level. This is shown by solving equation 8.5 for ψ and remembering equation 8.1:

$$\psi = \alpha / [(\lambda_{ideal} / \lambda_{base}) - 1] = [(t / \lambda_{base}) - 1)] / [(\lambda_{ideal} / \lambda_{base}) - 1]. \quad (8.6)$$

The weight placed on the ideal rate varies positively with the payer's expected rate of improvement, α, assuming $(\lambda_{ideal} / \lambda_{base}) > 1$, but it also varies inversely with the relative difference between the ideal and base rates. If the payer envisions a 95 percent ideal rate but sets a target rate of 85 percent on a base rate of 60 percent, then its implicit improvement rate $\alpha = 0.85 / 0.60 - 1 = 0.42$. In setting an 85 percent target, the payer implicitly assumes an ideal weight $\psi = (0.85 / 0.60 - 1) / (0.95 / 0.60 - 1) = 72$ percent. Hence, an 85 percent target rate requires closing only 72 percent of the gap between the base and ideal rates. Payers should be aware of the implications of setting α in terms of the percentage gap (ψ) they expect to close between the base and ideal quality rates.

Paying only for positive improvement (i.e., $\rho > 0$), instead of an actual-to-target ratio, provides much stronger financial incentives to improve quality. For example, suppose $\lambda_{base} = 0.50$ and a payer paid no bonus if the managed care or provider organization failed to achieve a 25 percent improvement $t = 1.25(0.50) = 0.625$. Under this scenario, if the organization raised the quality indicator by one-half of the required 25 percent, or 0.5625, it would receive no bonus in the rate-of-improvement scenario. By contrast, under the continuous unconstrained arrangement, it would receive 90 percent of its bonus (i.e., $\lambda / t = [1 / 1.25] + [0.25 / 1.25][0.50]$). We simulate the financial gains (losses) involved in paying on rates vs. levels of improvement later in the chapter.

Number and Interdependence of Quality Indicators (N)

Because physicians see a variety of patients every day, we need several quality indicators to capture even a modest share of their caseload. Spreading bonuses and penalties across more indicators reduces the variance of the expected gain (Research & Education Association, 1996, p. 266). Big bonuses or penalties are less likely as the number of indicators increases. Assuming that managers are risk averse and seek to reduce the likelihood of a zero bonus (or a large penalty) across all indicators, they should prefer more indicators. Yet diversifying their risk across more indicators may not be optimal if managers have negotiated "easily attainable" targets on one or two indicators. Moreover, because any positive correlation among indicators raises bonus (or penalty) variance, we also simulate the risk effects of varying degrees of indicator interdependence.

Quality Indicator Weights (ω)

Uncertainty surrounds not only the success of interventions that improve care processes but also the responsiveness of metrics such as life-years saved to better processes of care (Landon et al., 2007; Siu et al., 2006; Werner & Bradlow, 2006). This is why payers and clinicians are inclined to give some quality indicators more weight than others. Using simulation methods, we explore whether using dramatically different weights for some indicators substantially raises the rewards and risks associated with bonus payouts.

Simulation Methods

To determine the variation in indicator-specific expected bonus fractions under alternative payment algorithms, we simulated performance by using 500 random trials from a normal distribution of an organization's actual improvement rates. Results were essentially identical using 1,000 trials. If organizations are risk-averse, the likelihood function should be right-skewed and more weight given to below-target performance. We adjusted for risk aversion by simulating expected performance below target, which should give results similar to those from a log-normal or similar uncertainty distribution. We assumed no feedback loop of bonus payments on an organization's investment in raising quality, which should produce a downward bias in expected bonus payments.[5]

---

[5] We assumed no feedback loop of bonus payments on an organization's investment in raising quality, which should produce a downward bias in expected bonus payments. It is reasonable to assume that organizations facing low expected bonuses would invest more to raise their payments—at least up to a point—to increase their bonuses or minimize their penalties.

An important unknown is the effectiveness of quality-improvement interventions when baseline levels are very low, which is why we simulated some expected bonus impacts at a low baseline level of 20 percent (see Simulation 4b later in this section). We simulated an actual rate of improvement $\rho_{ipd} = E[\rho_{ip}] \pm r_{ipd}SE_\rho$ for 500 trials for each of five indicators (i) for the p-th payment arrangement; r_{ipd} is a random normal variable around the expected rate of improvement as a consequence of any single trial.

In the baseline simulation, we assume that the organization expects to meet the payer's required improvement rate, $E[\rho] = \alpha = 0.25$, with a medium level of uncertainty $\sigma_\rho = 0.125$ (coefficient of variation [CV] = 0.50). Thus, if a single draw from a random normal distribution was 1.96, then the organization's expected rate of quality improvement for an indicator would be $E[\rho] = 0.25 + 1.96(0.125) = 0.495$, and $E[\lambda] / t = 1.495 / 1.25 = 1.196$ above target. A random draw of only 0.20 would give a simulated success-to-target rate of 0.98 (1.225 / 1.25). The resulting relative performance ratios are then converted to individual indicator payout percentages using the bonus algorithms described in Table 8-1. We determined a final overall bonus percentage by aggregating across five indicators using equal weights.

We simulated the impacts of the six payment algorithms on the level and variability in gains (paybacks) that organizations face, while varying key elements in the final payout structure listed below:

1. **Organizational uncertainty (ρ) about achieving target growth rate ($\alpha = 0.25$):**

   1a. Low uncertainty: standard deviation of 0.051 and CV = 0.051 / 0.25 = 0.20

   1b. High uncertainty: standard deviation of 0.165 (CV = 0.66)

   Greater organizational uncertainty about an intervention's success should result in a greater expectation of smaller (no) bonuses or larger penalties, depending on payment arrangement.

2. **Number and correlation of quality indicators**

   2a. 10 indicators (all equally weighted)

   2b. Two pairs of indicators correlated 50 percent; a fifth uncorrelated

   Having more indicators reduces the likelihood of very small bonuses (or large penalties). Greater correlation among indicators works in the opposite direction to raise the likelihood of bigger gains or losses.

3. **Unequal indicator weights:** one indicator weighted 50 percent and four of five each weighted 12.5 percent

   Weighting one or more indicators disproportionately can increase the risk of small bonuses or large penalties.

4. **Expected levels arrangements (λ / t)**

   4a. Organization's expected performance level, λ, is two-thirds of payer's required improvement rate $\alpha = 0.25$ on $\lambda_{base} = 53.3$ percent: $(\lambda / t) = 0.622 / 0.666 = 0.934 = 53.3[1+0.67(0.25)]/53.3(1.25)$

   4b. Organization's expected performance level is one-half of payer's required improvement rate $\alpha = 1.5$ on a low $\lambda_{base} = 20$ percent: $(\lambda / t) = 0.35 / 0.50 = 0.70$

   4c. Organization's expected performance level is 1.5 times payer's required improvement rate $\alpha = 1.96(SE_\lambda = 0.01054)$ on $\lambda_{base} = 53.3$ percent using a 95 percent confidence interval: $(\lambda / t) = 0.564 / 0.533 = 1.019$

   Organizations that expect to exceed the target and its implicit rate of improvement over baseline will expect a higher total bonus or smaller penalties.

5. **Expected rate of improvement arrangements, $E[\rho] / \alpha$**

   5a. Neutral expected-to-target growth: $E[\rho] / \alpha = 1$

   5b. Robust required target growth: $E[\rho] / \alpha = 0.67$

   Paying bonuses on rates of improvement over baseline and not on actual-to-target levels requires organizations to substantially improve quality to receive any bonuses. Neutral expected target growth assumes that the organization's expected improvement just equals the payer's required target rate of improvement. Robust target growth assumes that the organization's expectation of improvement falls one-third short of the payer's "ambitious" rate of improvement.

Results

Table 8-2 presents the mean and first quartile threshold bonus percentages that would be paid out under scenarios that vary by several key parameters. The first four payout algorithms in Table 8-1 are shown as columns in Table 8-2: all-or-nothing, continuous unconstrained, continuous constrained (LL = 90 percent; UL = 100 percent; θ = 50 percent bonuses between limits), and composite. The fifth and sixth P4P payment algorithms in Table 8-1 are simulated as rows 4c and 5a,b in Table 8-2. The baseline and simulations 1a through 4c base final bonus percentages on relative success-to-target quality *levels*, λ / t. Simulations 5a and 5b pay bonuses under each of the four P4P columns only if a provider or managed care organization improves on some or all of the quality indicators.

Baseline Simulation

Using the baseline simulation parameters, Table 8-2, top row, the all-or-nothing payment arrangement (columns 1 and 2), has an expected baseline bonus payout of 0.50, or 50 percent, averaged across the five indicators, with a lower first-quarter threshold of 40 percent. Although an organization has a 50:50 chance of no bonus on any particular indicator in the all-or-nothing arrangement, it has only a 25 percent chance of receiving 40 percent or less of its maximum bonus because success on some indicators offsets failure on others. Of 500 baseline all-or-nothing trials, only 15 resulted in no overall bonus payout at all, whereas another 250 trials had one or more failures out of five indicators. The rest of the trials enjoyed successful payouts on all five indicators.

Organizations paid on a continuous unconstrained algorithm (columns 3 and 4, top row) could expect to receive 96 percent of their overall bonus percentage, on average, under the baseline scenario. Such a high percentage results from the payer's making minimum bonus payments of $1 / (1 + \alpha)$ = 1 / 1.25 = 80 percent or more—even when the organization simply achieves the baseline rate with no improvement.

When baseline bonuses are constrained to just 50 percent when success-to-target ratios are between 90 percent and 100 percent with no bonus below 0.90 of the target (columns 5 and 6, top row), the expected bonus percentage falls from 96 percent to 67 percent. The high first-quartile threshold of 60 percent implies a low likelihood of a very small bonus payout, even with a highly constrained bonus structure.

Under the baseline composite payment algorithm (last two columns, top row), an organization that expected to achieve the required improvement rate of 25 percent could expect to receive 100 percent of its overall bonus. The "composite" payment expectation is even higher than under the continuous unconstrained algorithm because it allows indicator-specific bonus payments in excess of 100 percent to offset lower bonus percentages on some indicators.

Table 8-2. P4P simulation results

Parameter	Bonus Payout Percentages[a]							
	All-or-Nothing		Continuous Unconstrained		Continuous Constrained[b]		Composite	
	Mean	25%ile	Mean	25%ile	Mean	25%ile	Mean	25%ile
Baseline Simulation[c]	0.50	0.40	0.96	0.95	0.67	0.60	1.00	0.97
1. Uncertainty[d]								
1a. Low: $\sigma(\rho) = 0.051$	0.50	0.40	0.98	0.98	0.75	0.70	1.00	0.99
1b. High: $\sigma(\rho) = 0.165$	0.50	0.40	0.95	0.93	0.64	0.50	1.00	0.96
2. Number/Correlation of Indicators								
2a. 10 indicators	0.51	0.40	1.00	0.95	0.68	0.60	1.00	0.98
2b. 2sets-of-5 correlated 50%[e]	0.48	0.25	0.96	0.94	0.66	0.50	1.00	0.96
3. Weights[f]								
1 $\omega = 0.50$; 4$\omega = 0.125$	0.48	0.25	0.96	0.94	0.66	0.50	1.00	0.96
4. Expected Levels Arrangements $E[\lambda] / t$								
4a. $E[\Delta\lambda] = 0.67(\Delta t = 0.133)$[g]	0.15	0.00	0.89	0.86	0.33	0.20	0.90	0.87
4b. $E[\Delta\lambda] = 0.50(\Delta t = 0.30)$[h]	0.00	0.00	0.70	0.68	0.00	0.00	0.70	0.68
4c. $E[\Delta\lambda] = 1.50(1.96 SE_\lambda)$[i]	0.59	0.40	0.97	0.95	0.73	0.60	1.00	0.96
5. Expected Improvement Arrangements (ρ / α)[j]								
5a. $E[\rho] / \alpha = 1.00$	0.50	0.40	0.81	0.73	0.54	0.40	0.99	0.84
5b. $E[\rho] / \alpha = 0.67$	0.15	0.00	0.50	0.39	0.18	0.00	0.49	0.34

[a] Percentages based on 500 random normal trials.
[b] Bonuses of 50 percent for $0.90 < \lambda / t < 1.0$, and 0 or 1.0 at lower level/upper level.
[c] Based on five equally weighted, uncorrelated indicators, $\alpha = 0.25$ target improvement rate, $\sigma(\rho) = 0.125$, baseline rate $\lambda_{base} = 53.3$ percent.
[d] $\sigma(\rho)$ = the standard deviation of an organization's own expected intervention effectiveness over baseline.
[e] Two pairs of five indicators correlated 50 percent; fifth indicator uncorrelated.
[f] One indicator weighted 50 percent; remaining four equally weighted.
[g] Organization expects to achieve only two-thirds of payer's targeted 25 percent improvement rate on $\lambda_{base} = 53.3$ percent, $\lambda / t = 0.622 / 0.666 = 0.934$.
[h] Organization expects to achieve only one-half of payer's targeted 150 percent improvement rate on $\lambda_{base} = 20$ percent, or $\lambda / t = 0.35 / 0.50 = 0.70$.
[i] Organization expects to achieve 50 percent above 5% confidence threshold = 1.96 times SE based on 1,000 patients on $\lambda_{base} = 53.3$ percent, or $\lambda / t = 0.564 / (0.533 + 0.01045) = 1.038$.
[j] Bonuses based on actual vs. target rates of improvement, not absolute levels.

Levels-Based Simulations

Financial losses from varying the degree of organizational uncertainty, the number of indicators, how indicators are weighted, and the correlation among indicators (simulations 1a–3) differ little from the baseline simulation as long as an organization (1) believes it has at least a 50:50 chance of just achieving the target growth rate, and (2) is paid on its actual-to-target rate of success. The type of payment arrangement—*not* the parameters—determines expected bonuses when conditions (1) and (2) exist.

Bonuses (or penalties) can change radically if an organization's expectation of success is less than 50:50 (simulations 4a and 4b). The expected bonus percentage under an all-or-nothing algorithm falls from 50 percent (baseline simulation) to 15 percent if an organization's expected improvement rate was only two-thirds of the target growth rate (simulation 4a, column 1). A constrained algorithm with no bonus below 90 percent of the target (4a, column 5) produces an expected bonus of only 33 percent. Unconstrained and composite bonuses are much less sensitive to robust (ambitious) target growth rates relative to organizations' expectations. This is because *any* level of quality relative to the target generates substantial bonuses that would not be paid at all in an all-or-nothing payment arrangement or only in a limited fashion in a constrained payment scenario.

Consider, next, simulation 4b. It may be unrealistic for a payer to assume that an organization has a fair chance to raise the baseline rate by 30 percentage points to 50 percent from a very low 20 percent baseline over a short demonstration period. If an organization that faced a 30 percentage point required increase (i.e., $\Delta t = 1.5[20\%] = 0.30$) felt that it could achieve only one-half of that rate of improvement over baseline (simulation 4b), then expected bonuses in the all-or-nothing and constrained payment arrangements fall to zero. This is because of the organization's relatively narrow (assumed) range of medium uncertainty, $\sigma(\rho) = 0.125$, which is around its lower expected rate of improvement. Continuous and composite bonus arrangements would continue to pay out 70 percent on average (columns 3 and 7), even when an organization expected to achieve only one-half the target rate of 50 percent on a baseline quality level of 20 percent. Again, these two arrangements continue to pay a high percentage of bonuses by always rewarding an organization for achieving a fraction of the target.

As shown in simulation 4c, setting the target at only the upper limit of a 95% confidence level around the baseline level requires little in the way of quality improvement—especially for large intervention populations. With 1,000 patients, 1.96 times the standard error around the mean baseline rate of 53.3 percent produces a higher quality target of only 56.4 percent (3.1 percentage points higher). All-or-nothing expected bonuses increase from 50 percent in the original baseline simulation to 59 percent if the organization believes its intervention's effectiveness would be 1.5 times as high as the targeted increase of just 3.1 percentage points. Constrained bonuses increase from 67 percent to 73 percent (column 5). Continuous and composite bonuses remain at nearly 100 percent because of the high baseline floor and an overall ceiling on the full bonus.

Improvement Rate Simulations

Even when an organization's expected rate of improvement is equal to the payer's required rate (simulation 5a), bonuses in the continuous unconstrained and constrained models decline from 96 percent and 67 percent in the baseline simulation to 81 percent and 54 percent, respectively. This is because paying only for positive growth *rates*, unlike levels, does not reward organizations if they achieve zero improvement over baseline. All-or-nothing bonuses are unaffected by paying on rates vs. levels because the arrangement never pays anything when failing to meet the target. In stark contrast, composite payment arrangements treat levels and rates of improvement the same when organizations expect to achieve the required rate of improvement—again because of offsetting large bonus percentages for some indicators.

Because paying on quality improvement factors out the baseline bias inherent in paying on levels, average expected bonuses generally fall to their lowest levels if an organization expects to achieve only two-thirds of the targeted improvement rate (simulation 5b). The continuous constrained arrangement pays only 18 percent on average. Even the generous composite arrangement has an average expected bonus percentage of only about 50 percent.

Discussion

Payers naturally seek the most cost-effective way to reward managed care organizations and provider groups when they improve quality. This requires that quality bonuses be neither too easy nor too difficult to achieve. Based on our simulation results, their strategy should be to

- select process quality indicators that are closely linked to patient outcomes,

- set challenging target levels over baseline performance levels and not targets that are only statistically greater than baseline, and
- base bonuses (or penalties) on rates of improvement over baseline (e.g., 20 percent improvement on a 50 percent baseline, or 10 percentage points) and not on target levels that pay bonuses even if the intervention only reproduces the baseline (50 percent) level.

Payers should think of "challenging" targets as a weighted average of the baseline and the ideal levels. Setting the ideal weight too high will produce unreachable targets that can discourage any serious investment in quality improvement. All-or-nothing or tightly constrained payment methods are particularly punitive if targets are not reasonably achievable over short periods. At the other extreme, simply requiring organizations to achieve a target that is only statistically different from the baseline rate implicitly assumes very little true intervention effect—especially for large patient populations—and guarantees sizable bonus payouts.

Payers should avoid unconstrained and composite P4P arrangements if they choose to pay on performance levels. Both essentially *guarantee* organizations a very high percentage of their total bonus (or very little payback of management fees). Constrained continuous and composite payment arrangements can produce more stringent, efficient bonus payouts (or more meaningful penalties) when based on *rates* of improvement instead of intervention vs. baseline levels.

References

Agency for Healthcare Research and Quality. (2006). *National healthcare quality report.* Rockville, MD: US Department of Health and Human Services.

Bokhour, B. G., Burgess, J. F., Jr., Hook, J. M., White, B., Berlowitz, D., Guldin, M. R., et al. (2006). Incentive implementation in physician practices: A qualitative study of practice executive perspectives on pay for performance. *Medical Care Research and Review, 63*(1 Suppl), 73S–95S.

Chassin, M. R., & Galvin, R. W. (1998). The urgent need to improve health care quality. Institute of Medicine National Roundtable on Health Care Quality. *JAMA, 280*(11), 1000–1005.

Cromwell, J., Drozd, E., Smith, K., & Trisolini, M. (2007, Fall). Financial gains and risks in pay-for-performance bonus algorithms. *Health Care Financing Review, 29*(1), 5–14.

Epstein, A. M. (2006). Paying for performance in the United States and abroad. *New England Journal of Medicine, 355*(4), 406–408.

Institute of Medicine, Board on Health Care Services. (2001). *Crossing the quality chasm: A new health system for the 21st century.* Washington, DC: National Academies Press.

Institute of Medicine, Board on Health Care Services. (2006). *Performance measurement: Accelerating improvement* (Pathways to Quality Health Care Series, 1). Committee on Redesigning Health Insurance Performance Measures, Payment, and Performance Improvement Programs. Washington, DC: National Academies Press.

Landon, B. E., Hicks, L. S., O'Malley, A. J., Lieu, T. A., Keegan, T., McNeil, B. J., et al. (2007). Improving the management of chronic disease at community health centers. *New England Journal of Medicine, 356*(9), 921–934.

LLanos, K., Rothstein, J., Dyer, M. B., & Bailit, M. (2007). Physician pay-for-performance in Medicaid: A guide for states. Hamilton, NJ: Center for Health Care Strategies. Available from http://www.chcs.org/publications3960/publications_show.htm?doc_id=471272

McCall, N. T., Cromwell, J., Urato, C., & Rabiner, D. (2008). *Evaluation of Phase I of the Medicare Health Support Pilot Program under traditional fee-for-service Medicare: 18-month interim analysis.* Report to Congress. Available from http://www.cms.hhs.gov/reports/downloads/MHS_Second_Report_to_Congress_October_2008.pdf

National Committee for Quality Assurance. (2006). *Health Plan Employer Data and Information Set (HEDIS).* Washington, DC: National Committee for Quality Assurance.

National Quality Forum. (2006). *National Voluntary Consensus Standards for Ambulatory Care: An initial physician-focused performance measure set.* Washington, DC: National Quality Forum.

Research & Education Association. (1996). *The statistics problem solver.* Piscataway, NJ: Research & Education Association.

Rosenthal, M. B., & Dudley, R. A. (2007). Pay-for-performance: Will the latest payment trend improve care? *JAMA, 297*(7), 740–744.

Siu, A. L., Boockvar, K. S., Penrod, J. D., Morrison, R. S., Halm, E. A., Litke, A., et al. (2006). Effect of inpatient quality of care on functional outcomes in patients with hip fracture. *Medical Care, 44*(9), 862–869.

Trude, S., Au, M., & Christianson, J. B. (2006). Health plan pay-for-performance strategies. *American Journal of Managed Care, 12*(9), 537–542.

Weinstein, J. N., Bronner, K. K., Morgan, T. S., & Wennberg, J. E. (2004). Trends and geographic variations in major surgery for degenerative diseases of the hip, knee, and spine. *Health Affairs (Millwood), Supplemental Web Exclusives,* VAR81–VAR89.

Weinstein, J. N., Lurie, J. D., Olson, P. R., Bronner, K. K., & Fisher, E. S. (2006). United States' trends and regional variations in lumbar spine surgery: 1992–2003. *Spine, 31*(23), 2707–2714.

Wennberg, J. E., Fisher, E. S., Sharp, S., McAndrew, M., & Bronner, K. K. (2006). The care of patients with severe chronic illness: An online report on the Medicare program by the Dartmouth Atlas Project. Available from http://www.dartmouthatlas.org/downloads/atlases/2006_Chronic_Care_Atlas.pdf

Werner, R. M., & Bradlow, E. T. (2006). Relationship between Medicare's Hospital Compare performance measures and mortality rates. *JAMA, 296*(22), 2694–2702.

Williams, T. R. (2006). Practical design and implementation considerations in pay-for-performance programs. *American Journal of Managed Care, 12*(2), 77–80.

CHAPTER 9

Overview of Selected Medicare Pay for Performance Demonstrations

Leslie M. Greenwald

Several current pay for performance (P4P) initiatives began as Medicare pilot projects, or demonstrations, that test both the administrative feasibility and outcomes-defined "success" of the individual performance models. This approach of pilot testing P4P initiatives allows Medicare policy makers to determine the models that best meet their intended goals and can be operationalized at an acceptable level of administrative cost and burden to physicians and health care provider organizations, insurers, and other stakeholders. Reliance on testing through demonstrations also allows policy makers to identify lessons learned and opportunities for improvement, and to adapt aspects of new initiatives that do not work—all on a manageable scale not possible with full implementation through a program the size of Medicare. Demonstrations also identify the most successful variants within a general type of innovation—such as P4P—for replication, expansion, and possible national application.

As one of the largest public insurers in the world, Medicare has played a special role in pilot testing a wide range of health care programs, in addition to P4P. The Medicare program has several advantages in testing health care innovation. First, because Medicare is a major publicly funded program, Congress often makes funding available both to support technical development of P4P and other innovations and for comprehensive independent evaluations of the pilot programs. Second, the Medicare program operates in a way that makes large amounts of administrative data available for development of a variety of P4P models, supports their implementation, and allows for relatively efficient evaluation options. Finally, because of Medicare's size and importance in the clinician and provider marketplace, it is often more feasible for this public program to gather practitioners and providers and other organizations willing to engage in demonstration projects to develop and evaluate P4P demonstration options (as well as other policy pilot projects). Thus, complex new initiatives such as proposed Medicare P4P models start out as demonstrations, with national implementation an implicit future goal (although national implementation of a demonstration is rare, a topic discussed in further detail in Chapter 11 of this book).

Medicare has a rich history of demonstration projects for even as relatively recent a policy initiative as physician or provider P4P. The dozens of new Medicare P4P and other related demonstrations mandated under the Affordable Care Act continue policy makers' reliance on the Medicare program to test new ideas for health care reform.

This chapter summarizes a range of the Medicare P4P demonstrations currently completed or near implementation. The demonstrations described here are not exhaustive of all the P4P demonstrations the Medicare program has considered, designed, or implemented. As a result of health care reform under the Affordable Care Act, this list will expand significantly. Rather, this selection of demonstrations is intended to give the reader a sense of the kinds of P4P projects that have been tried under Medicare and, when the information is available, whether they were successful in improving health care efficiency and quality of care. As a group, they may give some signals as to the possible success of P4P models in future years under reform.

Table 9-1. Overview of Medicare P4P demonstrations

Demonstration Name	Summary Description
Care Management Pay for Performance Demonstrations	
Medicare Coordinated Care Demonstration	Demonstration's goal was to identify intervention components that save the government money while maintaining quality of care or possibly improving the quality through better coordination of the chronically ill—without net increase in Medicare spending.
Medicare Health Support Pilot Program	The pilot is testing a P4P third-party non–health care provider contracting model. MHSOs aimed to improve clinical quality, increase beneficiary and clinician/provider satisfaction, and achieve Medicare program savings for chronically ill Medicare FFS beneficiaries with targeted conditions of heart failure and/or diabetes.

The demonstration projects described in this chapter are organized into three categories:

- Care management P4P demonstrations—projects that use a third-party care management organization or other strategies to coordinate Medicare beneficiary care
- Physician-focused P4P demonstrations—projects that base P4P models around outpatient and ambulatory care and/or use the physician group as the primary responsible organization
- Hospital-focused P4P demonstrations—projects that base P4P around hospital-based care and use the hospital as the primary responsible organization

This chapter provides an overview of each P4P demonstration, describes the key features of the initiative, and summarizes the status of each project. When evaluation findings to date are publicly available, they are presented here.

Some readers may not be interested in the full demonstration details provided here and may choose to refer to the detailed descriptions only to supplement points or references made in other chapters of this book. Therefore, Table 9-1 summarizes the P4P demonstration projects described in this chapter.

Demonstration Status and Available Findings

- Implemented in 2002
- Of 15 programs, only 1 had statistically significant reduction in hospitalizations. All programs saw increases in Medicare expenditures for care for intervention population between baseline and demonstration period. None of the 15 produced statistical savings in Medicare outlays on services relative to control group, but 2 had higher costs. Clinical measures showed few, scattered effects of self-reported flu and pneumococcal vaccinations, mammography, or other routine diabetic and CAD tests. No pattern of patient responses suggested that preventable hospitalizations had been reduced.

- Implemented in 2005/2006
- Only limited positive impacts achieved on positive improvements in patient overall satisfaction. No statistically significant findings for clinical interventions relative to comparison group. Limited Medicare savings achieved in first 18 months, but none of the gains were statistically significant.

(continued)

Table 9-1. Overview of Medicare P4P demonstrations *(continued)*

Demonstration Name	Summary Description
Care Management for High-Cost Beneficiaries Demonstration	Demonstration's principal objective was to test care management models for Medicare beneficiaries who are high cost and have complex chronic conditions, with goals of reducing future costs, improving the quality of care, and improving beneficiary and clinician/provider satisfaction.
Cancer Prevention and Treatment Demonstration	Demonstrations were aimed at reducing disparities in cancer screening, diagnosis, and treatment among racial and ethnic minority Medicare beneficiaries through use of peer navigators. Peer navigators help steer Medicare beneficiaries through health care system.
Physician-Focused Pay for Performance Demonstrations	
Medicare Physician Group Practice Demonstration	Medicare's first physician P4P initiative. PGP demonstration establishes incentives for quality improvement and cost efficiency at level of physician group practice. Goals included (1) encouraging coordination of health care furnished under Medicare Parts A and B, (2) encouraging investment in administrative structures and processes for efficient service delivery, and (3) rewarding physicians for improving health care processes and outcomes.
Medicare Medical Home Demonstration	A medical home is a physician-directed practice that provides care that is accessible, continuous, comprehensive, and coordinated and is delivered in context of family and community. Some variants combine use of health information technology and/or electronic medical records as a care-coordination tool.
Hospital-Focused Pay for Performance Demonstrations	
Medicare Participating Heart Bypass Center Demonstration	Under this demonstration, government paid a single negotiated global price for all Parts A & B inpatient hospital and physician care associated with bypass surgery. Demonstration was to encourage regionalization of procedure in higher-volume hospitals and to align physician with hospital incentives under bundled prospective payment. Hospitals shared global payment with surgeons and cardiologists based on cost savings. CMS allowed participants to market a CoE demonstration imprimatur referring to themselves as a "Medicare Participating Heart Bypass Center." Medicare patients were not restricted to demonstration hospitals for their surgery.
Expanded Medicare Heart and Orthopedics Centers of Excellence Demonstration	Developed as follow-on to Medicare Participating Heart Bypass Center Demonstration. Expanded demonstrations were to include more cardiovascular procedures and major orthopedic procedures such as hip and knee replacement.

Demonstration Status and Available Findings

- Implemented in 2006
- No evaluation findings publicly available.

- Implemented in 2006/2007
- Publicly available evaluation results focus on *implementation* issues. Based on available results, five of six demonstration sites encountered difficulty in identifying eligible beneficiaries and enrolling them in a demonstration, resulting in substantially fewer participants than initially projected.

- Implemented in 2005
- CMS has publicly reported evaluation of results through second demonstration year. In the second performance year, 4 of the 10 participating physician groups earned a total of $13.8 million in performance payments for improving quality and cost efficiency of care as their share of a total of $17.4 million in Medicare savings. When adjusted for predemonstration expenditure trends, reduction in expenditures was $58 per person, or 0.6% less than the target, and not statistically different from zero. Between base year and second demonstration year, 4 of 7 claims-based quality indicators showed greater improvement among PGP-assigned beneficiaries than among comparison beneficiaries. This improvement was statistically significant at 5% level.
- Implementation pending coordination with medical home mandates in Affordable Care Act health care reform legislation.

- Implemented in 1991
- Over the demonstration's 5 years, Medicare program saved $42.3 million on the 13,180 bypass patients treated in the seven demonstration hospitals. About 85% of savings came from demonstration discounts, another 9% from volume shifts to lower-cost demonstration hospitals, and 5% from lower post-discharge utilization.

- Not implemented due to health care provider resistance.

(continued)

Table 9-1. Overview of Medicare P4P demonstrations *(continued)*

Demonstration Name	Summary Description
Medicare Acute Care Episode Demonstration	Most recent iteration of CoE P4P model. Demonstration offers bundled payments and increased flexibility in financial arrangements between participating hospital-physician consortia. Will also focus on methods for improved quality of care for bundles of heart and orthopedic hospital-based procedures. Approved demonstration sites will be allowed to use term "Value-Based Care Centers" in approved marketing programs.
Premier Hospital Quality Incentive Demonstration	Demonstration recognizes and provides financial rewards to hospitals that demonstrate high-quality performance in hospital acute care. Conducted by Medicare in collaboration with Premier, Inc., nationwide organization of not-for-profit hospitals. Top-performing hospital participants rewarded with increased payment for Medicare patients.
Medicare Hospital Gainsharing Demonstration and Physician–Hospital Collaboration Demonstration	Both demonstrations test similar a gainsharing model. Overall concept is intended to allow hospitals to share efficiency savings with physicians under controlled setting in which quality of care standards are maintained or improved.

CAD = coronary artery disease; CMS = Centers for Medicare & Medicaid Services; CoE = Center of Excellence; FFS = fee-for-service; MHSO = Medicare health support organization; P4P = pay for performance; PGP = Physician Group Practice.

Note: This table describes the demonstrations discussed in this chapter only and is not an overview of all Medicare P4P demonstrations.

Care Management P4P Demonstrations

A large group of P4P demonstration projects center on the concept of disease and chronic care management: that by implementing specifically targeted chronic care/disease management interventions, we can improve beneficiaries' adherence to self-care and other preventative approaches that can potentially reduce overall costs of acute care. Under these demonstrations, the Medicare program pays disease management organizations (sometimes on a risk basis) for managing patients with specific target conditions such as diabetes and congestive heart failure (CHF). Medicare pays the organizations based on a per beneficiary per month (PBPM) fee. Under many of these models, disease management firms forfeit some or all of their fees if they fail to achieve savings targets.

Demonstration Status and Available Findings
• Implemented in 2009 • No evaluation findings publicly available.
• Implemented in 2003. Phase II projects operated between 2007 and 2009. • Findings from initial years of demonstration are publicly available. Over initial 2 years, both nonparticipating (those only reporting data) and hospitals participating in P4P program, showed quality improvements. In 7 of 10 quality indicators, P4P hospitals showed greater improvements. After adjusting for baseline differences in study and control groups, incremental increases in quality attributed to P4P incentives declined. Preliminary results from first 4 years suggest participating hospitals raised overall quality by average of 17 points over 4 years, based on their performance on more than 30 nationally standardized care measures for patients in five clinical areas.
• Medicare Hospital Gainsharing Demonstration implemented in 2008. • Medicare Physician–Hospital Collaboration Demonstration implemented in 2009. • No evaluation findings publicly available.

Medicare Coordinated Care Demonstration

Project Overview

The Balanced Budget Act of 1997 instructed the Secretary of Health and Human Services to conduct and evaluate care coordination programs in Medicare's fee-for-service (FFS) setting (Peikes et al., 2009). In 2002, Centers for Medicare & Medicaid Services (CMS) selected 15 demonstration programs of various sizes and intervention strategies as part of the Medicare Coordinated Care Demonstration (MCCD). The demonstration's goal was to identify intervention components that save the government money while maintaining quality of care or possibly improving the quality of care through better coordination of health care the chronically ill—without any net increase in Medicare spending. The MCCD used a randomized intent-to-treat (ITT) design. Eligible beneficiaries in areas served by the 15 programs were randomized on a 1:1 basis to the intervention and control groups. Four programs requested a stratified randomization process.

Project Status

Programs began enrolling beneficiaries in the intervention group over summer 2002, followed by a 3-year evaluation period. Beneficiary participation was voluntary. CMS paid a negotiated monthly management fee that ranged from $80 to $444. The average fee across the 15 programs was $235 (Peikes et al., 2009). Fees were limited to 20 percent of the historical average monthly PBPM costs of the chronically ill, given that savings on Medicare outlays were unlikely to be greater. After the 6-month enrollment period, CMS paid no fees on intervention beneficiaries who were not enrolled or had decided to drop out. Programs had to be budget neutral and were at financial risk if savings in Medicare outlays on intervention beneficiaries were less, on a monthly basis, than the monthly fee. Calculations of savings also included Medicare expenditures incurred by intervention beneficiaries who dropped out of the demonstration, thereby putting programs at risk for lower enrollment rates.

None of the programs charged beneficiaries to participate. Three types of quality measures were used in evaluating the programs: (1) Medicare claims were used to identify six disease-specific and preventive process-of-care indicators; (2) claims data were also used to track hospitalizations of eight ambulatory care sensitive conditions thought to be avoidable through improved care management; and (3) a beneficiary survey collected responses related to health education received from the programs, functional status, knowledge and adherence to medication and other protocols, and perceived quality of life.

The participating sites were a broad mix of disease management organizations, including commercial ones, academic medical centers, and community hospitals (an integrated delivery system, a long-term care facility, and a retirement community). The selection provided an opportunity to compare cost-effectiveness between two competing disease management models, one relying on commercial vendors and another grounded in physician practices. Programs served beneficiaries in diverse geographic areas, including Maine (statewide), southern Florida, South Dakota, Phoenix, and central California.

The programs targeted Medicare-aged and disabled beneficiaries with coronary artery disease (CAD), CHF, diabetes, chronic obstructive pulmonary disease (COPD), and a few minor chronic conditions. In identifying eligibles, 10 programs required at least one hospitalization (6 stipulated that the hospitalization be related to a target chronic condition), 4 excluded the nonelderly, 13 excluded end-stage renal disease (ESRD) beneficiaries, 9 excluded long-term nursing home residents, and all but 1 program excluded patients who were terminally ill, had AIDS, or had similarly complex conditions.

The number of beneficiaries in each program was generally small. The largest 3 programs had between 2,289 and 2,657 total beneficiaries and had only roughly as many in the intervention group. Three programs had between 90 and 115 intervention patients and fewer than 250 including the control group. Overall, 18,402 beneficiaries were spread across 15 programs. Consequently, the study's power to detect significant differences was low, although the evaluators generally had more than 90 percent power to detect a 20 percent or greater gain in outcomes and cost savings in the intervention over the control group. None of the programs appear to have had 80 percent statistical power to detect intervention gains of 10 percent or less. (Peikes et al., 2009, p. 608).

Participants varied widely across programs by geographic area (Peikes et al., 2009). A few sites had no minorities, whereas Georgetown University had 63 percent African American and Hispanic enrollees. Medicaid eligibility ranged from 0 percent to 28 percent. CAD and CHF generally were the dominant diagnoses, with significant numbers (>20 percent) of beneficiaries who had COPD, cancer, or stroke. Jewish Home & Hospital was exceptional with 33 percent of enrolled patients having dementia.

All of the programs assigned enrollees to a registered nurse care coordinator. Eleven programs contacted patients 1 to 1.5 times on average per month by telephone, and 3 contacted patients 4 to 8 times per month. All but 1 educated the patients regarding diet, medications, exercise, and self-care management. The University of Maryland did not educate patients but simply tested the effect of home monitoring of vital signs. One-half used transtheoretical or motivational interviewing approaches to behavior change. Most taught patients how to better communicate with their physicians using role playing. Only 4 programs concentrated on improving physicians' adherence to evidence-based practice guidelines. To avoid costly readmissions, 10 programs kept timely information on hospitalizations and emergency room visits that would allow them to intervene quickly post-discharge.

Findings to Date

Peikes and colleagues have already published findings for this project (2009). Similar to the Medicare Health Support Pilot Program's disappointing results, this demonstration found no statistically significant improvements in clinical outcomes or savings to Medicare. Of the 15 programs, only 1 (Mercy) had a statistically significant reduction in hospitalizations relative to its control group, controlling for patient characteristics. All of the programs saw increases in Medicare expenditures for care for the intervention population between baseline and the demonstration period. None of the 15 programs produced any statistical savings in Medicare outlays on services relative to the control group, but 2 had higher costs. Peikes and colleagues based these findings on regressions controlling for age, gender, race, disabled/aged entitlement, Medicaid coverage, and whether beneficiaries used skilled nursing facility or hospital services prior to the demonstration.

Once they added monthly fees to estimate savings net cost, 9 out of 15 programs had statistically higher costs to the Medicare program than did their control group (Peikes et al., 2009, p. 612). The one site with a reduction in hospitalizations had a large management fee that overwhelmed its (statistically insignificant) $112 in PBPM savings, resulting in higher net total Medicare costs.

Treatment beneficiaries were more likely to report having received education on diet, exercise, and disease warning signs than their corresponding control group. However, the "treatment group members were no more likely than control group members to say they understood proper diet and exercise" or that they were adhering better to prescribed diet, exercise, and medication regimens (Peikes et al., 2009, p. 613). Clinical measures showed few, scattered effects of self-reported flu and pneumococcal vaccinations, mammography, or other routine diabetic and CAD tests. No pattern of patient responses suggested that preventable hospitalizations had been reduced.

Care coordination activities, as practiced in the 15 varied interventions in this study, "hold little promise of reducing total Medicare expenditures" for the Medicare chronically ill (Peikes et al., 2009, p. 613). Two programs did show some promise in reducing hospitalizations and costs, however, suggesting that care coordination might be at least cost neutral.

The demonstration's main limitation was the small sample size and lack of statistical power to detect smaller savings rates. The study was unable to confirm a statistically significant savings rate of 9 percent at the 10 percent confidence level for the most successful site. This program also had one of the highest average monthly management fees, due in part to extensive registered nurse face-to-face contact with patients. A possible major reason for the lack of success in both Medicare savings and better health outcomes is the absence of a true transitional care model in which patients are enrolled during their hospitalization. Studies have shown the approach to significantly reduce admissions within 30/60 days post-discharge when the patient is at high risk of being readmitted (Coleman et al., 2006; Naylor et al., 1999; Rich et al., 1995). "By providing close links between the patient's nurse coordinator and physician, [with] substantial in-person contact between the patient and the care coordinator, . . . the medical home model may be able to replicate or exceed the success of the most effective MCCD programs" (Peikes et al., 2009, p. 617).

Medicare Health Support Pilot Program

Project Overview

Section 721 of the Medicare Prescription Drug, Improvement, and Modernization Act of 2003 (Pub. L. 108–173, also called the Medicare Modernization Act, or MMA), required the Secretary of Health and Human Services to provide for the phased-in development, testing, evaluation, and implementation of chronic care improvement programs (McCall et al., 2007). CMS selected eight Medicare Health Support (MHS) pilot programs under Phase I. The MHS initiative's principal objectives were as follows: to test a P4P contracting model and MHS intervention strategies that may be adapted nationally to improve clinical quality, increase beneficiary and clinician and provider satisfaction, and achieve Medicare program savings for chronically ill Medicare FFS beneficiaries with targeted conditions of heart failure and/or diabetes.

This initiative provides the opportunity to evaluate the success of the "fee at risk," P4P, model. MHS disease management organizations enjoy flexibility in their operations, coupled with strong incentives to expand outreach and refine intervention strategies to improve population outcomes. The MHS pilot program is distinct, legislatively, from most demonstration programs. A congressionally mandated pilot can be expanded easily into a national program if it reports positive results during the pilot phase; no additional legislation is required.

The MHS pilot's overall design follows an ITT model (McCall et al., 2007). Medical health support organizations (MHSOs) are held at risk for up-front monthly management fees based on the performance of the entire eligible Medicare population randomized to the intervention group and as compared with all eligible beneficiaries randomized to the comparison group. Beneficiary participation in the MHS programs is voluntary and does not change the scope, duration, or amount of Medicare FFS benefits that beneficiaries currently receive. The traditional Medicare FFS program continues to cover, administer, and pay for all Medicare FFS benefits, and beneficiaries do not pay any charge to receive MHS program services.

After the initial 6-month outreach period, the MHSOs accrue management fees for only those beneficiaries who verbally consent to participate and only during participation periods. Participation continues until a beneficiary becomes ineligible for the MHS program or opts out of services provided by the MHSO. To retain any monthly fees, MHSOs originally had to achieve 5 percent savings relative to the comparison group. Savings are defined as the difference in mean Medicare PBPM spending on services between the entire intervention and comparison groups, multiplied by the total number of eligible months in the intervention group. CMS subsequently dropped the 5 percent minimum savings requirements.

To retain all of its accrued fees, an MHSO had only to reduce average monthly payments equivalent to the monthly management fee. Because small differences remained in Medicare PBPM payments between intervention and comparison groups, CMS made an actuarial adjustment in the intervention PBPM for any difference from the comparison group in the 12 months just prior to each MHSO's start date. The MHSOs must also meet quality and satisfaction improvement thresholds or pay back negotiated percentages of their fees.

Project Status

Eight MHSOs launched their programs between August 1, 2005, and January 16, 2006. Several programs serve urban and suburban populations, whereas others target metropolitan and rural communities. Among the populations served are significant minority populations of African American, Native American, and Hispanic beneficiaries. During the second year of operations, three organizations requested early termination of their programs, primarily, they stated, out of concern that the 5 percent savings requirement plus savings covering accrued fees was too ambitious a goal. The MHS pilot targets beneficiaries with the threshold condition(s) of heart failure and/or diabetes from among the diagnoses listed on Medicare claims.

CMS prospectively identified 30,000 eligible beneficiaries from each MHSO area and randomly assigned them to intervention and comparison groups in a ratio of 2:1 under an ITT evaluation model. With 240,000 pilot beneficiaries, it is the largest disease management randomized trial ever conducted. Randomization produced statistically equivalent demographic, disease, Hierarchical Condition Category (HCC) risk score, and economic burden profiles between the intervention and comparison groups.

All programs provide MHS participants with telephonic care management services, including nurse-based health advice for the management and monitoring of symptoms, health education (via health information, videos, online information), health coaching to encourage self-care and management of chronic health conditions, medication management, and health promotion and disease prevention coaching. Only a few of the MHSOs actively serve an institutionally based population. Most of the MHS programs have an end-of-life intervention. Several of the MHSOs rely on sophisticated predictive models using proprietary logic with more than 100 variables to identify gaps in care, create risk strata scores, and achieve operational efficiency. MHSOs that found that their own stratification models did not adequately discriminate among different risk groups have relied on Medicare's HCC scores to target their MHS populations.

Findings to Date

Results available at this writing include the first 18 (of 36) pilot months (McCall et al., 2008a). Beneficiary participation averaged 84 percent across the eight MHSOs and ranged from a high of 95 percent to a low of 74 percent. Refusals explain nearly 0.4 percent of the 16 percent average nonparticipation rate. Defining *active engagement* as having five or more calls or two or more home visits over 18 months, MHSOs worked actively with two in three intervention beneficiaries (65 percent). Only two (of seven reporting) MHSOs achieved positive improvements in patient overall satisfaction, although a majority increased the number of beneficiaries who had received help to set goals for self-care management. None of the MHSOs demonstrated consistent positive intervention effects across six physical and mental health functioning indicators relative to the comparison group.

Out of the 40 evidence-based process of care tests (eight MHSOs, five process rates), 16 were statistically significant, all in the positive direction; however, the absolute rate of change was very small (perhaps not an unexpected finding given the relatively short period of time elapsed during the intervention). MHSOs had the greatest success in improving cholesterol screening among heart failure and diabetes beneficiaries: 9 gains out of 16 were statistically significant (McCall et al., 2008a). MHSOs did less well in improving urine protein screening and eye exams. Only one MHSO significantly improved on all five concordant care processes, and a second MHSO improved on four of five. Despite gains in several process measures, none of the MHSOs were able to reduce the mortality rate among intervention compared with comparison group beneficiaries.

During the pilot, all-cause admission rates ranged from a low of 767 to 1,078 per 1,000 intervention beneficiaries (McCall et al., 2008a). Heart failure and diabetes together were minor reasons for Medicare admissions (16–19 percent; roughly one in six). None of the eight MHSOs succeeded in statistically reducing hospitalization rates among intervention compared with comparison group beneficiaries. Although four of the eight MHSOs achieved Medicare savings during the pilot's first 18 months, none of the gains were statistically significant at the 95 percent confidence level. McCall and colleagues found no significant differences within disease cohort. Although savings among intervention beneficiaries willing to participate were somewhat greater, none were statistically significant. Savings rates between 1.0 percent and 2.1 percent fell far short of the MHSO budget neutrality criterion that ranged from 4.7 percent to 9.3 percent for the same MHSO. Sample sizes were large enough to detect savings rates as low as 3.5 percent to 4.5 percent of average PBPM costs. Medicare savings net of fees were negative for all eight MHSOs through 18 months, implying negative returns on investment. All MHSOs experienced substantial regression-to-the-mean PBPM growth across both intervention and comparison groups.

With 16 successes out of 40 possible gains in evidence-based process-of-care measures, the cost per successful improvement was approximately $15 million, based on $235 million in Medicare fees through 18 months (McCall et al., 2008a). The cost would be $6.6 million per percentage point quality improvement. There did not appear to be any correlation between MHSOs that "saved" money and their quality of care improvements.

Taken together, the findings from this demonstration were disappointing in terms of both clinical and cost impact. Results from this project show that third-party care management is a difficult model under which to achieve measurable clinical improvement and net savings.

Care Management for High-Cost Beneficiaries Demonstration

Project Overview

Medicare beneficiaries with multiple progressive chronic diseases are a large and costly subgroup of the Medicare population. The Congressional Budget Office estimated that in 2001, high-cost beneficiaries in the top 25 percent of spending accounted for 85 percent of annual Medicare expenditures (Congressional Budget Office, 2005). Beneficiaries who had multiple chronic conditions, were hospitalized, or had high total costs had expenditures that were twice as high as those for a reference group. Further, these beneficiaries currently must navigate a health care system that has been structured and financed to manage their acute, rather than chronic, health problems. When older patients seek medical care, their problems are typically treated in discrete settings rather than managed in a holistic fashion (Anderson, 2002; Todd et al., 2001). Because Medicare beneficiaries have multiple conditions, see a variety of clinicians and providers, and often receive conflicting advice, policy makers are concerned about the care that beneficiaries actually receive (Jencks et al., 2003; McGlynn et al., 2003).

Congress mandated the Care Management for High-Cost Beneficiaries (CMHCB) Demonstration to address current failings of the health care system for chronically ill Medicare FFS beneficiaries. In July 2005, CMS announced the selection of six care management organizations (CMOs) to operate programs in the CMHCB Demonstration (McCall et al., 2008c). The demonstration's principal objective was to test new models of care for Medicare beneficiaries who are high cost and have complex chronic conditions, with the goals of reducing future costs, improving quality of care, and improving beneficiary and clinician/provider satisfaction.

The CMHCB initiative employs a mixed-mode experimental design (McCall et al., 2008c). Two interventions are population based, whereas the other four are provider-based and provider-care services to a "loyal" patient population (Piantadosi, 1997). As a trial, it is unusual in employing a "pre-randomized" scheme, assigning eligible beneficiaries to an intervention or comparison group before gaining consent to participate. The Medicare program pays CMHCB organizations a monthly administrative fee per participant, and the organizations may participate in a gainsharing arrangement with the government contingent on improvements in quality, beneficiary and clinician/provider satisfaction, and savings to the Medicare program over a 3-year period. Participating organizations are held at risk for all fees based on the performance of the full population of eligible beneficiaries assigned to the intervention group (an ITT model). CMS developed the CMHCB Demonstration with considerable administrative risk as an incentive to reach targeted beneficiaries and their providers and to improve care management (i.e., 5 percent savings requirement).

Beneficiary participation in the CMHCB Demonstration is voluntary and does not change the scope, duration, or amount of Medicare benefits they currently receive. Beneficiaries do not pay a charge to receive CMHCB Demonstration program services. After the initial 6-month outreach period, the MHSOs accrue management fees for only those beneficiaries who verbally consent to participate and only during participation periods. Participation continues until a beneficiary becomes ineligible for the MHS program or opts out of services that the MHSO provides. Beneficiaries who become ineligible during the demonstration program are removed from the intervention and comparison groups for the total number of months following loss of eligibility for purposes of assessing cost savings and quality, outcomes, and satisfaction improvement.

Project Status

The participating sites implemented this demonstration with some differences. Among the six CMO programs, CMS assigned the two community-based programs—Care Level Management and Key to Better Health—approximately 15,000 and 5,000 intervention beneficiaries, respectively, in Southern California and New York City (McCall et al., 2008c). In contrast, for the four remaining programs, which are integrated delivery systems, CMS chose their intervention population based on a minimum number, or plurality, of visits to participating physicians and hospitals. The four provider-based organizations were Massachusetts General Hospital, Montefiore Medical Center, Texas Senior Trails, and the Health Buddy Consortium. Each CMO worked collaboratively with CMS to finalize its intervention population definition for the demonstration. All programs include high-cost beneficiaries and/or beneficiaries with high HCC risk scores. The definition for *high cost* and *cut-off* of the HCC score varies by program.

CMS awarded contracts under this initiative to CMOs offering approaches that blend features of the chronic care management, disease management, and case management models. Their approaches rely, albeit to varying degrees, on engaging both physicians and beneficiaries and supporting the care processes with additional systems and staff. They proposed to improve chronic illness care by providing the resources and support directly to beneficiaries, using their existing relationships with insurers, physicians, and communities in their efforts.

Although each of the CMOs has unique program characteristics, they share some common features (McCall et al., 2008c), which include educating beneficiaries and their families on improving self-management skills; teaching beneficiaries how to respond to adverse symptoms and problems; and providing care plans and goals, ongoing monitoring of beneficiary health status and progress, and a range of resources and support for self-management.

Findings to Date

No evaluation results of this demonstration are publicly available to date.

Cancer Prevention and Treatment Demonstration

Project Overview

Racial/ethnic disparities in cancer screening and treatment have been well documented. Minority populations are less likely to receive cancer screening tests than are white populations and, as a result, are more likely to be diagnosed with late-stage cancer (Agency for Healthcare Quality and Research [AHQR], 2004; National Institutes of Health & National Cancer Institute, 2001). For those with a positive test result, racial/ethnic minorities are more likely to experience delays in receiving the diagnostic tests needed to confirm a cancer diagnosis (Battaglia et al., 2007; Ries et al., 2003). Similarly, differences in primary cancer treatment, as well as appropriate adjuvant therapy, have been shown to exist between white and minority populations (AHQR, 2004). Although ability to pay is one of the explanatory factors, researchers have found similar disparities among Medicare beneficiaries.

To address this problem, Congress mandated that the US Department of Health and Human Services conduct demonstrations aimed at reducing disparities in screening, diagnosis, and treatment of cancer among racial and ethnic minority Medicare-insured beneficiaries (Section 122 of the Medicare, Medicaid, and State Children's Health Insurance Program [SCHIP] Benefits Improvement and Protection Act of 2000).

CMS decided to assess the use of patient navigators in reducing racial disparities. Patient navigators are individuals who help steer, or "navigate," Medicare beneficiaries through the health care system (Brandeis University Schneider Institute for Health Policy, 2003). Patient navigators primarily have helped cancer patients (Dohan et al., 2005; Hede, 2006); their use for cancer screening and diagnosis is more limited, although some recent studies are promising (Battaglia et al., 2007).

Project Status

CMS issued an announcement on December 23, 2004, soliciting cooperative agreement proposals for the Cancer Prevention and Treatment Demonstration (CPTD) for Racial and Ethnic Minorities. In particular, the announcement sought demonstration projects that targeted four legislatively mandated minority populations: American Indians, Asian Pacific Islanders, African Americans, and Hispanics. Following review of all applications and negotiations with individual sites, CMS announced the selection of six CPTD sites on April 3, 2006.

Each site has two study arms: screening and treatment. Both study arms have one intervention group and one control group. CMS assigned to the treatment arm participants with a diagnosis of breast, cervical, colorectal, lung, or prostate cancer who have received some form of treatment within the past 5 years; it excluded from the study those who have received treatment in the past 5 years for another type of cancer care. All other participants are assigned to the screening arm. The study uses a randomized ITT design; therefore, participants enrolled in the screening arm remain in that arm, even if they are diagnosed with cancer over the course of the study.

Each site developed its own navigation model to ensure that the intervention was culturally sensitive to the needs of each minority community. Three of the sites adopted a nurse/lay navigation model in which nurses play a leadership and oversight role, supported by lay navigators from the community. The other three sites rely almost entirely on lay navigators (community health workers) who provide the bulk of services to intervention group participants. Sites using the nurse/navigator model have more thoroughly developed patient-flow algorithms that may result in better monitoring of care over time. This model also includes more direct interaction with primary care providers in the community, thus allowing them greater influence over screening rates. Control groups in each arm receive relevant educational materials.

Each demonstration project has three sources of funding: (1) start-up payments, (2) payment for administration of CMS-mandated participant surveys, and (3) capitated payments for navigation services (Centers for Medicare & Medicaid Services [CMS], 2008a). The first source was a one-time $50,000 payment at the beginning of each project. As part of the second source, the sites received a fixed payment for each baseline survey they completed on participants in both the intervention and control groups, as well as for an exit survey administered at the end of the demonstration period for all participants. Sites also received payments for administering an annual survey to all intervention group participants. The third source was a capitated monthly payment to each site for all intervention group participants, which covered the cost of navigation services and varied across sites. The normal Medicare claims process handled billing and payment for all clinical screening, diagnosis, and treatment services.

Each site focuses on Medicare beneficiaries from a single racial/ethnic minority group. This substantially strengthens the experimental design, because intervention and control participants share the same racial/ethnic background and are drawn from the same community.

The screening intervention group received navigation services to help ensure that participants undergo the appropriate screenings for breast, cervical, colorectal, and prostate cancer in accordance with Medicare coverage policy for preventive services (CMS, 2009b), as well as clinical practice guidelines. Intervention participants received navigation services to ensure completion of all primary and secondary cancer treatments and all necessary follow-up and monitoring.

Findings to Date

Findings to date, based on site visits and CMS enrollment data, focus on implementation issues (Mitchell et al., 2008); Medicare will not assess demonstration impacts until the demonstrations end in late 2010. Five of the six sites (all but Josephine Ford Cancer Center) encountered difficulty in identifying eligible beneficiaries and enrolling them in the demonstration, resulting in substantially fewer participants than initially projected. At the end of year 1, projected enrollment was 6,484 in the screening arm. After 15 months, the number of screening participants totaled 4,138, more than half of whom were enrolled at Josephine Ford.

Enrollment in the treatment arm fared even worse, with none of the sites meeting their year 1 goals. After 15 months, only 300 treatment participants were enrolled, compared with the originally projected 1,276 for year 1. (The majority of treatment participants also are at Josephine Ford.) Challenges included a larger-than-expected proportion of the population enrolled in managed care (an exclusion criteria for CPTD); limited electronic medical record systems or linkages between existing systems; a lack of current partnerships with community agencies serving their targeted minority population; and lack of identification, recruitment, and retention of qualified staff. For some sites, actual implementation did not begin until well after the October 1, 2006, start date because of delays in institutional review board approval and staff recruitment.

Because staffing and other costs were not quickly offset by capitation payments owing to slower-than-expected enrollments, CMS increased capitation and lump sum payments for debt relief. In some instances, CMS also renegotiated total enrollment goals. Total CMS spending on the CPTD remains unchanged, however (i.e., not to exceed the $25 million obligated by Congress).

Physician-Focused P4P Demonstrations

Medicare has also experimented through demonstrations with physician-focused P4P. The rationale behind this group of projects is that, regardless of the institutional site of care, physicians are the primary drivers behind care treatment decisions, influencing both costs and outcomes. Therefore, initiatives that improve the incentives for physicians to improve quality and efficiency of care, in theory, could have a powerful impact on health care systems performance.

Physician Group Practice Demonstration

Project Overview

The Medicare Physician Group Practice (PGP) Demonstration, Medicare's first physician P4P initiative, establishes incentives for quality improvement and cost efficiency at the level of the PGP. The Medicare, Medicaid, and SCHIP Benefits Improvement and Protection Act of 2000 included a legislative mandate for the PGP Demonstration.

The premise of the PGP Demonstration is that PGPs can achieve higher quality and greater cost efficiency by managing and coordinating patient care. The physician groups participating in the PGP Demonstration engaged in a wide variety of care management interventions to improve the cost efficiency and quality of health care for Medicare FFS patients (RTI International, 2006). These interventions include chronic disease management programs, high-risk/high-cost care management, transitional care management, end-of-life/palliative care programs, practice standardization, and quality improvement programs. In addition, PGP participants use information technology, such as electronic medical records, patient disease registries, and patient monitoring systems, to improve practice efficiency and quality of care delivered to patients, and to better understand the utilization of services by the Medicare FFS population.

The PGP Demonstration tests whether care management initiatives generate cost savings by reducing avoidable hospital admissions, readmissions, and emergency department visits, while at the same time improving the quality of care for Medicare beneficiaries. This demonstration is a shared-savings clinician and provider-payment model in which participating physician groups and the Medicare program share savings in Medicare expenditures. In effect, this model is a hybrid between the FFS and capitation payment methods (Wallack & Tompkins, 2003). Medicare continues to pay physicians and provider organizations under FFS rules, and beneficiaries are not enrolled (i.e., they retain complete freedom of provider choice). However, participating physician groups are able to retain—through annual performance payments in addition to their FFS revenues—part of any savings in Medicare expenditures that they generate for their patients.

This shared-savings payment model gives participating clinicians and providers a financial incentive to control the volume and intensity of medical services, such as what exists under capitated payment. Moreover, physician groups retain a higher portion of savings as their measured quality of care increases. In this way, incentives for both cost efficiency and quality improvement are introduced into FFS payment. Because participating clinicians and providers retain only part of the savings generated by reducing expenditures, incentives for underservice and risk selection are lower than under full capitated payment. Another difference from capitation is that the Medicare program shares in any savings, benefiting from cost-efficiency improvements and lowering government expenditures.

As a Medicare FFS innovation, the PGP Demonstration does not have an enrollment process whereby beneficiaries accept or reject involvement. Therefore, CMS employs a methodology to assign beneficiaries to participating PGPs based on utilization of Medicare-covered services. CMS assigns beneficiaries to a participating PGP if the PGP provided the largest share (i.e., the plurality) of outpatient evaluation and management (E&M) visits to the beneficiary during a year. A beneficiary is assigned to the PGP for the entire year even if the visit occurred late in the year. The assignment methodology incorporates outpatient E&M services provided by specialists as well as by primary care physicians. Beneficiary assignment is redetermined after each year based on that year's utilization patterns. This algorithm assigns beneficiaries uniquely to a single PGP, obviating issues of shared responsibility or rewards among multiple PGPs serving overlapping patient populations. Approximately 50 percent of beneficiaries who were provided at least one Medicare Part B physician service by the PGP during a year are assigned to the PGP; groups with greater primary care orientation have more patients assigned (Kautter et al., 2007). PGPs generally retain approximately two-thirds of their assigned beneficiaries from one year to the next.

Local Medicare beneficiaries not assigned to the participating PGP serve as the comparison population. A PGP's comparison group resides in its service area, which is defined as counties in which at least 1 percent of a PGP's assigned beneficiaries reside. These counties typically include 80 to 90 percent or more of a PGP's assigned beneficiaries. Each participating PGP's service area may differ across years to reflect changes in the location of the PGP's assigned beneficiaries.

Demonstration savings are computed as the difference between the expenditure target and the PGP's expenditures in the performance year. A PGP's annual expenditure target is calculated as PGP's Base Year Expenditures × (1 + Comparison Group Growth Rate). Both the PGP base year expenditures and the comparison group-expenditure growth rate are adjusted for case-mix change between the base and performance years.

If the participating PGP holds the expenditures for its assigned beneficiaries to more than 2 percent below its target, it is eligible to earn a performance payment for that performance year (Kautter et al., 2007). The net savings are calculated as the amount of annual savings that exceeds the 2 percent threshold. The net savings are divided, with 80 percent going to the PGP performance payment pool and Medicare retaining 20 percent as program savings. The PGP performance payment pool is then itself divided between a cost-performance payment and a maximum-quality performance payment. The shares of the cost and maximum-quality performance payment change from 70 percent/30 percent in performance year 1 to 50 percent/50 percent in performance year 3 and after. The Medicare program determines the actual quality performance payment based on the percentage of the PGP Demonstration's quality targets that the PGP met in the performance year. Performance payments are capped at 5 percent of the PGP's target expenditures.

The PGP demonstration includes 32 quality measures covering five modules: (1) diabetes mellitus, (2) heart failure, (3) coronary artery disease,

(4) hypertension, and (5) preventive care. The 32 quality measures are a subset of those developed by CMS's Quality Measurement and Health Assessment Group for the Doctors Office Quality Project (CMS, 2005).

PGP participants are eligible to earn quality performance payments if they achieve at least one of three targets. The first two are threshold targets and the third is an improvement target:

- The higher of 75 percent compliance or the Medicare Health Plan Employer Data and Information Set (HEDIS) mean for the measure (for those measures where HEDIS indicators are also available).

- The 70th percentile Medicare HEDIS level (for those measures where HEDIS indicators are also available).

- A 10 percent or greater reduction in the gap between the baseline performance and 100 percent compliance (e.g., if a PGP achieves 40 percent compliance for a quality measure in the base year, its quality improvement target is 40 percent + (100-40)*10 percent = 46 percent).

Including both threshold and improvement targets gives participating groups positive incentives for quality whether they start out at high or low levels of performance. Groups starting at low levels of quality might view threshold targets as unachievable.

CMS uses claims data to calculate 7 of the 32 quality measures; it uses medical record abstraction or other internal PGP data systems for the other 25 measures. Claims measures receive a weight of four points compared with one point for medical records measures, reflecting the larger sample size of beneficiaries used in calculating claims measures. To calculate a PGP's quality performance payment for a demonstration year, we sum the points for each quality measure where at least one of the three targets was attained, then divide this sum by the total possible points for all quality improvements and apply the resulting ratio to the maximum quality performance payment.

Project Status

The PGP Demonstration began April 1, 2005, and has continued to run for more than 5 years. Calendar year 2004 is used as a baseline for cost and quality performance assessment.

Ten large multispecialty physician groups participated in the PGP Demonstration. CMS selected them through a competitive process based on organizational structure, operational feasibility, geographic location, and implementation strategy. Large PGPs were selected to ensure that participants would have the administrative and clinical capabilities necessary to respond to the PGP demonstration's incentives. The participating PGPs all had at least 200 physicians and together represented more than 5,000 physicians. They included freestanding group practices, components of integrated delivery systems, faculty group practices, and physician network organizations. The number of Medicare FFS patients assigned to the 10 participating physician groups ranged from 8,383 to 44,609, and totaled 223,203. Overall for the 10 physician groups, the percentage of assigned patients that were female was 57.5 percent, dually eligible for Medicare/Medicaid was 13.3 percent, and aged 85 or older was 10.3 percent. These distributions were broadly similar to the Medicare FFS population (CMS, 2006).

Findings to Date

CMS has reported the evaluation of results through the second demonstration year (CMS, 2008b; Sebelius, 2009). In the second performance year, 4 of the 10 participating physician groups earned $13.8 million in performance payments for improving the quality and cost efficiency of care as their share of a total of $17.4 million in Medicare savings. This compares to two physician groups that earned $7.3 million in performance payments as their share of $9.5 million in Medicare savings in the first year of the demonstration. In the first demonstration year, two PGPs accrued "negative savings" of $1.5 million combined. In the second demonstration year, one PGP accrued "negative savings" of $2.0 million. Subtracting the incentive payments to the PGPs and negative savings from Medicare savings, the net savings to the Medicare Trust Fund was $1.6 million in the second demonstration year and $0.7 million in the first.

Medicare expenditures were $120 per person, or 1.2 percent less than target (expected) expenditures per beneficiary for the combined 10 PGPs in the second demonstration year. This reduction was statistically significant ($p < .01$). However, when adjusted for predemonstration expenditure trends, the reduction in expenditures was $58 per person, or 0.6 percent less than the target, and not statistically different from zero. The majority of the second year demonstration savings occurred in outpatient, not inpatient, services. On average, outpatient expenditures were $83 per person year less than expected, whereas inpatient expenditures were $25 per person year less than expected and not statistically significant. Across the 10 PGPs, actual expenditures were lower than target expenditures for beneficiaries with diabetes mellitus ($224 per person year lower), CAD ($555 per person year lower), and COPD ($423 per person year lower). No statistically significant cost reductions were observed for beneficiaries with CHF, cancer, stroke, vascular disease, or heart arrhythmias.

All 10 groups achieved target performance on at least 25 of 27 quality measures applicable in the second performance year. Five of the 10 participating groups achieved target performance on all 27 quality measures for diabetes, CHF, and CAD, compared with 2 that achieved benchmark performance on all 10 measures used in the first demonstration year. Between the base year and the second demonstration year, the PGP groups showed improvement by increasing their quality scores an average of 9 percentage points on the diabetes mellitus measures, 11 percentage points on the heart failure measures, and 5 on the CAD measures.

Between the base year and second demonstration year, four of seven claims-based quality indicators (lipid measurement, urine protein testing, left ventricular ejection fraction testing, and lipid profile) showed greater improvement among PGP-assigned beneficiaries than among comparison beneficiaries. This improvement was statistically significant at the 5 percent level. The differences in the three other indicators (HbA1c management, eye exam, and breast cancer screening) between the PGP and comparison group beneficiaries were not statistically significant. The finding that participating PGPs improved their claims-based quality process indicators more than did their comparison group remained true even after adjusting for predemonstration trends in the claims-based quality indicators.

The PGP Demonstration shared-savings model changes payments to clinicians and providers, not the insurance arrangements of Medicare beneficiaries, who remain enrolled in the traditional FFS program with complete freedom of provider choice. The innovation of the PGP Demonstration model is that participating physicians and provider groups have the opportunity to earn additional performance payments for providing high-quality and cost-efficient care. The financial risk to clinicians and providers is mitigated by the continuation of FFS payment, the use of clinician- and provider-specific base costs as a starting point for measuring savings, and the lack of penalties for underperformance. However, like all payment innovations, the PGP Demonstration shared-savings model faces some challenges. For example, it remains to be seen how much control a physician or provider group can exert over its assigned beneficiaries when they retain freedom of provider choice and have limited incentives to restrain their use of services. This issue of "attribution" is discussed in Chapter 7 of this book.

Medicare Medical Home Demonstration

Project Overview

Policy makers are promoting the patient-centered medical home concept as a potentially transformative health system innovation. A medical home, in broad terms, is a physician-directed practice that provides care that is accessible, continuous, comprehensive, and coordinated and delivered in the context of family and community. Current interest in the medical home as the anchor for a patient's interaction with the health care system stems from growing recognition that even patients with insurance coverage may not have an established access to basic care services and that care fragmentation affects the quality and cost of care that patients experience. Studies (e.g., Rittenhouse et al., 2009; Reid et al., 2010) suggest that the medical home might be a component of health care reform, particularly useful for patients with chronic conditions who typically receive care from many physicians, prescriptions for several medications, and, generally, face unique problems related to redundant, or, worse, inconsistent care that compromises quality and increases spending.

The Tax Relief and Health Care Act of 2006 (TRHCA) mandated that CMS establish a medical home demonstration project to provide patient centered care to "high-need populations." The legislation has targeted the medical home demonstration to a "high-need population," defined as individuals with multiple chronic illnesses that require regular monitoring, advising, or treatment. CMS has decided to adopt a broad definition of the target population to include more than 80 percent of Medicare beneficiaries to broaden the scope and reach of the demonstration. The demonstration legislation provides that care management fees and incentive payments be paid to physicians rather than to practices per se, although qualifying physicians must be in practices that provide medical home services. To qualify, physicians must implement an interdisciplinary plan of care in partnership with patients, use clinical decision support tools to support practice of evidence-based medicine, rely on health information technology, and promote patient self-management skills. Additionally, the medical home itself is responsible for targeting eligible beneficiaries and for promoting patient access to personal health information, developing a health assessment tool for targeted individuals, and providing training for personnel involved in care coordination.

Project Status

CMS has completed work toward a solicitation and final design for the demonstration, and sites were originally projected to be operational sometime in 2010. However, the Affordable Care Act health care reform legislation also includes a mandate for a Medicare Medical Home Demonstration. Therefore, CMS put the TRHCA-mandated demonstration on hold until the outcome of the health care reform legislation made clear the specific parameters for a congressionally mandated Medicare Medical Home Demonstration. At this writing it is unclear whether this originally mandated Medicare Medical Home Demonstration will be implemented or combined with an Affordable Care Act–mandated demonstration.

Medicare Hospital-Focused P4P Demonstrations

A large proportion of Medicare expenditures goes to provide inpatient hospital services. As a result, Medicare has devoted significant attention to improving both the efficiency and quality of hospital care on behalf of its beneficiaries. Current demonstrations in the planning and development stage include projects aimed at implementing a new round of bundled payment/improved quality of care hospital-focused demonstration projects.

Medicare Hospital Heart Bypass Demonstration

Project Overview

Since the implementation of Medicare's inpatient prospective payment system (IPPS) in 1983, the annual update in allowed charges nationally has capped Part A hospital payments per discharge for bypass surgery. Both hospital managers and policy makers have expressed major concern about the asymmetric Medicare financial incentives facing hospitals compared with physicians. Unlike hospitals (and surgeons paid a global payment), other physicians seeing a patient are paid for every additional service they provide. Surgeons are also paid more for more complex bypass surgeries. Moreover, all hospital support services (e.g., nursing) are essentially "free" to physicians, who bear none of the financial risk of higher use of these services as a result of longer hospital stays, more tests, and higher utilization of other hospital-based services. Misaligned physician incentives were thought to raise the cost of an admission.

An alternative strategy focused on the structural characteristics of clinicians and provider organizations that set them apart as Centers of Excellence (CoEs).

In this strategy, payers "reward" both hospitals and physicians in an indirect way by allowing them to market a CoE imprimatur to potential patients in their plan. The CoE concept is straightforward: a payer (such as Medicare) solicits applicants that are then thoroughly reviewed according to a set of structure, process, and outcome measures. The payer then authorizes those meeting high standards to market an imprimatur to subscribers or beneficiaries as a CoE for inpatient surgery. Payers, like Medicare, may also request discounts off the usual payment rates—particularly if the payer believes that its seal of approval is highly valuable to a physician or a provider organization. The approach is a win-win-win for the payer, the payers' beneficiaries, and the hospitals and their medical staffs.

Project Status

In 1988, CMS solicited proposals from more than 40 hospital and physician groups to participate in the Medicare Participating Heart Bypass Demonstration (Cromwell et al., 1998). In the demonstration, the government paid a single negotiated global price for all Parts A and B inpatient hospital and physician care associated with bypass surgery (diagnosis-related-groups [DRGs] 106 and 107, bypass with and without cardiac catheterization). The intent of the demonstration was to encourage regionalization of the procedure in higher-volume hospitals and to align physician with hospital incentives under a bundled prospective payment. Hospitals shared the global payment with surgeons and cardiologists based on cost savings. CMS allowed participants to market a demonstration imprimatur as a "Medicare Participating Heart Bypass Center." Medicare patients were not restricted to demonstration hospitals for their surgery.

In May 1991, after extensive evaluation of 27 final applicants, CMS began paying four provider groups, later expanded to seven. Initial discounts averaged 13 to 15 percent, depending on DRG (Cromwell et al., 1998). Discounts were substantial considering that CMS could not offer exclusive contracting to sites, nor did CMS allow the sites the right to market a true Centers of Medicare Excellence imprimatur. All participants said that they would have offered even deeper discounts had they been allowed to market a CoE imprimatur.

Findings to Date

Over the demonstration's 5 years, the Medicare program saved $42.3 million on the 13,180 bypass patients treated in the seven demonstration hospitals (Cromwell et al., 1998). About 85 percent of the savings came from demonstration discounts, another 9 percent from volume shifts to lower-cost demonstration hospitals, and 5 percent from lower post-discharge utilization. In addition, beneficiaries (primarily their supplemental insurers) saved another $8 million, resulting in $50 million in overall demonstration savings. Total savings were $3,794 per bypass admission. Micro-cost analyses showed that three of the four initial sites experienced 10 to 40 percent declines in direct intensive care units and routine nursing expenses resulting in rising profit margins in spite of substantial discounts. Fewer surgeon requests for specialist consultations also produced Medicare savings (Cromwell et al., 1997b).

One-third of demonstration patients surveyed were aware of the hospital's demonstration status when choosing their site of surgery, and only one-third of knowledgeable patients said it had affected their hospital choice (Cromwell et al., 1998). Two-thirds of referring physicians were aware of the demonstration hospital's status, but this knowledge reportedly had little effect on their referral recommendation compared with the general reputation and their own familiarity with the hospital's staff. That the marketing of the imprimatur influenced only one in nine patients raises questions about the effectiveness of "consumer-driven" health care based on more information, given the government's goal of regionalizing bypass surgery to improve community-wide outcomes.

Controlling for risk factors (e.g., age, gender, ejection fraction, comorbid illnesses), demonstration hospitals exhibited a statistically significant decline in annual inpatient mortality (one-half of a percentage point from a mean of 4.6 percent). One-year post-discharge mortality exhibited the same rate of decline. The two sites with above-average mortality achieved statistically significant declines in mortality during the demonstration. The CMS-funded evaluation found a small, positive trend in complication rates that did not result in greater mortality and no significant trend in the appropriateness rating of bypass patients when angioplasty was an alternative (Cromwell et al., 1998).

Expanded Medicare Heart and Orthopedics CoE Demonstration

Project Overview
The first Medicare Hospital Heart Bypass Demonstration illustrated the potential of using the CoE imprimatur to self-finance higher quality care. Having proof of concept, CMS developed a follow-on demonstration with more cardiovascular procedures and a few major orthopedic procedures, such

as hip and knee replacement. The demonstration also was intended to provide a true test of the value of the CoE imprimatur to applicants.

Project Status
In 1997, CMS initiated a two-stage process that began with a pre-application form to nearly 1,000 hospitals seeking Medicare's CoE imprimatur in the San Francisco and Chicago regions. CMS received 538 pre-applications and invited 160 heart and orthopedics hospitals to submit full applications (Cromwell et al., 1997a). (Most pre-applicants did not meet the minimum-volume criteria.) Eventually, 123 (75 percent) submitted full applications. CMS then convened 10 government panels comprising expert clinicians from inside and outside the agency to conduct in-depth reviews of the applications. At the end of an intensive 3-month period, the panels recommended 31 (of 70 invited) cardiovascular and 42 (of 53) orthopedic applicants for final approval. The 73 winners represented 14 percent of the original 538 submitting pre-applications, suggesting a very select group of high-quality hospitals.

Discounts from the accepted applicants ranged widely from zero percent to 35 percent. Excluding 9 zero-discount applicants (of the 70 eligible applicants), the mean heart bypass discount was 9.3 percent (Cromwell & Dayhoff, 1998; Cromwell et al., 1997a). Two-thirds of the proposed discounts ranged between 5 and 14 percent. Part B physician discounts averaged 17 percent less than hospital Part A discounts. Four out of 10 applicants (including 8 monopolists) were considered dominant in their market and submitted discounts a full 3 percentage points lower than nondominant applicants (significant at the 1 percent level). However, another 25 percent of dominant providers offered discounts of 13.6 percent or more. Applicants operating in duopoly markets offered discounts more than twice as great (10.7 percent) as monopolists. High-cost (to Medicare) providers offered substantially greater discounts. The 18 applicants in very high health maintenance organization (HMO) penetration areas (>40 percent) offered discounts nearly 6 percentage points lower than those in low HMO penetration markets, a highly significant difference. This finding supports other research indicating that competitive pressures on prices may have already reduced costs with less financial leeway for further discounts (Hadley et al., 1996).

Project Findings to Date
Ultimately, CMS never implemented the expanded CoE demonstrations because of opposition on the part of the health care provider community in

addition to other logistical complications internal to CMS. Any P4P approach will encounter opposition from some clinicians and provider organizations. The CoE approach was particularly contentious because rejected (or ineligible) clinicians and providers argued that patients would perceive them as being less qualified. Since 1997, CMS has failed in three attempts to implement a CoE imprimatur P4P demonstration.

Acute Care Episode Value-Based Purchasing Demonstration

Project Overview

The Acute Care Episode (ACE) Demonstration is the most recent iteration of the CoE P4P model. The demonstration, implemented in late 2009, offers bundled payments and increased flexibility in financial arrangements between participating hospital-physician consortia (CMS, 2009c). Under the demonstration, a bundled payment is a single payment for both Part A and Part B Medicare services furnished during an inpatient stay (McCall et al., 2008b). Currently, under Medicare Part A, CMS reimburses a hospital a single prospectively determined amount under the IPPS for all the care it furnishes to the patient during an inpatient stay. Physicians who care for the patient during the hospital stay are paid separately under the Medicare Part B Physician Fee Schedule for each service they perform. The demonstration will also focus on methods for improved quality of care for bundles of heart and orthopedic hospital-based procedures.

The Medicare program will permit approved demonstration sites to use the term "Value-Based Care Centers" in approved marketing programs. This demonstration is intended to provide an opportunity for Value-Based Care Centers to develop efficiencies in the care they provide to beneficiaries through quality improvement in clinical pathways, improved coordination of care among specialists, and gainsharing. This demonstration also provides an opportunity for Medicare to share savings achieved through the demonstration with beneficiaries who, based on quality and cost, choose to receive care from participating demonstration providers (CMS, 2009a).

Project Status

CMS selected six sites for ACE demonstration participation: Baptist Health System in San Antonio, Tex.; Oklahoma Heart Hospital LLC in Oklahoma City, Okla.; Exempla Saint Joseph Hospital in Denver, Colo.; Hillcrest Medical Center in Tulsa, Okla.; and the Lovelace Health System in Albuquerque, N.M. Under this version of the CoE-type model, the bundled payment demonstration includes 28 cardiac and 9 orthopedic inpatient surgical services and procedures. CMS selected these elective procedures because volume for them has historically been high, and there is also sufficient marketplace competition and existing quality metrics. The ACE demonstration sites began implementation in 2009, with some procedures in some sites beginning implementation in 2010.

Findings to Date

No publicly available findings are ready yet.

Premier Hospital Quality Incentive Demonstration

Project Overview

In the Deficit Reduction Act of 2005 (DRA), Congress mandated CMS to develop initiatives for hospital value-based purchasing by 2009 (Lindenauer et al., 2007). Likely driving this mandate was interest in the earlier Hospital Quality Alliance (HQA) initiative, launched in December 2002 by the American Hospital Association, the Federation of American [proprietary] Hospitals, and the Association of American Medical Colleges. The Alliance was intended to build a collaborative relationship between private hospitals and the government to improve quality of care. The Alliance invited all hospitals to participate and report data on at least 10 quality indicators for clinical conditions such as heart failure and pneumonia. Building on this initiative, CMS tied Medicare hospital payment updates to reporting quality indicators, ultimately achieving a 98 percent participation rate among hospitals (Lindenauer et al., 2007, p. 487). CMS made hospital quality indicators available on its Hospital Compare Web site. In March 2003 CMS invited hospitals providing the quality indicator data to participate in its Medicare Premier Hospital Quality Incentive Demonstration (HQID), a P4P demonstration managed by Premier Healthcare Informatics. Nonparticipating hospitals could still report quality data but could not participate in the P4P program.

The Medicare Premier HQID project recognizes and provides financial rewards to hospitals that demonstrate high-quality performance in areas of hospital acute care. CMS conducts the Medicare demonstration in collaboration with Premier, Inc., a nationwide organization of not-for-profit hospitals. Under the demonstration, top-performing participating hospitals receive increased payment for Medicare patients.

Project Status

The Premier HQID phase one operated initially from 2003 through 2006. HQID paid bonuses for superior quality performance based on a limited set of 33 indicators, which spanned five clinical conditions: heart failure, acute myocardial infarction (heart attack), pneumonia, bypass surgery, and hip and knee replacement. Example indicators included the following:

- Heart attack: Percentage of patients given aspirin or beta blocker on arrival
- Heart failure: Percentage of patients assessed for left ventricular function
- Pneumonia: Percentage of patients assessed for oxygenation or given antibiotics within 4 hours of arrival.

To be eligible in any year, practitioners and hospitals needed a minimum of 30 cases per condition. For each clinical condition, hospitals performing in the top two deciles of all participants received a 2 percent or 1 percent bonus payment per Medicare patient along with their regular Medicare prospective payment. Bonuses were expected to be paid for by 1 to 2 percent payment penalties on Medicare payments for participants falling into the lowest two performance deciles. Thus, the demonstration design is budget neutral, reallocating Medicare payments away from poor performing to high-performing hospitals based on a limited set of quality measures. Hospitals qualified for bonuses based only on whether their absolute level of performance was superior and not by their rate of improvement. Multihospital groups submitted bills and quality data as a single entity, thereby sharing in financial gains and possible losses.

The primary metric in evaluating hospital performance was the improvement in their quality scores, even though financial incentives were based solely on absolute scores each year. CMS benchmarked performance improvements several ways. First, it developed a comparison group by matching each participating hospital with one or two HQA hospitals that agreed to participate in the HQID based on number of beds, teaching status, region, urban/rural, and ownership status (for-profit vs. nonprofit). Second, CMS benchmarked participant quality improvements against all HQA facilities, using linear regression methods with change in overall quality as the dependent variable. Third, to address a potential volunteer bias, CMS repeated multivariate analyses of performance by including all HQID hospitals in the intervention group following a clinical trial, ITT experimental design.

Of the 421 hospitals invited to participate in the P4P program, 63 percent accepted and began providing data on 33 quality indicators (Lindenauer et al., 2007, p. 488); 11 facilities eventually withdrew. Patient admission was the unit of observation for quantifying changes in process outcomes. CMS based approximately 117,000 P4P patients and 192,000 control patients for statistical testing with no apparent adjustment for clustering effects on variance in the 207 participating and 406 nonparticipating hospitals.

A second phase of the Premier Hospital demonstration is continuing, allowing for an additional 3 years of implementation and testing of new incentive models. Currently, about 230 hospitals continue to participate in this phase of the demonstration.

Findings to Date

Lindenauer and colleagues (2007) have published initial findings for this demonstration. Over 2 years, both nonparticipating hospitals (those only reporting data) and hospitals participating in the P4P program, showed quality improvements. In 7 of 10 quality indicators, P4P hospitals showed greater improvements (Lindenauer et al., 2007, p. 489). In these findings based on early years of the project, performance increases varied inversely with baseline rates. For example, among acute myocardial infarction (AMI) patients, the highest-performing quintile showed increases in composite process scores of 7.5 percentage points above the control group beginning from a baseline score of only 73 percent. The poorest-performing quintile saw a relative increase in its composite AMI score of only 2.4 percentage points (and a 1.1 percentage point decline from 97.9 percent to 96.8 percent). After adjusting for baseline differences in study and control groups, the incremental increases in quality attributed to P4P incentives declined.

Bonus payouts to hospitals participating in the P4P program averaged $71,960 per year per hospital, but they ranged widely from $914 to $847,227. Similar "losses" among the lowest-performing hospitals offset these bonus payments. Bonuses and penalties, however, were not based on rates of improvement over baseline but on absolute levels during the demonstration period. Hospitals in the lowest two deciles (or quintile) in terms of rates of improvement during the demonstration had the highest average baseline scores and tended to receive most of the bonuses. Hospitals in the highest demonstration quintile based on rate of improvement still had the lowest scores by demonstration's end and paid a disproportionate percentage of the bonuses.

With very small P4P financial incentives, this demonstration found relatively small improvements in several quality process indicators. Because of the large sample sizes, the analysis could detect and accept very small quality improvements of less than 1 percentage point. For example, baseline composite process scores for AMI increased from 88.7 percent to 94.8 percent. After adjusting for study-control differences in patient characteristics and volunteerism, the P4P effect fell to 1.8 percentage points, or an improvement from 88.7 percent to 90.5 percent. Baseline process scores for the other two conditions averaged roughly 80 percent for AMI, suggesting high adherence levels as well. We do not know from this demonstration how effective a larger financial incentive (and penalty) might be for another group of hospitals with much lower adherence rates.

According to Lindenauer and colleagues (2007), hospitals that already had the highest average baseline performance received the majority of performance bonuses. In fact, many of the hospitals with the greatest improvements in quality incurred payment penalties because their scores remained in the lowest quintile by the end of the demonstration. Another concern in using process measures at the hospital level was the narrow range of indicators. Using just 33 indicators to track quality in a few broad reasons for admission may be too narrow to accurately represent differences in absolute quality or rates of improvement in quality. Also of concern was the limiting of payment reallocations between just the bottom and top quintiles based on quality scores. In the longer run, this could discourage hospitals unable to achieve the highest quintile from continuing to strive (at high internal cost) to further raise quality.

The evaluation of the Premier demonstration is ongoing. Preliminary results from the first 4 years of the demonstration suggest that participating hospitals raised overall quality by an average of 17 points over 4 years, based on their performance on more than 30 nationally standardized care measures for patients in five clinical areas (heart attack, coronary bypass graft, heart failure, pneumonia, and hip and knee replacements (CMS, 2009d).

Medicare Gainsharing and Physician Hospital Collaboration Demonstrations

Project Overview

Ever since CMS implemented hospital prospective per case payments using DRGs (through the IPPS) in 1984, hospital managers and researchers have raised concerns about the misalignment of hospital and physician incentives. At the time, per case DRG payment represented an unprecedented bundling of facility services in a single Part A payment, including routine and intensive care unit nursing, operating room, and other ancillary services. Physicians, by contrast, remained under a fractionated Current Procedural Terminology billing system with thousands of codes that encouraged them to continue providing separate services with no incentive to conserve health care costs. The overall concept of gainsharing is intended to allow hospitals to share efficiency savings with physicians in a controlled setting in which quality of care standards are simultaneously maintained (or improved).

To test the gainsharing concept under Medicare, Congress mandated two separate but very similar demonstrations. Under Section 5007 of the DRA, Congress required CMS to conduct a qualified gainsharing program that tests alternative ways that hospitals and physicians can share in efficiency gains. Similarly, under Section 646 of the MMA, Congress authorized the Secretary of Health and Human Services to conduct a Physician Hospital Collaboration demonstration as part of the larger Medicare Health Care Quality Initiative (CMS, 2010b). Like the Gainsharing Demonstration, the purpose of this Physician Hospital Collaboration Demonstration is to test gainsharing models that facilitate collaborations between physicians and hospitals to improve quality and efficiency.

Under both demonstrations, incentive payments made to physicians under the Physician-Hospital Demonstration must tie directly to improvements in quality and/or efficiency, and cannot be based on other standards (such as volume or patient referrals). Physician payments are limited to 25 percent of Medicare payments made to physicians for other similar patients. Payments must also be based on a methodology that is replicable and auditable, and the demonstration must—at a minimum—be budget neutral.

However, unlike the Gainsharing Demonstration, which has a distinct hospital-based focus, the Physician Hospital Collaboration project places particular emphasis on participation of integrated delivery systems and coalitions of physicians in collaboration with hospitals. The project also places a greater emphasis on improved efficiency and quality of care over a longer episode of care, including post-acute services, beyond the acute-care stay.

Both of the current Medicare gainsharing demonstration initiatives are modeled on an earlier project that, because of legal challenges, was never fully implemented. In 2001, the New Jersey Health Association (NJHA) submitted an application to CMS to run an eight-hospital all-payer refined DRG (APR-DRG) Demonstration of gainsharing in its state (NJHA, 2001). Introducing all the facets that other gainsharing proposals are likely to include, this gainsharing methodology was likely the most complex ever proposed. The New Jersey plan was to establish maximum pools of Part A hospital savings for each APR-DRG in the hospital to be shared with the medical staff. These pools were limited to 25 percent of total Part B outlays. Next the pools were converted to a per discharge basis for each APR-DRG based on average costs of the lowest 90 percent of cases (i.e., so-called Best Practice Norms).

Excluding the most expensive cases from the target baseline cost per discharge was the primary mechanism to achieve reductions in hospital costs. Once the demonstration site identified responsible physicians, they became eligible for gainsharing, depending on how the average cost of their cases related to the mean cost of the 90 percent baseline group of cases. The demonstration standardized baseline and demonstration cases for case severity and inflation. In the early demonstration years, responsible physicians could participate in gainsharing, even if they failed the Best Practice Norms, as long as they showed reductions in their Part A costs per case. The demonstration carved out gainsharing pools for hospital-based and consulting physicians to partially shelter them from lost billings associated with shorter stays and less testing.

The demonstration used process and outcome indicators to restrict gainsharing to physicians maintaining high-quality standards. Physicians in the NJHA project were put at risk for excessive post-acute Medicare outlays from any source (including outpatient physician services: "any absolute increase in Medicare PAC [post-acute care] payments per discharge [must] be smaller than any absolute decrease in Part B inpatient physician payments per discharge" [Cromwell & Adamache, 2004]). The two demonstrations also differed in that CMS negotiated up-front discounts in its cardiac DRG global Part A and B rates, whereas New Jersey hospitals had to reduce baseline Part A and B inpatient outlays by 2 percent after adjusting for inflation and case-mix changes.

Project Status

CMS solicited volunteer participating sites for the Gainsharing Demonstration in fall 2006 (CMS, 2010a), with applications due November 17, 2006. CMS initially selected five sites from this solicitation for participation but also issued a new announcement to resolicit for rural demonstration sites. CMS designated five sites as potential Medicare Gainsharing Demonstration participants. Two sites signed terms and conditions and initially participated in the demonstration:

- Beth Israel Medical Center (BIMC), New York, New York
- Charleston Area Medical Center (CAMC), Charleston, West Virginia

These two demonstration sites began the implementation process as of October 1, 2008. Charleston Area Medical Center withdrew from the demonstration effective December 31, 2009.

The BIMC site includes all DRGs in its demonstration. Enrollment is voluntary for physicians. A pool of bonus funds will be prospectively estimated from hospital savings based on variances from best practices. If no hospital savings are realized, no bonuses will be allocated to participating physicians. In the BIMC model, each patient is assigned to one practitioner who will take financial responsibility for the care of the patient. For medical patients, the "responsible physician" is the attending physician. For surgical patients, the responsible physician is the surgeon. The actual bonus paid to physicians is called the performance incentive, which is calculated as a percentage of the maximum performance incentive, based on performance. Gainsharing payments are capped according to CMS policy at 25 percent of the physician's affiliated Part B reimbursements. BIMC proposes a range of physician quality standards, which, if not met by individual physicians, would make them ineligible for the gainsharing bonus (Greenwald et al., 2010a).

The CAMC gainsharing model focused on cardiac care. Each cardiac-related DRG included in the demonstration had established savings initiatives. CAMC measured participating physicians on several grounds to ensure that quality of patient care remained the same. Worse performance on any of the following standards for an individual physician made that physician ineligible to receive the gainsharing bonus (Greenwald et al., 2010a).

CMS has solicited participants for the Physician Hospital Collaboration Demonstration in this project and selected the NJHA/New Jersey Care Integration (NJCI) Consortium, Princeton, N.J. (with 12 hospitals), targeting all inpatient Medicare beneficiaries, to participate in the demonstration (CMS, 2010c). The 12 hospitals participating in the NJCI Consortium began implementing the demonstration in July 2009.

The NJCI Consortium sites will include all DRGs in their demonstration. Enrollment is voluntary for physicians. Physicians must have at least 10 admissions at the consortium member to be eligible for incentive payments.

In the NJCI model, each patient is assigned to one practitioner who will take financial responsibility for the patient's care. For medical patients, the "responsible physician" is the attending physician. For surgical patients, the responsible physician is the surgeon. Up to 12.5 percent of internal hospital savings will be available for incentive payments (Greenwald et al., 2010b).

Physician incentive payments will consist of two parts: a performance incentive and an improvement incentive. In the initial year, the improvement incentive will be two-thirds of the gainsharing payment, and the performance incentive will be one-third. In year 2, the maximum improvement incentive is reduced to one-third, and by year 3, the improvement incentive will be eliminated, with all funds directed to the performance incentives. A physician's peer performance incentive is based on his or her average cost per case relative to the best practice cost per case of a cost-efficient peer group. The NJCI Consortium proposes a range of physician quality standards to ensure that patient safety and quality of care. In addition, the consortium proposes to track and review several parameters for any unusual or exceptional changes (Greenwald et al., 2010b).

Findings to Date

No publicly available evaluation findings are ready for either the Medicare Gainsharing or the Physician Hospital Collaboration demonstrations.

Medicare Demonstrations and the Future of Pay for Performance

The examples cited previously in this chapter and in earlier chapters make up only a partial list of Medicare demonstrations related in some way to P4P. Previous chapters (especially Chapters 1 and 2) also discuss private-sector P4P initiatives implemented by a range of sponsors (see Table 1-1 in Chapter 1 for a complete list of all demonstrations). Because the Affordable Care Act health care reform legislation mandates dozens more P4P, accountable care organization and other value-based purchasing projects and demonstrations, the range of models, provider types, payment incentives, and other variations will only expand in the next 5 years.

Conspicuously missing from these lists of P4P initiatives is a nationally implemented program for P4P. Of course, P4P initiatives sponsored by regional employers and insurers will logically remain focused on the issues and needs of these regional sponsors. Resources to fund implementation, evaluation, and refinement of P4P models may be scarce. In contrast, the Medicare program presents a very likely candidate for eventual national implementation of P4P initiatives. Medicare is the largest US insurer and sponsor of a national program with access to implementation and evaluation funding from Congress. It is curious then that given the extent of Medicare P4P demonstrations currently completed or ongoing, no serious move toward national implementation of any of the existing P4P models is currently under serious consideration. Chapter 11 of this book discusses this issue and explores the challenges of implementing Medicare P4P on a national level.

References

Agency for Healthcare Quality and Research. (2004). *National healthcare disparities report, 2003: Summary*. Retrieved March 24, 2010, from http://www.ahrq.gov/qual/nhdr03/nhdrsum03.htm

Anderson, G. (2002). *Chronic conditions: Making the case for ongoing care*. Baltimore, MD: Partnership for Solutions, Johns Hopkins University, and the Robert Wood Johnson Foundation.

Battaglia, T. A., Roloff, K., Posner, M. A., & Freund, K. M. (2007). Improving follow-up to abnormal breast cancer screening in an urban population. A patient navigation intervention. *Cancer, 109*(2 Suppl), 359–367.

Brandeis University Schneider Institute for Health Policy. (2003). *Cancer Prevention and Treatment Demonstration for Ethnic and Racial Minorities*. Retrieved March 24, 2010, from http://www.cms.hhs.gov/DemoProjectsEvalRpts/downloads/CPTD_Brandeis_Report.pdf

Centers for Medicare & Medicaid Services. (2005, November). *CMS Physician Focus Quality Initiative: Chronic disease and prevention measures*. Baltimore, MD: CMS.

Centers for Medicare & Medicaid Services. (2006). *Health Care Financing Review, statistical supplement* (Publication No. 03477). Available from http://www.cms.hhs.gov/review/default.asp.

Centers for Medicare & Medicaid Services. (2008a, October 1). *Cancer Prevention and Treatment Demonstration for Ethnic and Racial Minorities: Fact sheet*. Retrieved November 12, 2009, from http://www.cms.hhs.gov/DemoProjectsEvalRpts/downloads/CPTD_FactSheet.pdf

Centers for Medicare & Medicaid Services. (2008b, August 14). *Physician groups earn performance payments for improving quality of care for patients with chronic illnesses*. Retrieved November 12, 2009, from http://www.cms.hhs.gov/DemoProjectsEvalRpts/downloads/PGP_Press_Release.pdf

Centers for Medicare & Medicaid Services (2009a). *CMS announces sites for a demonstration to encourage greater collaboration and improve quality using bundled hospital payments.* Retrieved November 12, 2009, from http://www.cms.hhs.gov/DemoProjectsEvalRpts/downloads/ACEPressRelease.pdf

Centers for Medicare & Medicaid Services. (2009b). *Quick reference information: Medicare preventive services.* Retrieved November 12, 2009, from http://www.cms.hhs.gov/MLNProducts/downloads/MPS_QuickReferenceChart_1.pdf

Centers for Medicare & Medicaid Services. (2009c). *Acute Care Episode Demonstration.* Retrieved March 9, 2010, from http://www.cms.hhs.gov/DemoProjectsEvalRpts/downloads/ACEFactSheet.pdf

Centers for Medicare &Medicaid Services. (2009d). *Medicare demonstrations show paying for health care quality pays off.* Retrieved July 13, 2010, from http://www.cms.gov/HospitalQualityInits/downloads/HospitalPremierPressReleases20090817.pdf

Centers for Medicare & Medicaid Services. (2010a). *DRA 5007 Medicare Hospital Gainsharing Demonstration solicitation.* Retrieved March 9, 2010, from http://www.cms.hhs.gov/DemoProjectsEvalRpts/downloads/DRA5007_Solicitation.pdf

Centers for Medicare & Medicaid Services (2010b). *Physician-Hospital Collaboration Demonstration solicitation.* Retrieved March 9, 2010, from http://www.cms.hhs.gov/DemoProjectsEvalRpts/downloads/PHCD_646_Solicitation.pdf

Centers for Medicare & Medicaid Services. (2010c). *Press release: Medicare demonstrations show paying for quality health care pays off.* Retrieved March 9, 2010, from http://www.cms.hhs.gov/apps/media/press/release.asp?Counter=3495

Coleman, E. A., Parry, C., Chalmers, S., & Min, S. J. (2006). The care transitions intervention: Results of a randomized controlled trial. *Archives of Internal Medicine, 166*(17), 1822–1828.

Congressional Budget Office. (2005). *High-cost Medicare beneficiaries.* Retrieved March 24, 2010, from http://www.cbo.gov/ftpdocs/63xx/doc6332/05-03-MediSpending.pdf

Cromwell, J., & Adamache, W. (2004). *Rates and savings report: Final design report.* CMS Contract No. 500-92-0013. Waltham, MA: RTI International.

Cromwell, J., & Dayhoff, D. A. (1998). *An analysis of proposed discounts under the Medicare Participating Cardiovascular and Orthopedic Centers of Excellence Demonstration: Draft report.* Health Care Financing Administration Contract No. 500-95-0058-004. Waltham, MA: Health Economics Research, Inc.

Cromwell, J., Dayhoff, D. A., McCall, N., Subramanian, S., Freitas, R., Hart, R., et al. (1998). *Medicare Participating Heart Bypass Center Demonstration: Final report.* Health Care Financing Administration Contract No. 500-92-0013. Retrieved January 11, 2011, from http://www.cms.hhs.gov/reports/downloads/CromwellExecSum.pdf; http://www.cms.hhs.gov/reports/downloads/CromwellVol1.pdf; http://www.cms.hhs.gov/reports/downloads/CromwellVol2.pdf; http://www.cms.hhs.gov/reports/downloads/CromwellVol3.pdf

Cromwell, J., Dayhoff, D. A., Subramanian, S., & Hart, R. (1997a). *Medicare Negotiated Bundled Payment Demonstration: Full application banking report.* HCFA Contract No. 500-92-0013. Waltham, MA: Health Economics Research, Inc.

Cromwell, J., Dayhoff, D. A., & Thoumaian, A. H. (1997b). Cost savings and physician responses to global bundled payments for Medicare heart bypass surgery. *Health Care Financing Review, 19*(1), 41–57.

Dohan, D., & Schrag, D. (2005). Using navigators to improve care of underserved patients: Current practices and approaches. *Cancer, 104*(4), 848–855.

Greenwald, L., Cromwell, J., Adamache, W., Healy, D., Halpern, M., & West, N. (2010a). *Evaluation of the Medicare Gainsharing Demonstration: Design report.* CMS Contract Number HHSM-500-2005-00029I; Task 3. Waltham, MA: RTI International.

Greenwald, L., Cromwell, J., Healy, D., Adamache, W., Drozd, E., Bernard, S., et al. (2010b). *Evaluation of the Medicare Physician Hospital Collaboration Demonstration: Design report.* CMS Contract Number HHSM-500-2005-00029I; Task 17. Waltham, MA: RTI International.

Hadley, J., Zuckerman, S., & Iezzoni, L. I. (1996). Financial pressure and competition. Changes in hospital efficiency and cost-shifting behavior. *Medical Care, 34*(3), 205–219.

Hede, K. (2006). Agencies look to patient navigators to reduce cancer care disparities. *Journal of the National Cancer Institute, 98*(3), 157–159.

Jencks, S. F., Huff, E. D., & Cuerdon, T. (2003). Change in the quality of care delivered to Medicare beneficiaries, 1998-1999 to 2000-2001. *JAMA, 289*(3), 305–312.

Kautter, J., Pope, G. C., Trisolini, M., & Grund, S. (2007). Medicare Physician Group Practice Demonstration design: Quality and efficiency pay-for-performance. *Health Care Financing Review, 29*(1), 15–29.

Lindenauer, P. K., Remus, D., Roman, S., Rothberg, M. B., Benjamin, E. M., Ma, A., et al. (2007). Public reporting and pay for performance in hospital quality improvement. *New England Journal of Medicine, 356*(5), 486–496.

McCall, N. T., Cromwell, J., & Bernard, S. (2007). *Evaluation of Phase I of Medicare Health Support (formerly Voluntary Chronic Care Improvement) Pilot Program under traditional fee-for-service Medicare: Report to Congress.* CMS Contract No. 500-00-0022. Retrieved March 24, 2010, from http://www.cms.hhs.gov/reports/downloads/McCall.pdf

McCall, N. T., Cromwell, J., Urato, C., & Rabiner, D. (2008a). *Evaluation of Phase I of the Medicare Health Support Pilot Program under traditional fee-for-service Medicare: 18-month interim analysis. Report to Congress.* CMS Contract No. 500-00-0022. Available from http://www.cms.hhs.gov/reports/downloads/MHS_Second_Report_to_Congress_October_2008.pdf

McCall, N. T., Dalton, K., Cromwell, J., Greenwald, L., Freeman, S., & Bernard, S. (2008b). *Medicare Acute Care Episode Demonstration: Design, implementation, and management: Design report.* CMS Contract No. 500-2005-00291. Waltham, MA: RTI International.

McCall, N. T., Urato, C., Spain, P., Smith, K., & Bernard, S. (2008c). *Evaluation of Medicare Care Management for High Cost Beneficiaries (CMHCB) Demonstration: Second annual report.* CMS Contract No. 500-00-0024, T.O. # 25. Waltham, MA: RTI International.

McGlynn, E. A., Asch, S. M., Adams, J., Keesey, J., Hicks, J., DeCristofaro, A., et al. (2003). The quality of health care delivered to adults in the United States. *New England Journal of Medicine, 348*(26), 2635–2645.

Mitchell, J. B., Holden, D. J., & Hoover, S. (2008). *Evaluation of the Cancer Prevention and Treatment Demonstration for racial and ethnic minorities: Report to Congress.* Available from http://www.cms.hhs.gov/reports/downloads/Mitchell_CPTD.pdf

National Institutes of Health & National Cancer Institute. (2001). Voices of a broken system: Real people, real problems. President's Cancer Panel. Report of the Chairman, 2000–2001. Available from http://deainfo.nci.nih.gov/advisory/pcp/video-summary.htm

Naylor, M. D., Brooten, D., Campbell, R., Jacobsen, B. S., Mezey, M. D., Pauly, M. V., et al. (1999). Comprehensive discharge planning and home follow-up of hospitalized elders: A randomized clinical trial. *JAMA, 281*(7), 613–620.

New Jersey Hospital Association. (2001). *Medicare demonstration performance-based incentives.* Prepared for the Centers for Medicare & Medicaid Services. Princeton, NJ: New Jersey Hospital Association.

Peikes, D., Chen, A., Schore, J., & Brown, R. (2009). Effects of care coordination on hospitalization, quality of care, and health care expenditures among Medicare beneficiaries: 15 randomized trials. *JAMA, 301*(6), 603–618.

Piantadosi, S. (1997). *Clinical trials: A methodologic perspective.* New York: Wiley-Interscience.

Reid, R. J., Coleman, K., Johnson, E. A., Fishman, P. A., Hsu, C., Soman, M. P., et al. (2010). The Group Health Medical Home at year two: Cost savings, higher patient satisfaction, and less burnout for providers. *Health Affairs, 29*(5), 835–843.

Rich, M. W., Beckham, V., Wittenberg, C., Leven, C. L., Freedland, K. E., & Carney, R. M. (1995). A multidisciplinary intervention to prevent the readmission of elderly patients with congestive heart failure. *New England Journal of Medicine, 333*(18), 1190–1195.

Ries, L. E., Eisner, M. P., Kosary, C. L., Hankey, B. F., Miller, B. A., Clegg, L., et al. (2003). *SEER Cancer Statistics Review, 1975–2000.* Available from http://seer.cancer.gov/csr/1975_2000/.

Rittenhouse, D. R., & Shortell, S. M. (2009). The patient-centered medical home: Will it stand the test of health care reform? *JAMA, 301*(19), 2038–2040.

RTI International. (2006). *PGP site visit reports.* Retrieved November 12, 2009, from http://www.cms.hhs.gov/DemoProjectsEvalRpts/dowloads/PGP_Site_Visit_Reports.zip

Sebelius, K. (2009). *Report to Congress: Physician Group Practice Demonstration evaluation report.* Retrieved January 31, 2010, from http://www.cms.hhs.gov/reports/downloads/RTC_Sebelius_09_2009.pdf.

Todd, W., & Nash, T. (Eds.). (2001). *Disease management, a systems approach to improving patient outcomes.* New York: Jossey-Bass Publishers.

Wallack, S. S., & Tompkins, C. P. (2003). Realigning incentives in fee-for-service Medicare. *Health Affairs (Millwood), 22*(4), 59–70.

CHAPTER 10

Evaluating Pay for Performance Interventions

Jerry Cromwell and Kevin W. Smith

Current and potential approaches to paying for performance vary widely. As payers implement new models for pay for performance (P4P), they must test these models to determine whether they do improve performance. Interventions differ considerably in their design and implementation. A few evaluations of interventions are based on randomized trials, although most are observational studies. A lack of rigorously constructed comparison groups hampers much of the existing evaluation literature on P4P, particularly in the private sector. In their systematic review of the effects of financial incentives on health care quality, Petersen and colleagues (2006) found that nearly half of the eligible studies neither involved a concurrent comparison group nor compared quality indicators at baseline.

In this chapter, we explore many of the technical challenges of deriving scientifically rigorous estimates of P4P impacts. We begin by reviewing common threats to the internal validity of findings that introduce positive or negative bias in the quantitative estimate of P4P effects. We suggest using comparison groups to isolate a true P4P effect in an unbiased manner. After reviewing internal threats, we review the theory and approaches underlying the selection of comparison groups. We emphasize the importance of establishing a balance between intervention and comparison groups on the baseline level and in the rate of growth for any outcome variable of interest. We also consider the need to isolate the true P4P effect from exogenous shocks that may be contemporaneous with the intervention and suggest two approaches: first, one can isolate the true P4P effect by forming a comparison group ex ante, or before the demonstration begins (called "propensity score matching"); second, one can statistically correct imbalances ex post, or after the demonstration is over (called "ex post regression matching"). Having considered alternative ways to form the comparison group, we then introduce two external threats to valid findings that are quite common in P4P demonstrations. These threats undermine the generalizability or replicability of P4P effects to a national program. In the last section of the chapter, we summarize how five Medicare P4P demonstrations formed their comparison groups, and we critique their success in avoiding the various threats discussed earlier in the chapter. Where appropriate, we note the inherent limitations that enabling legislation places on Medicare demonstrations.

Internal Threats to Validity

Internal threats pertain to the validity of estimated intervention effects exclusively among the demonstration population. External threats arise from extrapolating intervention effects to other populations. We organize internal threats into six broad categories:

1. changes over time (or history)
2. differential selection of study and control groups
3. statistical regression
4. statistical significance
5. differential mortality
6. instrumentation

Changes Over Time

"History" threats result from changes in the experimental setting that may explain (confound) intervention performance. In P4P evaluations, changes in medical technology, clinical practice, and payment methods may affect intervention outcomes in positive or negative ways. For example, minimally invasive heart surgery, endoscopic vascular surgery and diagnostic testing, new cancer and psychotropic drugs, laser surgery, and numerous other technologies have completely changed the context of costs and quality of care. Relatively recent changes in clinical practice include the dramatic shift to outpatient surgery, greater reliance on antipsychotic drugs, and evidence-based treatment protocols. In addition, third-party insurers like Medicare constantly change the way they pay for services by altering the basis of payment (per admission, per procedure), accounting for local input cost differences, and adjusting for cost inflation.

Differential Selection

Different study and control subjects can also invalidate estimates of intervention effects. Randomized trials are least subject to selection bias. However, beneficiaries in most P4P demonstrations must agree to participate or at least be informed about an intervention. For demonstrations involving explicit care management, beneficiaries must be officially contacted and agree to talk with case managers. Some beneficiaries refuse formal invitations to undergo active case management, although this is generally only a small percentage (e.g., 5–10 percent). Other beneficiaries are simply unreachable or uninterested. Often, they have moved out of the area, have new telephone numbers, are institutionalized, or are otherwise unreachable. Nonparticipants, not surprisingly, often are more expensive to reach, more costly to treat, and sicker than those who agree to participate. These tendencies can introduce a bias in population studies using an intent-to-treat evaluation design because intervention staff do not have the opportunity to interact with groups that may be more or less amenable to the intervention.

Further selection occurs when disease management care teams actively work with a smaller group of participants (i.e., engage in targeting; McCall et al., 2008). Success critically depends upon the accuracy of the algorithms for targeting participating patients at high risk of using health care services or having lifestyle and medication issues. To date, targeting strategies have not proved successful (McCall et al., 2008).

Provider-based demonstrations are more vulnerable to selection bias than are randomized trials because of their implicit hierarchical sampling method. Providers and their extended networks can work only with their own "loyal" patients and cannot manage patients who are seen primarily in other practices. Patients loyal to a particular provider group introduce two potential selection biases in evaluating group performance. They may be different from the local (potentially control) population with the same chronic condition who may not have a usual source of care. They may be better or less well educated, younger or older, and have more or less insurance. In addition, the care they receive within the intervention network may not be comparable to the care available to the average person outside of the network.

Statistical Regression

Medicare demonstrations commonly select unusual, or even extreme, populations to participate; selection is usually based on high costs or poorer health status. This tendency can expose demonstrations to potentially serious regression-to-the-mean (RtoM) effects.[1] RtoM effects also arise from repeated observations on the same set of subjects. In the subsequent demonstration period, however, costs among intervention patients are likely to gravitate toward mean annual costs of all patients with a similar disease. This tendency exacerbates the RtoM effects that result from random selection and adds to the estimated bias in the intervention effect. A simple pre/post experimental design could show much lower intervention costs that result simply from RtoM in studies conducted without matched control groups.[2]

RtoM creates problems even if subjects are randomly sampled. One way to think about it is to assume that subjects have been sampled at one point in time. Some will likely have very high costs and others will have low costs. In a regression of costs on an intervention indicator, the resulting coefficient, or intervention effect, will not necessarily be biased if the comparison group was well matched. However, the standard error of the estimated effect will be higher than if one had a more permanent measure of each subject's costliness (Greene, 2000, p. 277). RtoM effects reduce the likelihood of finding a statistically significant intervention effect due to the transitory nature of extreme changes in health care utilization in either direction. Over time, any initially high-cost outlier group quickly bifurcates into a smaller group of continually high-cost beneficiaries and a larger group whose costs fall (rise) rapidly from earlier extremely high (low) levels. Short-run, random costs obscure the underlying effect of the intervention that may not be discernable through the statistical "noise" from the churning of subjects from low to high cost and vice versa. RtoM that occurs between baseline and a subsequent

---

[1] RtoM was first explored by Francis Galton (1822–1911); Galton, Carl Friedrich Gauss, and Adrien Marie LeGendre are the fathers of the modern regression analysis that is so popular among social scientists (Stigler, 1986). In his famous study of sweet peas, Galton noticed a "reversion" in pea size in the second generation toward the overall mean size. He observed the same inverse correlation between the heights of parents and their children. "Reversion" later became "regression," and originally meant, in Galton's framework, regression to the mean in the natural order of things.

[2] Flawed designs that do not include control groups may explain why some commercial disease management organizations claim to have achieved large reductions in costs of younger workers who have only one chronic illness.

period, also called regression attenuation bias, is generally greater in truncated samples, as seen in equation 10.1 (Barnett et al., 2005, p. 217):

$$\text{RtoM} = \{\sigma^2_w / [\sigma^2_w + \sigma^2_b]^{.5}\}C(z), \tag{10.1}$$

where σ^2_w = the within-individual or within-group variance over time, and σ^2_b = the between (or across) individuals variance. $C(z)$ reflects the additional impact of taking a truncated sample of the entire population and is determined as

$$C(z) = \varphi(z) / \Phi(z), \tag{10.2}$$

where $\varphi(z)$ and $\Phi(z)$ are the probability and cumulative density functions of the random normal distribution of z underlying the truncated sample. If the demonstration has selected subjects because of particularly high costs, as in the Care Management for High-Cost Beneficiaries Demonstration described later in this chapter, $z = (c - \mu) / \sigma_t$, where c = the cutoff value, μ = the mean of the entire population, and σ_t = the overall variance in the population (i.e., the square root of $[\sigma^2_w + \sigma^2_b]$).

For example, if c = \$6,000, μ = \$3,000, $\sigma^2_w = 4 \times 10^6$, $\sigma^2_b = 2.25 \times 10^6$, and σ_t = \$2,500, then z = \$3,000 / \$2,500 = 1.20. The probability density $\varphi(1.2) = 0.194$, and the cumulative $\Phi(1.2) = 0.885$. The resulting RtoM effect = $(4 \times 10^6 / 2,500)[0.194 / 0.885]$ = \$3,504. Working with a cost outlier sample greater than a \$6,000 cutoff should reproduce a new mean that is \$3,504 lower in the subsequent period before including any intervention effects.

Equation 10.1 shows that RtoM is greater when the within-individual variance over time caused by RtoM is greater. Observing a particular value for an individual at a certain point in time will be farther from the individual's true average value if there is more variation in observations for each individual. In addition, the more extreme the cutoff point is in selecting the intervention sample from the general population, the greater the RtoM effects will be, thereby accentuating the underlying random variation within individuals.

Both sampling and analytic solutions can mitigate the effects of RtoM (Barnett et al., 2005). Random assignment of individuals to the intervention and comparison groups is one solution that can cancel out RtoM bias effects, but this solution does not eliminate statistical noise. Taking multiple measurements of the criterion variable in the base period, then applying the cutoff value to each person's mean baseline value, should also reduce random attenuation bias. Equation 10.1 is modified by dividing σ2w by the number of measurements per person. Once the within and between variance is known, we can use equation 10.1 to calculate RtoM effects. The result is subtracted from the overall change in the criterion variable, leaving the net effect of the intervention.

An alternative parametric approach to control for RtoM uses analysis of covariance regression to isolate RtoM effects. The regression specification is

$$Y_{pt} = \alpha + \beta I + \rho[Y_{pb} - Y_b^*] + \varepsilon \qquad (10.3)$$

$$Y_{pt} = (\alpha - \rho Y_b^*) + \beta I + \gamma Y_{pb} + \varepsilon \,;\, \gamma = 1 + \rho;\, \text{RtoM} = \rho = \gamma - 1, \qquad (10.4)$$

where Y_{pt} = the value of the dependent variable for the p-th subject in the current (t) period, I = 0,1 intervention indicator, Y_{pb} = the p-th subject's value in the base period, and Y_b^* = the mean baseline value of Y across all persons. The ρ coefficient captures the average RtoM effect that is purged from Y, leaving the true β effect of being in the intervention (I).

Statistical Significance

Statistical noise is not an issue in determining actuarial savings, bonuses, or fee paybacks in Medicare demonstrations. The Centers for Medicare & Medicaid Services (CMS), however, does not make recommendations to Congress based on actuarial findings. Rather, the agency uses an evaluation firm to determine whether any savings or health improvements are statistically robust, thereby justifying program expansion beyond the limited demonstration.

Conventionally, evaluators use a two-sided 95 percent confidence interval to determine gross savings or quality improvements.[3] A problem with two-sided 95 percent confidence intervals is the high bar set for statistical savings (i.e., the bottom 2.5 percent of the distribution of savings). Should policy makers reject intervention savings if the savings could happen by chance only 3–4 percent of the time? One would expect most P4P initiatives to *reduce*, not increase, Medicare outlays. This suggests that evaluators should use a one-sided *t*-test. If an evaluator actually found *greater* outlays in a demonstration, the evaluator generally would not attribute them to the intervention but rather to

---

[3] Pierre LaPlace originated the infatuation with such stringent levels of confidence in the early 1800s when he conducted tests of differential birth rates between London, Paris, and Italy (Stigler, 1986). For those analyses, LaPlace was working with hundreds of thousands of observations over many years and applied a 99 percent confidence level. Rarely do modern social scientists have the luxury of so many observations in making contentious inferences regarding social policies and programs.

some other factor. Even less likely is a disease management intervention that causes a decline in quality. Using a one-sided test would make it more likely that evaluators would find that an intervention had reduced costs or improved quality, thus avoiding a Type II error (i.e., rejecting a true intervention effect).

Potential Type II errors plague provider-based demonstrations because of their inherently small samples. Many provider groups have few physicians and Medicare patients—especially when Congress mandates that demonstrations occur in particular areas such as rural counties. Demonstration sample sizes are further reduced when the government and management groups impose narrow beneficiary eligibility criteria (e.g., only heart failure beneficiaries, elderly minority cancer beneficiaries in Hawaii). In the Medicare Coordinated Care Demonstration, none of the demonstration's 15 programs had even 80 percent power to detect a savings rate less than 10 percent, and the evaluator had to reject savings of 20 percent or more. Congress further limits the size of demonstrations by putting a dollar limit on total up-front management fees or Medicare outlays. Underpowering provider-based demonstrations can result in the government's rejecting potentially valuable interventions.[4]

On a technical level, many demonstrations involve repeated measurement, or longitudinal, designs in which the same beneficiaries are tested at different points in time. Using differences in each beneficiary's performance between the base and demonstration periods halves the number of observations and produces a more conservative level of confidence in any intervention impacts (Rosner, 2006, p. 296).

Differential Mortality

All Medicare demonstrations lose beneficiaries from various forms of attrition. Demonstrations involving the Medicare chronically ill population are especially prone to monthly attrition of 1 percent or more. As sicker, more costly beneficiaries die or become otherwise ineligible, the overall intervention population becomes healthier, on average, and makes the intervention appear more successful on a cost basis.

---

[4] Simple *t*-tests of mean differences in savings or quality improvement suffer from statistical noise created by differences in beneficiaries within intervention and comparison groups. Multivariate regression can filter out the variation created by age, race, Medicaid eligibility, disabled status, and the like, as discussed in ex post regression matching later in this chapter. The result is an outcome estimate adjusted for beneficiary characteristics with a lower standard error and detectable threshold. This method reduces detectable thresholds even in randomized demonstrations with equal percentages of key beneficiary characteristics. See the discussion of the Medicare Health Support Pilot Program later in this chapter.

Attrition, even with initial randomization, can also introduce an imbalance between the original study and control groups before the demonstration's start date. During the time between prerandomization and the beginning of recruitment, death or other factors produce small differences in the intervention and comparison populations.

Some demonstrations allow beneficiaries to enter and exit managed care during the intervention. This practice can produce spending gaps that bias estimates of performance between intervention and control groups. These gaps also create additional statistical noise and reduce the reliability of results.

Attrition in disease management demonstrations is substantial, and sites often request "refresh" samples. Usually analyzed separately, refresh populations suffer from smaller sample sizes and truncation in the number of months in which the intervention is active. Also, assuming that intervention staff have been learning how best to manage the original population, the refresh sample experiences a maturation effect that would not exist in a new national program with new disease groups.

Instrumentation

A constant threat to the validity of intervention success is how success is measured. Problems may stem from changes in data collection instruments or from imperfect measures of outcomes and from inconsistent data collection methods. Changes in outcome measurement for Medicare demonstrations are uncommon, but exclusion criteria in calculating costs and health outcomes can bias the performance of the intervention, even with an initially balanced comparison group.

Costs can be measured in dollars for both study and control groups but may be limited to Medicare outlays on health services and may exclude government and provider internal management costs and ignore savings to beneficiaries (e.g., the 20 percent coinsurance on Part B services). Excluding certain services (e.g., postdischarge utilization) provides an incentive for intervention providers to discharge their patients early if savings are based only on inpatient costs. Demonstrations may also exclude hospice and end-stage renal disease utilization because they are beyond the control of intervention providers. These exclusions encourage interventions to transfer demonstration eligibles to these services to reduce within-demonstration costs. In this case, both the demonstration and national programs overestimate total program savings.

Quality is an even more difficult outcome to measure than costs (see Chapter 5). Because quality is a latent, multidimensional construct that must be quantified to be evaluated, 1, 2, or even 10 measures may not adequately capture the true quality impacts of an intervention. If evaluators choose measures badly, an intervention can appear to have no quality impacts at all. To date, no P4P demonstrations have used rewards and penalties for outcomes such as mortality, functional status, or quality of life. An intervention could possibly score well on several process-of-care indicators and yet have no statistical impact on mortality, morbidity, and functional status.

P4P interventions that focus on particular diseases depend upon the consistent reporting of diagnostic codes from the International Classification of Diseases, Ninth Revision (ICD-9). Inconsistent disease reporting can lead to including patients who do not actually have a particular illness in an intervention for that illness. Moreover, if changes in the frequency of comorbid conditions and complications affect P4P bonuses or penalties, providers have a financial incentive to overreport the former and underreport the latter.

Determining the Counterfactual

Internal threats of history, differential selection, statistical regression, and experimental mortality raise the question of what would have happened in the intervention group without the intervention. This question is known as the "counterfactual" and is critical in avoiding most internal threats. Simple pre/post experimental designs that compare provider performance before and after the introduction of a new P4P program are open to criticism because other temporal changes (or exogenous shocks to previous trends) may be partially, if not totally, responsible for causing observed changes in performance. For example, a local economy might experience a cyclical downturn, or a new technology might revolutionize medical practice (e.g., angioplasty making heart bypasses unnecessary). These and other phenomena may explain changes in beneficiary or provider behaviors.

The common approach to inferring intervention effects that has gained widespread acceptance is the counterfactual model of causality—sometimes known as "Rubin's causal model" (Holland & Rubin, 1988; Rosenbaum & Rubin, 1983). The counterfactual model posits that each individual subject (e.g., a patient in a hospital or medical home) has two potential outcomes: one if the subject is treated by the intervention and another if the subject is untreated by the intervention. The difference between these two potential outcomes is considered an unbiased estimate of the effect of the intervention for that subject. Unfortunately, while we can directly observe an outcome for a subject participating in an intervention, we cannot observe their outcome without treatment. The untreated outcome is therefore counterfactual in the sense that we cannot simultaneously observe both what would happen under intervention and what would have occurred in the absence of the intervention.

There are three ways to form a counterfactual:

- Predict what outcomes would be using study baseline information.
- Form a control or comparison group and use its performance during the demonstration period to benchmark intervention performance.
- Use a comparison group and further adjust for any remaining imbalances in characteristics relative to the intervention group before benchmarking performance.

Consider Figure 10-1, which plots per beneficiary per month (PBPM) average costs across base and intervention periods. Assume that the intervention population exhibits trend line a'b'c' during the base year before the start of the intervention. A counterfactual estimate of performance using just intervention beneficiaries can be forecasted using regression modeling of the type

$$\text{PBPM Costs}_{pt} = f[\text{time}, Z_{pt}]. \tag{10.5}$$

PBPM Costs$_{pt}$ = PBPM costs of the p-th patient in time t; Z_{pt} = vector of causal factors (e.g., age, health status) tracked for the p-th beneficiary across

Figure 10-1. Trends in intervention and comparison groups with imbalanced cost levels

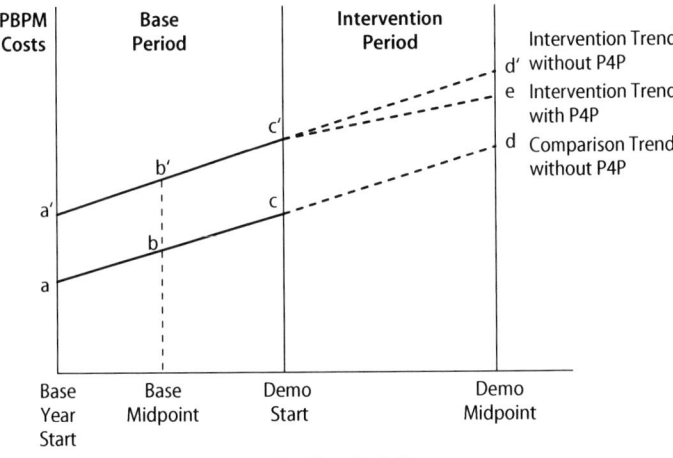

PMBM = per beneficiary per month.

t months; and time = a separate time indicator to capture inflation and other forces not captured in Z. If equation 10.5 predicted cost d′, midway through the demonstration, whereas the intervention group achieved cost level e, then the P4P cost-saving effect is (d′− e). Macroeconomic analysts of the national economy often use this method when no control group is available.

Two drawbacks to this approach are (1) the prediction errors surrounding point d′, based on short base periods and few causal variables, are highly correlated with cost increases; and (2) the baseline data used in the predicting equation may not capture exogenous shocks (e.g., payment changes, epidemics). Thus, it is more common to form a control or comparison group and use its performance during the intervention period as a benchmark. (*Comparison groups* apply to quasi-experimental designs with nonrandomized control groups. We use *comparison* rather than *control* hereafter because of the general lack of randomized designs in health payment demonstration research.) In Figure 10-1, a hypothetical comparison group shows a cost trend line of abc during the base period and a cost trend line of cd during the demonstration period. As drawn, the comparison group was not perfectly balanced on costs with the intervention group. Comparison PBPM costs were (a′ − a) dollars less than costs for the intervention group at the start of the base year. We also assume that no exogenous shocks occurred; hence, the comparison slope of cd = the slope of abc.

A naïve calculation of the intervention effect halfway through the demonstration is the observed difference (e − d) > 0, or dissavings, which we can decompose into an initial imbalance (d′ − d) and the true, unbiased intervention effect (e − d′):

$$(e - d') = (e - d) - (d' - d) \tag{10.6}$$

Because (d′ − d) = (b′ − b), the true intervention effect is recovered by simply subtracting out the initial cost imbalance:

$$(e - d') = (e - d) - (b' - b). \tag{10.7}$$

If (b′ − b) = $40 and (e − d) = $30, then the true intervention effect = $30 − $40 = -$10 and not (e − d′) = +$30. In Medicare demonstrations, independent actuarial contractors often use equation 10.7 to estimate the financial savings for a particular intervention. Making base period adjustments sometimes favors or disfavors a participating organization in a demonstration.

The adjustment is more complicated with exogenous shocks and imbalances—not only in base period cost levels, but in trends as well, as Figure 10-2 shows. The comparison group begins the base period with cost α. Through 12 base period months, comparison costs rise $\beta_2 > 0$ dollars per month. The intervention group starts with costs $\beta_1 > 0$ greater than the comparison group that rise monthly by $(\beta_2 + [\beta_3 > 0])$, resulting in beneficiary monthly costs of f dollars at the demonstration's go-live date. Thus, at the demonstration's start, intervention costs are even higher than comparison costs than at the beginning of the base period (i.e., $[f - g] > \beta_1$). A "shock" to the system is assumed to begin during the intervention period; this shock has the general effect of lowering both the comparison and intervention cost trends by $\beta_4 < 0$ dollars monthly. It could be exogenous to the system (a new cost-saving drug) or endogenous (RtoM from selecting high-cost beneficiaries).

Figure 10-2. Trends in intervention and comparison groups with imbalanced cost levels and trends

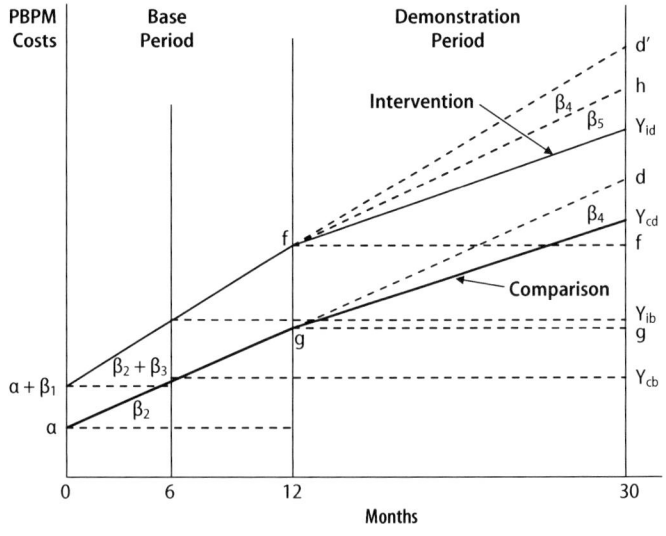

PMBM = per beneficiary per month.

The intervention is assumed to have an additional β_5 cost-reducing impact over and above the negative shock effect. β_5 is the true, unbiased, intervention effect. The mid-demonstration average costs are Y_{id} and Y_{cd} for the intervention and comparison groups, respectively. Evaluating intervention savings averaged at the demonstration midpoint produces dissavings of

$(Y_{id} - Y_{cd}) > 0$, instead of negative, lower costs. This calculation overestimates the intervention effect, which is negative. The standard difference-in-differences (D-in-D) method produces savings of $(Y_{id} - Y_{cd}) - (Y_{ib} - Y_{cb})$, which adjusts for cost differences at the midpoint of the base period. This D-in-D effect in a 36-month demonstration can be expressed using growth rates as

$$\text{D-in-D} = (30 - 6)\beta_3 + (30 - 12)\beta_5, \tag{10.8}$$

assuming no group differences in attrition or Z level or growth imbalances. β_1-level imbalances cancel out, but β_3 does not unless it is equal to zero, which is not the case in Figure 10-2 (the intervention base year slope is greater than the comparison slope). Suppose $\beta_3 = \$10$ and $\beta_5 = -\$5$. This would produce a D-in-D estimate of intervention dissavings equal to $(30 - 6) \times \$10 + (30 - 12) \times (-\$5) = \$240 - \$90 = \$150$. The D-in-D bias = $\$150 - (-\$90) = \$240$, or 2.5 times the true estimate in absolute terms. Such a large discrepancy has occurred because the baseline difference in growth rates was so large, and in the opposite direction, compared with the intervention effect. The baseline growth rate imbalance also ran a longer time (24 months) than the intervention effect did (18 months).

Another common way of controlling for unbalanced baseline values is to measure intervention performance exclusively in terms of growth rates, not just over the base year but from the midpoints of the base year and demonstration period. That is, intervention savings are determined by multiplying the intervention group's average base period costs by the differences in comparison and intervention growth rates:

$$\text{Savings} = Y_{ib}[\%\Delta Y_{cd} - \%\Delta Y_{id}]. \tag{10.9}$$

where $\%\Delta Y_{cd}$ and $\%\Delta Y_{id}$ = the percent change in comparison and intervention PBPM costs between the base and demonstration period midpoints. If comparison costs grow faster than intervention costs, then positive savings are generated; the opposite is true if intervention costs rise faster than comparison costs. Now, assume $\alpha = \$1,000$, $\beta_1 = \$50$, $\beta_2 = \$10$, $\beta_3 = \$5.25$, $\beta_4 = -\$2.50$, and $\beta_5 = -\$1.25$. Intervention costs start $50 higher and grow $5.25 per month faster than comparison costs, $10. A shock reduces the rate of cost increase by $2.50 per month in both groups during the demonstration period, but intervention costs grow an additional $1.25 slower. Dissavings based on equation 10.8 are $103.50, compared with -$27.00 per month of true savings by mid-demonstration, based on equation 10.9 (-$1.25).

In most instances, using the differential growth rate method will not produce such erroneous estimates of savings because comparison and base period imbalances are generally not as pronounced as in our example. The point, however, is that success in isolating P4P effects depends on how well matched control groups are on *both* levels and growth rates. Balancing the comparison group on average baseline costs is insufficient when using the D-in-D method. The comparison group must also be balanced on baseline trends. Whether the comparison group is, in fact, dually balanced at baseline is an empirical, hence statistical, question. If baseline intervention costs average 2 percent more than comparison group costs but the difference is not statistically significant, then no adjustment should be made for baseline imbalances in cost levels. If 2 percent were statistically significant, we would have to assume a consistent difference over time that then should be adjusted for when determining the counterfactual benchmark.

Statistically testing differences in baseline *trends* is much more difficult than in testing differences in *levels* because doing so requires a test of multiple observations over time. Using *t*-tests for beginning and end periods in the base period is a weak approach because one or both observations may not represent the overall trend. A better approach is to estimate the trend line through several subperiod points (e.g., monthly or quarterly average costs). For short trend lines, which are common in demonstration designs, few observations are available (e.g., 12 months or 4 quarters), resulting in relatively high standard errors of estimated growth rates. In these circumstances, rejecting the null hypothesis that baseline trends do not differ between the intervention and comparison groups is subject to Type II errors (i.e., rejecting true baseline differences in growth rates).

Regardless of the results of statistical tests of baseline differences in levels and trends, using an unbalanced comparison group can bias intervention effects. The gold standard of research designs is the randomized trial. With randomization, two groups should have no systematic differences, on average, in their baseline levels or trends. Netting out the performance of the control group leaves the intervention effect as a residual D-in-D. Nevertheless, balanced randomization requires large enough sample sizes to ensure that any meaningful imbalances disappear, usually within 30 cases (Rosner, 2006, p. 184). Very few Medicare demonstrations, unfortunately, have the ability to randomize subjects (later in this chapter, we describe how a few prominent demonstrations have selected comparison groups). In the next section, we review two methods for balancing study and comparison groups in quasi-experimental designs: (1) ex post regression matching and (2) propensity score matching. Both methods are becoming more common ways of balancing the two groups and raising the confidence that policy makers can have in demonstration findings.

Ex Post Regression Matching

In quasi-experimental demonstration designs, available data, short start-up periods, and preferences of participating sites for matching on only one or two characteristics (e.g., age, baseline costs) may make forming a balanced comparison group infeasible. Also, not being able to evaluate intervention subgroups, especially beneficiaries who agree to participate in a Medicare intervention, is a real drawback to policy makers and interveners, who often wish to have the program evaluated on the patients who are willing to undergo the intervention.

We can use ex post regression matching to address both of these drawbacks. Consider the following fully interacted model that uses a D-in-D regression approach to isolate the impact of a P4P intervention on the rate of growth in costs with one explicit group characteristic Z (e.g., percent disabled):

$$Y_{pt} = \alpha + \beta_1 I_p + \beta_2 t + \beta_3 I_p t + \beta_4 t D_t + \beta_5 t D_t I_p \qquad (10.10)$$
$$+ \mu Z_p + \gamma_1 Z_p I_p + \gamma_2 t Z_p + \gamma_3 t Z_p I_p + \gamma_4 t Z_p D_t$$
$$+ \gamma_5 t Z_p D_t I_p + \varepsilon_{pt},$$

where
- Y_{pt} = the cost of the p-th beneficiary in year t;
- I_p = an intervention indicator: 1 = in intervention; 0 = in comparison group;
- D_t = a demonstration period time indicator: 1 = during demonstration; 0 = base year;
- $I_p t$ = an interaction term tracking time changes (t) in Y for the intervention group;
- tD_t, $tD_t I_p$ = interaction terms tracking time changes only during the demonstration period for the control and intervention groups, respectively;

- Z_p = an unmatched, observable characteristic of p-th patient in the t-th period;
- $tZ_p \times D_t \times I_p$ = interaction term tracking changes in only the intervention group during the demonstration period, adjusted by the patients' Z value;
- α, β_x, μ, and γ_y = regression coefficients; and
- ε_{pt} = regression error term for costs unexplained by included variables.

The α, β, μ, γ and I, t, and D values together reproduce Figure 10-2. Table 10-1 presents the full-effects matrix. The sums of coefficients in each cell are an exact analog to a nested cross-tabulation of means of the two demonstration groups at the midpoints in each of two periods. Definitions are as follows:

- Z_{io}, Z_{co} = intervention and comparison average Z values at the start of the base period (e.g., percent female);
- $t_{ib} = (t_i - t_{ib})$, $t_{cb} = (t_c - t_{cb})$ = average number of months of eligibility, adjusted for attrition, for intervention and comparison beneficiaries halfway through the base period;
- t_i, t_c = average total number of months of eligibility of intervention and comparison beneficiaries;
- t_{id}, t_{cd} = average number of months of eligibility of intervention and comparison beneficiaries during the intervention; and
- $Z_{i,12}$, $Z_{c,12}$ = intervention and comparison average Z values in month 12 at the start of the demonstration period.

Excluding Z from the regression model produces beta coefficients that reflect all of the uncontrolled factors affecting levels and trends in costs in the intervention and comparison groups.

Beginning with the base period comparison group cell, $\alpha + \mu Z_{co}$ = the comparison group's mean cost at the beginning of the base period. μZ_{co} accounts for the initial comparison cost specific to a Z variable (e.g., poor health status), weighted by γ_1, or the marginal effect on cost of a higher percentage of the comparison group in initially poor health. β_2 is the comparison group's monthly change in cost unrelated to Z (Z-neutral). Multiplied by t_{cb}, the average number of attrition-adjusted baseline eligible months for comparison beneficiaries halfway through the base period, $t_{cb}\beta_2$, captures the growth effects of any other factors besides Z. $Z_{co}t_{cb}\gamma_2$ reflects the additional baseline cost growth contributed by the Z variable.

Table 10-1. Difference-in-differences mean predicted costs with imbalances

	Base Period (D = 0)	Demonstration Period (D = 1)	Difference
Intervention (I = 1)	$\alpha + \beta_1 + (\mu + \gamma_1) Z_{io}$ $+ t_{ib}(\beta_2 + \beta_3)$ $+ Z_{io}t_{ib}(\gamma_2 + \gamma_3)$	$\alpha + \beta_1 + t_i(\beta_2 + \beta_3)$ $+ t_{id}(\beta_4 + \beta_5)$ $+ Z_{io}[\mu + \gamma_1) + t_i(\gamma_2 + \gamma_3)]$ $+ t_{id}Z_{i,12}(\gamma_4 + \gamma_5)$	$(t_i - t_{ib})(\beta_2 + \beta_3)$ $+ t_{id}(\beta_4 + \beta_5)$ $+ (t_i - t_{ib})Z_{io}(\gamma_2 + \gamma_3)$ $+ t_{id}Z_{i,12}(\gamma_4 + \gamma_5)$
Comparison (I = 0)	$\alpha + \mu Z_{co}$ $+ t_{cb}\beta_2$ $+ Z_{co}t_{cb}\gamma_2$	$\alpha + t_c\beta_2 + t_{cd}\beta_4$ $+ Z_{co}(\mu + t_c\gamma_2)$ $+ t_{cd}Z_{c,12}\gamma_4$	$(t_c - t_{cb})\beta_2 + t_{cd}\beta_4$ $+ Z_{co}(t_c - t_{cb})\gamma_2$ $+ t_{cd}Z_{c,12}\gamma_4$
Difference	$\beta_1 + \beta_2(t_{ib} - t_{cb}) + \beta_3 t_{ib}$ $+ \mu(Z_{io} - Z_{co}) + \gamma_1 Z_{io}$ $+ \gamma_2(Z_{io}t_{ib} - Z_{co}t_{cb})$ $+ \gamma_3 Z_{io}t_{ib}$	$\beta_1 + \beta_2(t_i - t_c) + t_i\beta_3$ $+ \beta_4(tid - t_{cd}) + t_{id}\beta_5$ $+ \mu(Zio - Zco) + \gamma 1 Zio$ $+ \gamma_2(t_i Z_{io} - t_c Z_{co}) + \gamma_3 t_i Z_{io}$ $+ \gamma_4(t_{id}Z_{i,12} - t_{cd}Z_{c,12})$ $+ \gamma_5 t_{id}Z_{i,12}$	$[(t_i - t_c) - (t_{ib} - t_{cb})]\beta_2$ $+ (t_i - t_{ib})\beta_3$ $+ (t_{id} - t_{cd})\beta_4 + t_{id}\beta_5$ $+ \gamma_2[(t_i - t_{ib})Z_{io} - (t_c - t_{cb})Z_{co}]$ $+ \gamma_3(t_i - t_{ib})Z_{io}$ $+ \gamma_4(t_{id}Z_{i,12} - t_{cd}Z_{c,12})$ $+ \gamma_5 t_{id}Z_{i,12}$

Table 10-1 allows for several imbalance effects. The difference between the intervention and comparison groups' base period average costs is found in the difference row in the first column. $\beta_1 + \mu(Z_{io} - Z_{co}) + \gamma_1 Z_{io}$ is the difference in intervention versus comparison base period cost levels including the separate Z variable effects. $\mu(Z_{io} - Z_{co})$ is the group difference in initial base period Z values, weighted by the comparison group's marginal effect, μ, of Z. $\gamma_1 Z_{io}$ is the cost effect of the intervention Z value over and above μ. Similarly, $\gamma_2(Z_{io}t_{ib} - Z_{co}t_{cb})$ is the impact of the Z differences on midpoint baseline average costs, weighted by the comparison group's marginal trend factor, γ_2. $\gamma_3 Z_{io}t_{ib}$ is the additional intervention Z effect from having a different base period growth rate. $\beta_3 t_{ib}$ is any additional, Z-neutral cost impact of a different baseline growth rate in the intervention group. Demonstration period differences in mean intervention-comparison group costs build on initial period level and growth imbalances. β_4, β_5, γ_4, and γ_5 (second column) add potentially new differences resulting from shocks and true intervention effects.

The elements that make up the lower right cell in Table 10-1 are the most interesting. They represent an expanded version of the D-in-D intervention effect in equation 10.7. The eight terms capture all of the possible ways that the intervention may have altered the expected cost levels during the demonstration.

To illustrate the effects of the eight possible terms in the D-in-D calculation, we developed a simulation model to test the effects of various imbalances. The results are shown in Table 10-2. The first column assumes no imbalances from differences in intervention and comparison group costs at the beginning of the base period ($Z_{io} = Z_{co} = 0.1$, or 10 percent) or from any differences in base period growth rates (i.e., $\beta_1 = \beta_3 = \gamma_1 = \gamma_3 = 0$).

It also assumes no differences in the 1 percent attrition rates between study and control groups; that is, $(t_{id} - t_{cd})$ and $(t_{ib} - t_{cb}) = 0$. The only D-in-D effects are those due to nonzero β_5 and γ_5 values. The β_5 coefficient was assumed to be one-half as large as the decline in the comparison group's growth rate due to a shock after the base period, which, in turn, was assumed to be -25 percent of the 1 percent monthly comparison growth in costs during the base period, or -0.005. Starting base period costs were decomposed into a $1,000 mean PBPM

Table 10-2. Simulation effects of intervention, imbalances, and attrition on per beneficiary per month costs

Simulation Effects	Baseline Complete Balance	Low Z-Only Level Imbalance	High Z-Only Level Imbalance: 60% vs. 40%	High Level/Growth Imbalance	Baseline with 2% Demonstration l-attrition
$[(t_i - t_c) - (t_{ib} - t_{cb})]\beta_2$	0.00	0.00	0.00	0.00	-20.97
$(t_i - t_{ib})\beta_3$	0.00	0.00	0.00	69.93	0.00
$(t_{id} - t_{cd})\beta_4$	0.00	0.00	0.00	0.00	5.24
$t_{id}\beta_5$	-16.65	-16.65	-16.65	-25.39	-14.03
$\gamma_2[(t_i - t_{ib})Z_{io} - (t_c - t_{cb})Z_{co}]$	0.00	3.33	13.32	13.32	-1.05
$\gamma_3(t_i - t_{ib})Z_{io}$	0.00	0.00	0.00	20.98	0.00
$\gamma_4(t_{id}Z_{i,12} - t_{cd}Z_{c,12})$	0.00	-0.83	-3.33	-3.33	0.26
$\gamma_5 t_{id}Z_{i,12}$	-0.83	-1.25	-5.00	-7.62	-0.70
Total	**-17.48**	**-15.40**	**-11.66**	**67.89**	**-31.25**
Percent PBPM Cost	**-1.44**	**-1.27**	**-0.84**	**4.89**	**-2.57**

PBPM = per beneficiary per month.
Baseline Complete Balance Parameters:
$\alpha = \$1,000$; $\mu = \$500$; $\beta_1 = 0$; $\gamma_1 = 0$
$\beta_2 = 0.01\alpha$; $\beta_3 = 0.5(0.01)(\alpha + \beta_1)$; $\beta_4 = -0.25\beta_2$; $\beta_5 = 0.5(-0.25)(\beta_2 + \beta_3)$
$\gamma_2 = 0.01\mu$; $\gamma_3 = 0.5(0.01)(\mu + \gamma_1)$; $\gamma_4 = 0.25\gamma_2$; $\gamma_5 = 0.5(-0.25)(\gamma_2 + \gamma_3)$
$Z_{io} = Z_{co} = 0.1$; attrition = 0.01 per month
$t_{ib} = t_{cb} = 5.62$; $t_{id} = t_{cd} = 13.32$; $t_i = t_c = 18.94$

in Z-neutral costs plus another $500 in PBPM costs associated with a single Z variable (e.g., percent disabled).

These assumptions result in β_2 = $10 per month increase in Z-neutral comparison group costs, and another γ_2 = $5 monthly increase resulting from Z. Assuming the base period ran 12 months and the demonstration period was 36 months with a constant attrition of 1 percent per month, $t_{id}\beta_5$ = 13.32(-1.25) = -16.65. The intervention's cost effect via Z was $\gamma_5 t_{id} Z_{i,12}$ = (-1.25)(13.32) (0.1) = -0.83, assuming the percentage disabled did not change during the demonstration. The Z effect is only 5 percent of the Z-neutral β effect because of the compound assumption: μ = 0.5α and γ_1 = 0.10. The overall effect of the intervention on PBPM costs was -$17.48 in savings, or 1.4 percent of the comparison group's mean PBPM cost during the demonstration period. This amount is the average dollar savings per month per beneficiary that resulted from the intervention's effect being half as large (-0.125) as the 25 percent reduction caused by the negative shock in the comparison (and intervention) group's 1 percent monthly cost growth in the base (and demonstration) periods. (We assumed that RtoM would slow the rate of cost growth in both groups.)

In Table 10-2, the second and third columns show the intervention effects of a 50 percent imbalance only in the Z variable—first, when Z_{co} = 10 percent and then when Z_{co} = 40 percent ($\beta_1 = \beta_3 = \gamma_1 = \gamma_3 = 0$). The ($\gamma_2 + \gamma_4$) rows capture the effects of the Z imbalance on differential cost growth. The γ_2 effect of 3.33 in the second column is the higher growth that occurs in the intervention during the demonstration period caused by, say, a higher percentage disabled (i.e., 15 percent versus 10 percent). The γ_4 effect debits the γ_2 effect by -0.83 because the demonstration period's shock effect is working off a higher Z value at the start of the demonstration. The net Z imbalance effect is $2.08 higher costs. With a much higher percentage disabled in the base period (i.e., 60 percent intervention, 40 percent comparison group), the Z imbalance increases by a factor of 4 under the model's assumptions, resulting in higher mean intervention costs of $10 ($13.32 − $3.33) during the demonstration period. Note, too, that under our assumption, the intervention's γ_5 effect becomes more negative with a higher percentage disabled in the intervention group. This phenomenon occurs because we assume that the constant γ_5 intervention effect on Z is working on a higher Z-related cost at the beginning of the demonstration period. A disabled rate of 60 percent among the intervention group versus 40 percent among the comparison group results in intervention savings falling from -$17.48 to -$11.66, a one-third reduction.

Column 4 shows the impact when we assume that the intervention group begins with 5 percent higher Z-neutral and Z-related costs at the start of the base period ($\beta_1 = 50$, $\gamma_1 = 25$) and the intervention's percentage disabled is 60 percent versus 40 percent in the comparison group. This produces $69.93 in *higher* intervention group costs averaged halfway through the demonstration from higher base period Z-neutral growth on a base of $1,050, which is offset somewhat by -$25.39 because of the intervention's cost reduction effect during the demonstration. Z-specific level and growth imbalances add another $23.35 that offsets intervention-induced cost reductions. Thus, level and growth imbalances appear to make the intervention group more expensive than the comparison group when the true intervention effects, $\beta_5 + \gamma_5$, reduce costs by -$33 (-$25.39 + -$7.62). The standard D-in-D model that compares average changes in mean costs between the base and demonstration periods fails to adjust for differences in base period growth trends that build upon cost imbalances from the start of the base period.

The last column simulates the impact of a doubling of the 1 percent beneficiary attrition rate in the intervention group during the demonstration period. In this case, the β_2 effect is negative and substantial (-$20.97) given the other baseline parameters. The β_2 effect is negative because intervention beneficiaries, on average, remain in the demonstration period a shorter time than comparison beneficiaries and experience a censoring in their underlying monthly growth in costs. This negative effect is partially offset by $5.24 more in β_4-related costs because intervention beneficiaries are not experiencing the cost-saving effects as long as the comparison group is, over the full demonstration period. The net attrition effect of -$31.25 slightly overstates intervention cost savings of -$14.73 (-$14.03 + -$0.70). Note that this exercise assumes that intervention attrition is randomly distributed among beneficiaries, which is unlikely to be the case. If higher attrition occurred among the more costly, sicker beneficiaries, then apparent (but not true) savings would be greater. That is, early intervention mortality saves money.

Inserting additional Z characteristics into the regression model will help to purge the crucial β_3 coefficient of imbalances in levels and growth rates. Nonetheless, unobservable Z characteristics may still bias the results. In provider-based P4P interventions, these variables might include intervention-comparison group differences in risk taking, local competition, staffing quality, efficiency levels, and Hawthorne effects of simply being monitored. A few of these characteristics may be quantifiable with some effort, but staff quality, Hawthorne effects, and such will likely remain uncontrolled, although they are technically accounted for in the β_3 and β_5 coefficients. Because the missing characteristics are likely to contribute to the intervention's ostensible success in a simple D-in-D test, one should be cautious in interpreting intervention success even after using ex post regression matching.

Propensity Score Matching

A common alternative to using regression to statistically adjust for study-control imbalances ex post is to conform the comparison group as closely as possible to the intervention group ex ante. The objective of propensity score matching is to create a comparison group that is similar to the intervention group on the basis of observed characteristics. The mechanism for this matching is the propensity score—the probability that a subject would be selected for the intervention, conditional on measured characteristics (Rosenbaum & Rubin, 1983). Typically, one would identify a pool of untreated subjects and merge this pool with the intervention subjects. One would then estimate a logistic regression by regressing intervention status (1 = intervention group, 0 = comparison pool) on a set of characteristics that is available for all subjects. Using the logistic equation, one can then estimate the probability that each potential comparison subject would have been a member of the intervention group if he or she had been asked. This estimated probability is the propensity score, which condenses information from all the characteristics into a single score (Rubin, 1997). Propensity scores are particularly useful when many continuous variables are available for comparison.

One can plot the distribution of propensity scores in each group to examine the overlap in scores. Intervention group subjects will often have higher propensity scores than those in the comparison pool, as illustrated by Figure 10-3. One can infer causal effects only for regions of overlap in the scores because subjects with low propensity scores are very unlikely to have been selected for the intervention. Strictly speaking, one should also omit intervention patients outside the region of overlap from the analyses, although this may be objectionable in studies that initially have only small numbers of intervention participants.

Once all subjects have a propensity score, a common approach is to divide all subjects into quintiles of propensity scores and then to select equal numbers of intervention and comparison subjects from each quintile. If a large enough pool of comparison subjects is available, it may also be possible to select multiple comparisons for each intervention subject (called "many-to-one matching").

Figure 10-3. Hypothetical distribution of propensity scores by group, showing regions in which scores do not overlap

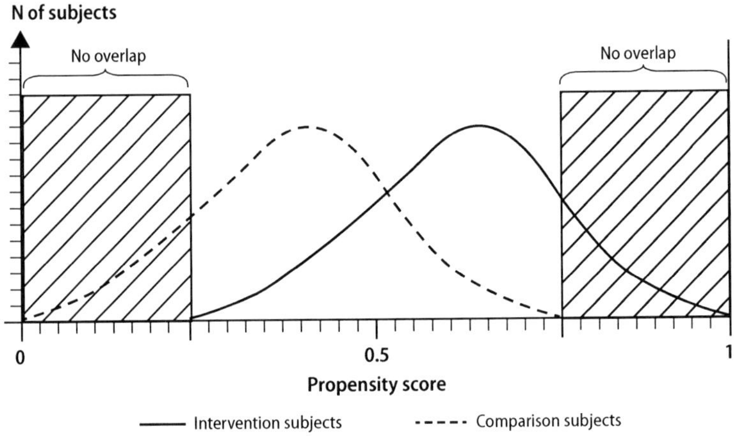

There are three main methods for estimating intervention effects using propensity scores. The first is simply to conduct comparative analyses using the propensity-matched comparison group because it serves as a counterfactual for inferring intervention effects without any further adjustment. If we observe all relevant confounding characteristics, then the difference between the outcomes for the intervention and comparison groups provides an unbiased estimate of the true effect of the intervention. This is a strong assumption, and in practice, the propensity score approach may remove much but not all of the potential selection bias in the estimate. The threat of bias is especially strong when the available observed characteristics include only demographic variables or administrative data that are not causally related to the actual selection mechanism.

The second method is to employ the propensity score as a covariate in the statistical model estimating intervention effects. The third method is to weight the results by the propensity score. All nonintervention subjects form the comparison group, and their outcomes are weighted by their propensity scores. Subjects with higher weights are more likely to be intervention subjects; hence, their performance justifies a higher weight. Recent simulation work, however, suggests that many propensities close to 0 or to 1 may compromise the validity of using a continuous weighting approach (Schafer & Kang, 2008).

In practice, the propensity score approach has several limitations when applied to P4P analyses. One problem is that policy makers and program developers may not be familiar with the propensity score approach. They may not trust the extent to which the propensity score actually adjusts for potential biases, although recently policy makers appear to be more open to the method (see the discussion later in this chapter on Medicare provider-based demonstrations). Traditionally, policy makers and participating sites have preferred a matching approach in which the matching factors and comparisons are more readily apparent.

A second drawback is that propensity scoring may be difficult because of the nature of P4P outcomes. The most straightforward application of propensity scores occurs when the outcome is a single measure that applies to all patients. However, most quality indicators are relevant only for select subgroups of patients (e.g., mammograms for women aged 50 or older, hemoglobin A1c tests for persons with diabetes). Technically, the propensity score approach requires a separate comparison group for each of these outcomes, which would each require separate propensity models as well. One way to address this complexity might be to use propensity scoring to identify a single overall comparison group. For example, patients from a set of physician group practices might be matched by propensity scores with patients from a comparison set of physician group practices. One could then proceed under the assumption that any subgroups within the intervention and comparison groups (e.g., diabetic patients) were comparable for analysis purposes.

External Threats to Validity

Even if comparison groups were well matched with intervention groups, and even if adequate adjustments produced statistically balanced groups ex post, policy makers would still have to extrapolate responses to P4P incentives from small samples of intervention organizations and beneficiaries to the whole population in a national program. Two particularly relevant threats to external validity, or "to what populations, settings, treatment variables, and measurement variables can [the intervention's] effect be generalized," (Campbell et al., 1963, p. 5) are as follows:

1. *interaction of selection and the experimental (intervention) variable*, or the selection of unique participants who respond differently to an intervention than most other groups; and

2. *reactive effects to experimental arrangements*, in which respondents and interveners respond in part to an intervention simply by being in an experiment.

Selection-Intervention Interactions

Under internal threats, differential selection occurs when the intervention selects subjects (e.g., beneficiaries) who differ in important ways, affecting outcomes from subjects in the comparison group. External interactive selection threats often result from systematic differences in the whole population of subjects in different geographic areas regardless of whether they end up in the intervention or control group. Threats to generalizability also depend on beneficiary eligibility criteria.

Cultural differences across areas may inhibit changes in eating habits. Getting exercise may also be more difficult in an inner-city environment than in other areas. Introducing the same disease management program in both rural Appalachia and Salt Lake City is likely to produce very different results simply because of differences in each area's underlying health needs. Areas with greater need for health care services may result in greater intervention gains in health outcomes and possibly lower costs from changed lifestyles than would occur across a nationwide program.

Population-based P4P demonstrations put sites at risk for intervention nonparticipants, in which case the overall probability of intervention success is inversely related to the beneficiary participation rate assuming nonparticipants perform more like the control group. Beneficiaries in some areas will be more or less willing or capable of participating than in other areas. Moreover, populations most in need may be the least likely to participate because of poorer health status, poverty, language, and other factors.

Success also depends upon participants' ability to access the health care system for key tests, check-ups, and timely treatment. Distance and scarcity of primary care physicians and hospitals with state-of-the-art equipment are both problems in rural areas. Some P4P interventions that extend a Centers of Excellence imprimatur to selected providers depend on local competition for their success (Cromwell et al., 1998). A Centers of Excellence imprimatur extended to a monopolistic hospital will not show the same shift in patient volumes as it would to a hospital in a competitive market.

Provider-based demonstrations present particularly serious interactive selection threats to generalizing findings. Because all provider applicants are volunteers, they automatically differ (often in unmeasurable ways) from other potentially qualified providers who decide not to apply. As volunteers, participants are likely to be more comfortable with and adaptable to incentives in the intervention. High-occupancy hospitals are likely to be more interested in avoiding unnecessary admissions. Teaching hospitals, which place a higher value on quality, research, and prestige than community hospitals do, are likely to be more inclined to participate in quality-enhancing demonstrations.

Often, demonstration applicants already have exceptional staff available to implement the intervention. In fact, CMS usually insists on this. Demonstration staff tend to have previous experience with similar interventions, have more extensive training, and be more innovative than clinicians elsewhere. They are likely to be more familiar with evidence-based practice and the research literature. Moreover, charismatic leaders with a personal interest in the success of their proposed intervention often are integral to demonstration success. Although national programs can disseminate intervention protocols to other practitioners in other settings, they cannot transfer the original staff and working environments.

Expenditures made by providers before the demonstration, called infrastructure sunk costs, lower their financial risk of participating in a novel payment reform. Providers tend to apply only if they have already made considerable investments, such as electronic medical records, in the intervention. They build on existing investments and simply spread their fixed costs over more patients. Far less infrastructure capacity would be available to other provider groups in a national program.

The more risk-averse the provider group, the less likely it is to participate in a demonstration when profits are at risk. Dominant market providers are often more risk-averse than smaller competitors and do not participate because they enjoy a secure financial position. P4P gainsharing demonstrations depend on the willingness of physicians to share financial risk and bonuses with hospitals. Physician-hospital organizations that agree to participate in these demonstrations are more willing than other groups to reduce inpatient consulting services. Gainsharing incentives are more attractive to providers who have been practicing *less* efficiently in the past. These providers are most likely to make improvements and capture whatever rewards are offered.

Reactive Effects to Experimental Arrangements

Three kinds of reactive effects to P4P interventions can limit the generalizability of a demonstration's findings to a national program. First, almost by definition, applicants to CMS demonstrations are innovative risk takers. They have fewer qualms about being closely monitored by the government, submitting the necessary data for evaluation, and permitting probing interviews regarding their operations, intervention strategies, and challenges. They are naturally more adaptive and flexible in structure and philosophy than most nonapplicants. Their management also tends to be more confident that they can bring the demonstration to a successful conclusion. Implementing the intervention in other sites will likely meet more internal resistance with less management support, less data support, and thus less success.

Second, CMS assigns each site a demonstration project officer who interacts with site management regularly and ensures that the site adheres to the terms and conditions of the demonstration contract. All sites also have a CMS research project officer who oversees the independent cost and quality evaluator. Consequently, providers and disease management organizations may spend more time contacting patients than they would without such intensive oversight (i.e., the Hawthorne effect). Moreover, in interviews, demonstration site staff may tend to report what evaluators want to hear, downplay problems, and generally overstate their successes. In any case, a national program is unlikely to have such intensive government oversight.

Third, CMS contractors provide nearly all P4P demonstration sites with periodic updates in their cost and quality performance. Although to some degree sites are able to track their performance using their own data, they need Medicare to provide them with utilization and spending statistics for services outside their network. Commercial disease management organizations depend to an even greater extent on the government for data because they have no immediate access to any Medicare claims. A large, geographically diverse, national program is unlikely to match the cost, rapidity, and depth of information that CMS provides to demonstration sites. Without such feedback, such groups would find it difficult to make midcourse corrections in their intervention implementation protocols, resulting in less satisfactory performance than among better-informed demonstration participants.

Examples of Demonstration Comparison Groups

Having reviewed the internal and external threats to valid P4P findings and ways that comparison groups can be balanced to ensure valid inferences, we next describe how a select number of Medicare P4P demonstrations constructed the counterfactual comparison group and the problems inherent in deriving valid, unbiased findings. To illustrate the different needs and kinds of comparison groups, we divide demonstrations into those that are population-based and those that are provider-based. We further stratify provider-based demonstrations by whether they cover essentially all Medicare services or just hospital inpatient services. We further stratify all-services demonstrations into those that cover all beneficiaries and those that cover a subset of high-cost and/or high-risk beneficiaries. The last subset includes most of the disease management interventions. Demonstrations we selected for special study are as follows:

1. Medicare Health Support Pilot Program (MHS): population-based, all services

2. Indiana Health Information Exchange Demonstration (IHIE): population-based, all services

3. Physician Group Practice (PGP) Demonstration: provider-based, all services, all beneficiaries

4. Care Management for High-Cost Beneficiaries Demonstration (CMHCB): provider-based, all services, high-cost and high-risk beneficiaries

5. Medicare Hospital Gainsharing Demonstration: provider-based, hospital services, all or selected diagnosis-related groups (DRGs)

Medicare Health Support Pilot Program

In randomized clinical trials, beneficiaries are usually asked to participate in a trial, then the investigators randomize them to the study or control (placebo) group. In population-based demonstrations such as the MHS pilot,[5] CMS first solicited applications from commercial disease management organizations to recruit and then manage the intervention group mostly using registered nurses in remote call centers. Once CMS selected applicants within their proposed geographic areas, the agency then prerandomized beneficiaries meeting heart failure and diabetes eligibility and minimum cost thresholds to intervention and control groups. We discuss threats from historical trends and differential selection three ways, involving the prerandomization process, the success in engaging beneficiaries, and the way CMS addressed discontinuities in eligibility.

---

[5] The Care Level Management and Key to Better Health interventions in the CMHCB Demonstration also involved prerandomization.

Prerandomization selection problems. In MHS, prerandomization took place 3–8 months before the disease management organization began to actively recruit beneficiaries. At each site, 30,000 beneficiaries were randomized on a 2:1 ratio to the intervention and comparison groups. Because beneficiaries had not been screened for their willingness to participate, participating sites had to recruit beneficiaries appearing on their assigned list of intervention beneficiaries. The demonstration's financial incentives followed a standard intent-to-treat design. Sites were financially responsible for the cost and outcome trends of all intervention beneficiaries and not just those they recruited. This strategy minimized any incentive to recruit only beneficiaries thought to be most successful in the intervention.

CMS expected such large sample sizes to balance intervention and comparison groups in terms of cost levels and baseline growth rates (i.e., minimize differences in history and differential selection). Subsequent analysis, however, showed a cost imbalance of 1–4 percentage points for or against the site. Most of this difference occurred during the 3- to 8-month hiatus between randomization and the start date. Although the differences were statistically insignificant, the fact that sites had to achieve at least 5 percent lower intervention costs during the demonstration or return all fees made such differences meaningful. During financial reconciliation, the program eventually made an actuarial adjustment to pilot period costs for any difference between intervention and comparison group costs caused by the lagged start date. Some organizations benefitted from the adjustment and others were penalized by it. Adjusting performance for only statistically significant base year differences may have been more justified. In a separate analysis, evaluators used ex post regression to control for minor imbalances in patient characteristics (e.g., age, gender, race, severity scores) that might cause different rates of cost growth in the intervention and comparison groups (McCall et al., 2008).

Engagement problems. The MHS intervention did not recruit all beneficiaries who were randomized to the intervention group and alive at the start date. The program was unable to reach some beneficiaries because they were in institutions, their addresses and telephone numbers had changed, or they had moved out of the area. A few beneficiaries were too cognitively impaired to participate. Others refused when asked; some refused because they had a terminal illness with short life expectancy and others did not want to be bothered. For whatever reason, the engaged intervention group was healthier than either the pool of intervention nonparticipants or the entire comparison group. This phenomenon produced an additional imbalance between the benchmark comparison group and those whom the sites were actually counseling. Failure to achieve the required level of savings may be out of the site's control to some degree. This failure also posed a political disadvantage with the prerandomization strategy when sites complained that they ought to be judged on the participants they did have rather than on nonparticipants. Evaluating success only on participants can be a persuasive argument with politicians who are unfamiliar with the intent-to-treat logic of a clinical trial.

Participation discontinuities. In clinical trials, all intervention subjects are followed for the length of the trial; beneficiaries in population-based Medicare demonstrations such as the MHS pilot, however, may drop out from the intervention within the first few months or even days.[6] Even if participants and nonparticipants incur similar costs, differential attrition can bias the estimate of intervention savings, as we showed when analyzing the D-in-D method (see Table 10-2). Shorter average participation in the intervention group can produce lower average monthly costs in that group than in the comparison group because the truncated time period censors the effects of positive cost increases in general.

Of particular concern is any differential mortality of beneficiaries who die only a few days or weeks after the start of the demonstration. Those beneficiaries remain in the analytic database, usually with a very high monthly cost, although the disease management organization had little chance to intervene in these cases. Although the frequency of such cases should be similar in the intervention and comparison groups if the groups are balanced at the start date, they can add substantial noise to any statistical test of intervention success.

The MHS used two approaches for dealing with dropouts in evaluating costs. The simplest approach is to drop beneficiaries with fewer than 1-3 months of eligibility, assuming that the intervention would have had no time to interact with them. The second approach weights the monthly costs by the fraction of time they were eligible in either the intervention or control group. Weighting mimics the actuarial approach of simply summing costs incurred regardless of a beneficiary's duration in the intervention and dividing by all eligible months.

---

[6] Techniques exist for adjusting patient dropouts and drop-ins in randomized clinical trials (Rosner, 2006, pp. 422–423). Drop-ins are patients who receive the treatment outside of the intervention.

Almost all Medicare demonstrations also experience intermittent eligibility that produces "involuntary noncompliance." In the MHS pilot, some beneficiaries experienced ineligible spells because they lost Medicare Part A or B coverage, joined a managed care plan, or experienced other disqualifying events. Beneficiaries were allowed to return to the intervention, and some did. Disease management organizations argue that they should not be responsible for any utilization or adverse events that occurred while beneficiaries were ineligible, and CMS agreed. Yet gaps in cost (and quality) measures can introduce bias if related in any way to intervention status.

Instrumentation problems. As in many other demonstrations, the MHS excludes certain services from its cost calculations. Usual exclusions include hospice, end-stage renal disease, and nursing home costs. The MHS pilot also allowed sites to choose up to 14 different quality process measures to demonstrate that quality of care had not been compromised by disease management. Some sites chose as few as 4 indicators that covered both heart failure and diabetes. Although evaluators applied the same exclusion criteria to the comparison group, the exclusions raise questions about the validity of intervention performance.

Statistical significance and regression-to-the-mean problems. Despite substantial variation in beneficiary cost levels and growth rates, the large sample sizes in the MHS pilot generated levels of statistical significance on the order of 3.5–4.0 percent of the comparison group's PBPM cost. Given the 5 percent minimum savings requirement, the demonstration required study and control sample sizes of 20,000 and 10,000, respectively, in each of eight sites. Both groups also exhibited substantial RtoM in demonstration period costs. The large cost increases experienced by beneficiaries with low costs in the base period made it particularly difficult for sites to target beneficiaries at high risk for cost increases. Most targeting algorithms focus on beneficiaries who experienced unusually high utilization in the base period.

Selection-experiment interactions. The eight MHS sites were distributed around the country. Some were more urban than others. Although none of the commercial interventions were successful in controlling costs (McCall et al., 2008), geography could have been responsible for some of the null findings. Local beneficiaries may have been less responsive to management counseling or less likely to participate and comply with recommendations. The local supply of primary care physicians and hospitals may have biased results against disease counseling. The fact that all eight interventions used registered nurses in remote call centers to counsel participants suggests, however, that geography did not play a critical role because the selection of area did not correlate with the key intervention activity. If several sites had been successful, then policy makers could have been confident that the success would generalize to other areas in a national program.

Reaction to experimental arrangements. The major concern with most CMS demonstrations is their replicability. Because commercial disease management organizations do not provide direct care to a predetermined set of their own patients, no concern exists about replicating the remote call center approach elsewhere. However, CMS staff perform a close oversight role on demonstration groups in tracking quarterly contacts and performing other such tasks. The agency also provides sites with quarterly cost information. Whether such support would exist in a national program is unclear. (See Chapter 11 for a longer discussion of this topic.)

Indiana Health Information Exchange Demonstration

IHIE is a population-based P4P initiative that includes all health insurers and physician practices treating the majority of patients in the greater Indianapolis area. The program makes incentive payments to provider groups and also provides them with claims and clinical information to improve quality. The demonstration is still in progress and no findings are available as of October 2010, but because of the need for regional approaches to solve problems in care discontinuities, reviewing the way the comparison group was formed is valuable.

For the IHIE Medicare demonstration, savings are based on the differences in cost trends between demonstration groups in Indianapolis and in a set of comparison cities. Nine similarly sized potential metropolitan statistical areas (MSAs) were identified in the Indiana region of the country. The program identified a final set of comparison MSAs based on 14 indicators (see Table 10-3). The nine candidate MSAs were ranked from 1 to 9, relative to Indianapolis, on the 14 characteristics using an absolute difference from the IHIE measure. In producing an overall ranking for each MSA, Medicare per capita spending levels and growth rates were weighted 5, mortality was weighted 3, and all other characteristics were weighted 1. Medicare spending was weighted 5 times more than most other characteristics because of the shared savings model used in the demonstration.

Table 10-3. Selection of comparison cities for Indianapolis in the Indiana Health Information Exchange Demonstration

Measure	MSA Code 17140: Cincinnati-Middletown, Ohio-Ky.-Ind., MSA	MSA Code 17460: Cleveland-Elyria-Mentor, Ohio, MSA	MSA Code 18140: Columbus, Ohio, MSA	MSA Code 19380: Dayton, Ohio, MSA
Medicare Expenditures per Beneficiary, 2005	7,303	8,428	7,415	7,192
Medicare Part A Share per Beneficiary, 2005	0.57	0.59	0.58	0.55
Percent Change in Medicare Expenditures per Beneficiary, 2000-2005	47.2	45.2	42.4	46.5
Hospital Beds per 10,000 Population	21.9	35.9	24.3	33.4
Patient Care Physicians per 10,000 Population	26.1	36.0	26.9	25.7
Advantage Share of Medicare Beneficiaries, 2003	0.143	0.180	0.118	0.130
Population, 2004	2,058,221	2,137,073	1,693,906	845,646
Population Density, 2004	468	1,066	425	495
Elderly Share of Population, 2002	0.116	0.144	0.100	0.135
Aged and Disabled Medicare Beneficiaries, 2003	277,165	337,978	192,187	132,487
Aged Share of Medicare Beneficiaries, 2003	0.842	0.883	0.847	0.851
Age-Adjusted Elderly Mortality Rate	0.057	0.054	0.058	0.056
Median Income, 2002	46,265	42,726	46,881	42,658
Poverty Rate, 2002	0.089	0.105	0.095	0.094
Weighted Average Rank	4.750	7.08	2.83	4.83

MSA Code 26900: Indianapolis, Ind., MSA	MSA Code 28140: Kansas City, Mo.-Kan., MSA	MSA Code 31140: Louisville, Ky.-Ind., MSA	MSA Code 33340: Milwaukee-Waukesha-West Allis, Wis., MSA	MSA Code 36540: Omaha-Council Bluffs, Neb.-Iowa, MSA	MSA Code 41180: St. Louis, Mo.-Ill., MSA	Weight
7,374	7,094	7,358	7,154	6,784	7,632	5
0.57	0.55	0.57	0.59	0.56	0.57	1
40.5	28.3	44.7	41.5	34.2	37.5	5
30.6	29.2	31.9	27.1	35.0	34.4	1
32.3	24.8	27.7	32.0	29.8	27.2	1
0.042	0.138	0.036	0.034	0.111	0.212	1
1,621,613	1,925,319	1,200,847	1,515,738	803,801	2,787,701	1
420	245	290	1,038	184	297	1
0.104	0.114	0.122	0.124	0.109	0.128	1
190,606	245,834	173,845	213,907	98,110	404,793	1
0.849	0.854	0.818	0.865	0.858	0.851	1
0.056	0.053	0.058	0.052	0.053	0.055	3
49,314	49,529	43,470	48,607	47,745	45,729	1
0.088	0.085	0.106	0.105	0.089	0.099	1
	6.42	4.83	4.38	6.13	4.83	

We selected Columbus, Ohio; Louisville, Kentucky; and Milwaukee, Wisconsin, as comparison market areas. They were also the cities best matched to Indianapolis when using equal weights for all characteristics. Simulation analysis using historical data indicated that Medicare weighted average expenditure growth in the three comparison cities was 1.72 percentage points lower than in Indianapolis. A minimum threshold of 1.75 percentage points (based on a statistical 95 percent confidence interval for savings) was recommended before paying bonuses to ensure real savings from the intervention.

History and Differential Selection. The CMS implementation contractor faced a challenge in forming the IHIE comparison group because beneficiaries could not be drawn from the immediate area in which the intervention was taking place. Balancing on costs requires balancing on both utilization rates and payer prices. Selecting similar beneficiaries in the same market area usually does this. The implementation contractor used population size and density as matching characteristics to approximate the area's capability of supporting expensive new technologies. Mortality served as a proxy for area quality of care and health status deriving from lifestyles and access to key services. Hospital and physician supply indicators also capture differences in utilization. The rate of growth in hospital spending approximated expected spending trends in general. Medicare spending and growth rates were also weighted more heavily than other variables to ensure a balance on both levels and trends.

Although matching comparison with intervention beneficiaries one-to-one is common practice, using this procedure is questionable when whole cities are the unit of analysis. Choosing a single city as a comparison risks failure to capture some unmeasured shocks to either group over the demonstration period. A more statistically sound approach would be to match multiple comparison cities with the intervention city. The IHIE contractor eventually selected three well-matched cities, but one could argue for selecting several more cities and then averaging performance over the larger group to minimize the bias from shocks isolated in one or two cities. The researcher must trade off the possible prediction error from using fewer comparison cities with the error when using less well-matched cities.

Statistical significance. The fact that evaluators will compare IHIE performance in Indianapolis with performance in three nearby cities raises a concern about the unit of observation. In each area, evaluators will evaluate all Medicare beneficiaries, thereby guaranteeing the statistical power to detect small changes in costs *in each city*. This, in turn, also guarantees the ability to detect small differences in cost trends between Indianapolis and the other three cities that, together, form the counterfactual comparison group. Evaluators will determine performance based on relative *trends* and not levels in base and demonstration period mean costs. This approach factors out differences in levels but not in *baseline* growth rates, which is why evaluators give a high weight to rates of cost growth when selecting the comparison group.

Selection-experiment interactions. The most important drawback to regionally grounded demonstrations is the possibility of a serious correlation between the intervention site and the experiment. If only one city participates in a regional demonstration with only three comparison cities, they will all look alike in terms of the matching characteristics (e.g., Medicare spending, medical supply, population). Consequently, policy makers will not know whether a demonstration that is successful in Indianapolis will be equally successful in Mobile, Alabama; Manchester, New Hampshire; Chicago, Illinois; or Boise, Idaho, which are cities with very different cost and utilization trends and Medicare penetration.

Reaction to experimental arrangements. Siting a demonstration based on sharing patient information across practices in Indianapolis is subject to serious threats of special experimental circumstances. The Regenstrief Institute at the University of Indiana is the primary group that supports the computer exchange of information. The institute has been a premier organization in conducting computerized analyses of local health care utilization and costs for more than 25 years. Very few other groups in the country (e.g., Intermountain in Utah) would be capable of implementing the IHIE model without major infrastructure and networking investments. Thinking of the IHIE demonstration as a proof of concept may make more sense. If the initial demonstration is successful, a larger demonstration with several information exchange groups would be necessary to justify a national program.

Physician Group Practice Demonstration

Most Medicare demonstrations do not involve the strict randomization found in randomized clinical trials or population-based disease management initiatives. CMS conducts most demonstrations with provider groups that use their own "loyal" patients (CMS's term) as intervention participants. The first, and largest, provider-based PGP demonstration currently involves 10 large physician practices with between 8,000 and 44,000 Medicare patients

(Kautter et al., 2007). Most of the practices are part of a larger integrated delivery system anchored by a major acute care hospital. The demonstration bases savings on risk adjusted differences in *increases* in intervention and comparison group costs per beneficiary and determines PGP bonuses in each year of the demonstration. Because the program determines savings and bonuses annually, this determination requires an ex post determination of beneficiary loyalty each year to each PGP. Beneficiaries have been assigned, or are loyal, to each demonstration PGP if they have multiple office-type visits to PGP physicians during the previous 12 months. (See Chapter 9 for details of the assignment process.)

Once intervention beneficiaries have been identified at year's end, a new comparison group is constructed each year based on revised PGP patient-flow information. CMS' implementation and evaluation contractor performs this elaborate, hierarchical, computer-intensive process. First, the evaluators identify comparison counties for each PGP that have at least 1 percent of the PGP's own eligible patients. Together, these counties include 80–90 percent of all PGP Medicare patients. Second, the evaluators identify all beneficiaries in each comparison county that had at least one ambulatory Medicare visit claim in the base year (i.e., beneficiaries had to be accessing the health care system). No loyalty constraints were placed on the eligibility of comparison group beneficiaries. Consequently, the evaluators test each PGP's performance against regular fee-for-service Medicare beneficiaries regardless of whether they received treatment from another large PGP outside of the demonstration. Third, evaluators average costs at the county level. Fourth, evaluators weight county averages by the county's share of PGP-loyal beneficiaries to produce an overall base and demonstration year average beneficiary cost. Fifth, evaluators use Medicare concurrent Hierarchical Condition Category (HCC) scores to risk-adjust beneficiary costs in both groups in the base year and each subsequent year. Evaluators calculate an aggregate average HCC score for the comparison group using the same PGP weighting procedure that was used to determine average costs and then use these scores to adjust for differential changes in case-mix costliness between the intervention and comparison groups.

History and differential selection. Because savings are based on differences in intervention and comparison cost *trends*, differences in average costs during the base period cancel out. Because some of the PGPs dominate their local market and draw from widely dispersed areas, evaluators believed that a patient-flow method of identifying comparison beneficiaries would be superior to using either a prespecified georadius around each PGP or administrative units such as nearby counties or metropolitan areas. Weighting each comparison county by its contributing share of PGP beneficiaries produces a single, synthetic market area believed to be identical in utilization patterns and unit costs of services for both intervention and comparison beneficiaries. How comparable the PGP approach to matching comparison with intervention beneficiaries is to a propensity score matching procedure is unknown. Nonloyal comparison and loyal intervention beneficiaries may also be on different growth curves during the base year. Any PGP cost savings, therefore, could be caused by previous PGP patient care patterns and not by offering additional bonuses for "bending the cost curve."

Differential mortality. The PGP demonstration is unique because it assigns beneficiaries at year's end to loyal PGPs ex post. Beneficiaries can rotate in and out of both groups each year rather than being a constant set of subjects in either group. The same is true of counties that succeed or fail to meet the 1 percent of PGP beneficiaries criterion. One advantage of this strategy is that it avoids the degradation in sample sizes that occurs in other Medicare demonstrations for the elderly. Its large sample sizes presumably guarantee that dropouts and drop-ins are random in intervention and comparison groups. The requirement that eligible comparison beneficiaries, like PGP beneficiaries, must access the health system each year reinforces this assumption.

An unexpected drawback to assigning beneficiaries to PGPs only after the year is over is that it prevents CMS from providing the PGP interim utilization and cost information against a predetermined comparison group. As a substitute, the evaluation contractor provides PGPs with interim feedback of utilization and costs against a "simulated" comparison group. The simulated comparison group overlaps substantially, but not perfectly, with the final ex post comparison group.

Instrumentation problems. Adjusting cost trends for differential changes in HCC risk scores raises concerns about upcoding comorbid conditions. If PGPs began to code for more, and more serious, comorbid health conditions than they had in the past, practices could deflate their risk-adjusted rate of cost increases relative to physicians and hospitals in comparison groups that did not have bonus incentives.

Statistical significance and regression-to-the-mean problems. CMS purposely solicited only PGPs with very large numbers of Medicare beneficiaries because the agency recognized the high degree of cost differences among elderly beneficiaries. Using an annual rather than a monthly cost indicator also avoids the additional variation that occurs when very short-duration beneficiaries have their costs adjusted upward to a full month. Because the program includes all beneficiaries rather than just the high-cost chronically ill, the program experiences much less cost churning of the population caused by regression-to-the-mean.

The PGP demonstration begins to cumulate bonuses only when intervention cost increases fall 2 percent below total comparison group costs. Originally, this criterion replaced any statistical determination of true savings. Later, CMS argued that no PGP should be able to accrue bonuses on very small rates of savings because of the sizable demonstration costs that the agency incurred. This requirement is not as stringent as the 5 percent criterion that the CMHCB demonstration used, requiring sites to return all management fees if cost increases were not at least 5 percent less than comparison group trends.

Selection-experiment interactions. Provider-based PGPs, almost by definition, operate within particular health and economic markets. For example, beneficiaries in some areas of the country may be more amenable to and more compliant with cost-saving initiatives than beneficiaries in other areas. Marshfield Clinic in central Wisconsin and Geisinger Clinic in central Pennsylvania are premier tertiary organizations that operate in essentially rural environments. Would these organizations be equally successful operating in the more economically and socially disadvantaged areas of large cities such as New York or Chicago? Could their approach to medicine be equally successful in rural McAllen, Texas, an area with one of the most entrepreneurial, high-cost provider systems in the country (Gawande, 2009)?

Reaction to experimental arrangements. The unique characteristics of the participating PGP pose a larger threat to generalizability. These sites tend to have charismatic leaders and be trailblazers in providing cost-effective care. Moreover, they tend to be research-oriented institutions with strong experimental capacity. Some of these sites have very high hospital occupancy rates and benefit from avoiding unnecessary admissions. Moreover, CMS has provided each site with a remarkable amount of statistical information to help guide performance and tweak their intervention activities; this information would be prohibitively expensive on a national level.

Care Management for High-Cost Beneficiaries Demonstration

The CMHCB demonstration includes PGPs usually associated with a major tertiary acute care hospital (e.g., Montefiore Medical Center in the Bronx, New York City, or Massachusetts General Hospital in Boston). The original PGP demonstration included all assigned beneficiaries; however, in the CMHCB demonstration, each site is responsible for providing disease management services to high-cost and/or high-severity Medicare beneficiaries. Another way in which the CMHCB demonstration differs from the original PGP demonstration is that although physicians share in bonuses for quantified savings beyond 2 percent, CMHCB sites receive substantial up-front fees to manage a negotiated set of beneficiaries over 3 demonstration years. If sites have not saved at least 5 percent of the comparison group's average PBPM cost by the end of the demonstration period, they must return all fees.

In each site, the design of the comparison group strives to replicate as closely as possible the unique cost and diagnostic characteristics of the intervention group (e.g., annual baseline costs greater than $5,000 for heart failure or diabetic beneficiaries). As in the original demonstration, the process required first identifying comparable geographic areas, but comparison beneficiaries in those areas then had to be loyal to a subset of nondemonstration physician groups.

Intervention PGPs tended to serve a large share of Medicare beneficiaries in their target areas. As a result, geographic comparison areas usually were counties or ZIP codes in other regions of a state that had demographic and health care utilization characteristics similar to those of the intervention area. CMS asked demonstration PGPs to identify other comparable PGPs in the designated comparison areas. Pilot tests indicated that their lists would not yield enough comparison beneficiaries and would require a claims-based approach to identify additional comparison PGPs. CMS identified several primary care PGPs in the comparison areas through their Tax Identification Numbers. Requiring a minimum of 20 percent of total PGP payments from office visits eliminated single-specialty practices. A final group of high-volume PGPs, similar to the intervention PGPs in their focus on primary care, was selected in each comparison area.

To enhance cost equivalence, the final step in the selection process was to match comparison group beneficiaries to intervention group members based on monthly beneficiary costs in the base year. CMS matched the comparison group to the intervention group by defining three to five cost ranges, determining the distribution of beneficiaries across these ranges in the intervention group, and then randomly selecting the same number of comparison beneficiaries in each category. This method produced equally sized comparison and intervention groups. As a final check, CMS also compared the two groups on a range of health status, payment category, and health care utilization variables.

History and differential selection. CMS closely matched CMHCB intervention beneficiaries on cost *levels* but not necessarily on baseline *trends* in costs. Actuarial reconciliation of cost differences raised (or lowered) the comparison group's average PBPM cost by the ratio of base year intervention to comparison costs, thereby factoring out differences in cost levels. Paybacks of management fees became contentious when sites failed the 5 percent criterion and complained about imbalances in patient characteristics that might create different cost trends. (An earlier section of this chapter deals with biases that one or more imbalanced Z variables can create.) In all cases, ex post regression matching found little difference in patient characteristics between the two groups, although matching focused on cost categories. For an imbalance to make a material difference in savings, an imbalanced variable (e.g., minority status) had to (1) substantially affect the growth (not just level) in costs and (2) be very different on a percentage point basis.

For example, a minority beneficiary, according to the regression coefficient, might have a $100 per month greater cost increase than a nonminority beneficiary—a meaningful difference. In addition, the intervention site may have served 10 percent minorities versus 5 percent in the comparison group, a 100 percent difference. Yet the impact of this imbalance would have been only $5 (0.05 × $100),[7] a trivial difference when compared with comparison PBPM costs that average $1,000 or more. As a practical matter, for an imbalance to have a meaningful effect on financial performance, the imbalance must be at least 20 percentage points or more and the Z-imbalance effect on cost growth, not levels, must be on the order of $200 or more. In demonstrations so far, imbalances rarely differ more than a few percentage points, even with 1,000 beneficiaries, and when they do, the regression weight is usually statistically insignificant.

Differential mortality. Attrition due to death presents a definite problem in most provider-based demonstrations—particularly in the CMHCB demonstration because of its strict eligibility criteria. Death rates for

---

[7] This calculation is analogous to $\gamma_4(t_{id}Z_{i,12} - t_{cd}Z_{c,12})$ in the D-in-D (bottom right) cell of Table 10-1.

Medicare heart failure and diabetes beneficiaries together average more than 1 percent per month. In negotiating inclusion and exclusion criteria with CMS, intervention sites had to choose between narrowing diagnostic characteristics that they believe are most amenable to intensive management and having a sufficient number of eligible beneficiaries to pay for the high fixed costs associated with the program. The problem could be exacerbated if the evaluator dropped beneficiaries who died in the first several months because the intervention could not have influenced their utilization and cost trajectories.

Statistical significance and regression-to-the-mean problems. Narrowly defined eligible groups, along with high attrition rates, have resulted in intervention and comparison samples of 2,000–4,000 each with little power to detect cost trends as small as 5 percent of PBPM average costs. Infusions of refresh beneficiaries have not helped because those populations are even smaller and are evaluated separately. When evaluators find a sizable point estimate of cost savings still to be statistically insignificant, interveners blame CMS for not having taken the limited nature of the eligible pool into consideration.

Selection-experiment interactions. All provider-based demonstrations face the same threats to generalizing findings because of the one-to-one link between the particular intervention and the local health and economic markets. CMS could not immediately generalize the results from running four different interventions in Lubbock, Texas; Bend, Oregon; the Bronx, New York City; and Boston, Massachusetts—unless all four were (un)successful, in which case we would be more confident that success, or the lack of, would extend to other areas of the country. If any single intervention were successful, the most prudent decision would be to replicate the intervention in several other different areas to test how robust the results are in those local environments.

Reaction to experimental arrangements. Unfortunately, transplanting provider-based interventions is vulnerable to this last threat to generalizing results. CMS can reproduce disease management and efficiency protocols and insert them into other practices; the original, successful, staff generally cannot, although some staff could be used as implementation consultants. Very little qualitative research has been done on either the transferability of P4P protocols to other sites or the effects of local markets and staffs on outcomes.

Medicare Hospital Gainsharing Demonstration

Another type of provider-based P4P demonstration more narrowly focuses on acute inpatient services with the goal of aligning physicians with hospital incentives to practice more efficiently. The Medicare Hospital Gainsharing Demonstration waived the Medicare prohibition against hospitals sharing savings from more efficient inpatient care with physicians. CMS allowed participating hospitals to share savings with their medical staff based on strict algorithms used to link individual physicians with the costs of inpatients for whom they were responsible. The program benchmarked trends in Medicare Part A and B inpatient costs per discharge against trends in a comparison group of similar hospitals.

The Charleston Area Medical Center (CAMC) in West Virginia presented a challenge in identifying a comparison group. CAMC is a 718-bed major teaching hospital with 70 percent of all beds in Charleston, a relatively modes-sized city. CAMC also entered a gainsharing arrangement with the medical staff only for a few major cardiac DRGs, including valve replacement, heart bypass, and angioplasty. CAMC had a 90–100 percent market share in these procedures, making it impossible to form a comparison group of local competitors. Selection of a comparison group focused on similar dominant hospitals in other smaller markets nationwide. To identify hospitals, the program used 11 criteria (see Table 10-4 on pages 44 and 45): bypass/valve and angioplasty volumes were weighted 6, given existing literature on the importance of high volumes in producing better outcomes; bypass/valve market shares were weighted 4 to reflect the unique effect that market dominance can have in setting local practice norms; acute care beds, Medicare discharge share, and number of residents were weighted 3 to capture the effects of size on cost and intensity. The rest of the indicators were weighted 1 when developing an overall similarity index with CAMC. The program derived the index as a weighted sum of a 1–10 ranking of absolute differences between CAMC and each hospital. Using these methods, CMS identified 35 comparison hospitals. The 10 hospitals most similar to CAMC formed the comparison group (see Table 10-4); all but two were also located in the South.

History and differential selection. This approach to comparison group selection is analogous to how cities were selected in the IHIE demonstration. The program encountered challenges typical in selecting comparison sites. First, determining whether comparison and intervention sites are on the same cost trend lines during a period prior to the demonstration is generally infeasible. This phenomenon is particularly true of individual provider groups whose claims histories are a very small part of an enormous database that is costly and time-consuming to manipulate. Second, although the literature has identified many variables affecting cost levels, it has paid much less attention to variables that affect cost trends or to the γ coefficients appearing in equation 10.10. Weighting bypass and valve volumes higher when selecting comparison hospitals seems logical, but these variables may not capture important differences in cost trends. Whether or not the hospital is a safety net provider in an inner city may be a better way of approximating the rate of cost growth. Teaching hospitals have generally exhibited higher rates of growth than other facilities because of their early adoption of new technologies, and thus residents per bed should have a higher weight.

The sites' perceptions of who would be a good match further complicate the selection process. Early in the CAMC matching process, CMS considered using several hospitals outside the South as candidates for the comparison group but eventually discarded those hospitals because CAMC did not believe the markets or local practice norms reflected their own. How important regional differences are in predicting cost trends is also unknown.

Statistical significance. A primary advantage of having several comparison hospitals is greater statistical power to test for smaller cost-saving effects. The intervention sample is necessarily limited by the size of the hospital (or physician practice), which is sometimes relatively small. For CAMC, evaluators will have fewer than 800 bypasses and valves and roughly 1,100 percutaneous transluminal coronary angioplasty (PTCA) cases to study yearly. The comparison group, however, will have more than 3,000 bypasses and valves and more than 6,000 PTCAs and will likely produce a much smaller standard error than for the intervention site by approximately the ratio of the square root of comparison to intervention beneficiaries. For bypasses and valves, the comparison standard error should be about one-half that of CAMC's, ignoring intersite variation. CMS expects roughly the same gain for PTCAs.

Selection-experiment interactions. Sites such as CAMC present some of the biggest challenges from interactions between the site and the environment in which they operate. CAMC is one of the largest heart hospitals in the country, but it is located in a city of modest size in one of the poorer states in the country. The medical center has a complete monopoly in heart surgery for hundreds of miles. Matching the hospital with the environment is very difficult. The hospitals that most closely resemble CAMC are located in large

Table 10-4. Selection of comparison hospitals for CAMC

Medicare Provider ID	Hospital Name	City	State	Mean Rank Score	Acute Care Beds	Medicare Discharges	
						Number	Share
		Weight:			3	1	3
510022	Charleston Area Medical Center	Charleston	WV	—	718	13,824	62%
490024	Carilion Medical Center	Roanoke	VA	7.6	664	13,381	66%
200009	Maine Medical Center	Portland	ME	8.5	581	11,033	47%
340002	Memorial Mission Hospital and Asheville Surgery Center	Asheville	NC	8.6	646	16,194	65%
440002	Jackson-Madison County General Hospital	Jackson	TN	9.6	558	12,635	82%
010039	Huntsville Hospital	Huntsville	AL	10.7	786	16,256	73%
340040	Pitt County Memorial Hospital	Greenville	NC	11.5	618	12,619	100%
110107	Medical Center of Central Georgia	Macon	GA	12.8	534	11,606	68%
440063	Johnson City Medical Center	Johnson City	TN	15.5	478	10,734	77%
200033	Eastern Maine Medical Center	Bangor	ME	15.6	302	8,388	76%
340141	New Hanover Regional Medical Center	Wilmington	NC	15.9	539	13,331	84%

ID = identification number; CABG = coronary artery bypass graft; PTCA = percutaneous transluminal coronary angioplasty; DSH = disproportionate share hospital.

Source: 2008 Medicare Impact File.

Medicare Volume							
CABGs/valves		PTCA and Stents					
Hospital Volume	Hospital Share	Hospital Volume	Hospital Share	DSH Adj Factor	Number of Residents	Residents per bed	Medicare Case-Mix Index
6	4	6	4	1	3	1	1
751	100%	1,120	89%	0.12	116	0.16	1.82
386	80%	1,066	81%	0.07	83	0.12	1.76
424	100%	896	94%	0.08	171	0.30	1.95
571	100%	750	100%	0.13	39	0.06	1.79
326	100%	1,315	97%	0.16	18	0.03	1.74
359	100%	684	93%	0.07	31	0.04	1.66
492	100%	749	100%	0.24	155	0.27	1.96
493	82%	1,323	77%	0.21	88	0.16	1.92
286	100%	755	100%	0.16	62	0.14	1.55
329	100%	658	100%	0.16	24	0.08	1.85
245	100%	563	100%	0.12	54	0.11	1.65

cities with major teaching programs. As a general rule, the more unique the hospital and surrounding market are, the more prudent it is to select several "somewhat matched" hospitals to cancel out any erroneous mismatches.

References

Barnett, A. G., van der Pols, J. C., & Dobson, A. J. (2005). Regression to the mean: What it is and how to deal with it. *International Journal of Epidemiology, 34*(1), 215–220.

Campbell, D. T., Stanley, J. C., & Gage, N. L. (1963). *Experimental and quasi-experimental designs for research.* Boston: Houghton Mifflin.

Cromwell, J., Dayhoff, D. A., McCall, N., Subramanian, S., Freitas, R., Hart, R., et al. (1998). *Medicare Participating Heart Bypass Center Demonstration: Final report.* Health Care Financing Administration Contract No. 500-92-0013. Retrieved January 11, 2011, from http://www.cms.hhs.gov/reports/downloads/CromwellExecSum.pdf; http://www.cms.hhs.gov/reports/downloads/CromwellVol1.pdf; http://www.cms.hhs.gov/reports/downloads/CromwellVol2.pdf; http://www.cms.hhs.gov/reports/downloads/CromwellVol3.pdf

Gawande, A. (2009, June 1). The cost conundrum: What a Texas town can teach us about health care. *The New Yorker*, pp. 36–44.

Greene, W. H. (2000). *Econometric analysis* (4th ed.). Upper Saddle River, NJ: Prentice Hall.

Holland, P. W., & Rubin, D. B. (1988). Causal inference in retrospective studies. *Evaluation Review, 12,* 203–231.

Kautter, J., Pope, G. C., Trisolini, M., & Grund, S. (2007). Medicare Physician Group Practice Demonstration design: Quality and efficiency pay-for-performance. *Health Care Financing Review, 29*(1), 15–29.

McCall, N. T., Cromwell, J., Urato, C., & Rabiner, D. (2008). Evaluation of phase I of the Medicare Health Support Pilot Program under traditional fee-for-service Medicare: 18-month interim analysis. Report to Congress. Retrieved January 11, 2011, from http://www.cms.hhs.gov/reports/downloads/MHS_Second_Report_to_Congress_October_2008.pdf

Petersen, L. A., Woodard, L. D., Urech, T., Daw, C., & Sookanan, S. (2006). Does pay-for-performance improve the quality of health care? *Annals of Internal Medicine,* 145(4), 265–272.

Rosenbaum, P. R., & Rubin, D. B. (1983). The central role of the propensity score in observational studies for causal effects. *Biometrika, 70*, 41–55.

Rosner, B. (2006). *Fundamentals of biostatistics* (6th ed.). Belmont, CA: Thomson-Brooks/Cole.

Rubin, D. B. (1997). Estimating causal effects from large data sets using propensity scores. *Annals of Internal Medicine, 127*(8 Pt 2), 757–763.

Schafer, J. L., & Kang, J. (2008). Average causal effects from nonrandomized studies: A practical guide and simulated example. *Psychological Methods, 13*(4), 279–313.

Stigler, S. M. (1986). *The history of statistics: The measurement of uncertainty before 1900.* Cambridge, MA: Belknap Press of Harvard University Press.

CHAPTER 11

Converting Successful Medicare Demonstrations into National Programs

Leslie M. Greenwald

Ever since Medicare was implemented in the mid-1960s, this public program has been a leader in health care payment development and innovation, including pay for performance (P4P) reforms. Many of the P4P projects currently operating are Medicare pilot projects, or demonstrations, that test both the administrative feasibility and success of various performance models.

Though policy makers sometimes use the terms *pilot testing* and *demonstration projects* interchangeably, there are key differences. Demonstrations operate under specific legislative authority that allows the Department of Health and Human Services (DHHS) Secretary to suspend (or waive) specific Medicare payment regulations for the purposes of testing policy alternatives. Pilot tests, likewise, test policy alternatives but do not operate under the Secretary's somewhat limited demonstration authority and, therefore, may sometimes be more expansive. Pilot tests typically operate under specific project-by-project congressional legislative directive and so are less common than demonstration initiatives.

P4P initiatives use both pilot tests and demonstrations to allow sponsors to identify models that best meet their intended goals and that can be operationalized at an acceptable level of cost and burden to health care organizations and clinicians, insurers, and other stakeholders. Pilot tests and demonstrations also enable sponsors to identify opportunities for improvement and modify aspects of new initiatives that do not work—all on a manageable scale. Despite this long history of demonstration programs particularly related to innovative payment approaches such as P4P, findings from major demonstrations rarely become part of permanent Medicare program policy. This chapter examines reasons that Medicare's significant experience in conducting demonstration projects to test program innovations has a less lasting impact on the current national program than might be expected.

Important Early Policy Changes

Medicare's demonstration programs yielded some important policy changes in its early years. In the late 1970s, Medicare granted demonstration waivers to novel hospital "prospective payment rate-setting" systems in several states; Medicare paid the hospitals set fees in advance for groups of services. In 1984 Medicare established its own national Inpatient Prospective Payment System, which bundled all hospital services into a single per-case rate for nearly 500 diagnosis-related groups (DRGs). The background research for such a revolutionary change in payment came from the New Jersey hospital DRG demonstration (Hsiao et al., 1986). Medicare eventually extended prospective payment systems to post-acute care, rehabilitation, and psychiatric hospitals, skilled nursing facilities, home health, and hospice services, in all cases beginning initial implementation with either pilot tests or demonstrations.

After freezing payments for high-cost procedures in the late 1980s, Medicare then designed a prospective payment system for physicians. The Medicare fee schedule put thousands of services on a common scale based on physician work effort. Medicare also pioneered capitated rates for its managed care population, setting separate rates for every county in the United States. All these systems changes divorced payment from the costs of individual clinicians and provider groups, and they supported implementation of Medicare's prospective payment systems.

Caring for more than 35 million elderly and disabled beneficiaries provided the number of patients and data needed to develop these systems. The Medicare program generates significant amounts of administrative data available for the development, implementation, and efficient evaluation of a range of P4P models. Because of its size and financial importance in the marketplace, Medicare is often able to recruit providers and other willing organizations to its projects demonstrating P4P options.

Of course, Medicare's large scale and economic importance also translate to downsides for innovative payment policy development. Because Medicare is the largest health insurance program in the United States, any potential changes to it face close scrutiny. In general, Congress sets out in statute almost all key Medicare program parameters: from the ways clinicians and provider organizations are paid, to the policies governing the covered benefits, the provider groups and clinicians who can participate in Medicare, claims reporting requirements, and other operational policies. Federal regulation then fills in the details required by legislative changes in policy direction, including modifications to payment methodologies and rates.

The agency that administers Medicare (the Centers for Medicare & Medicaid Services, or CMS) is responsible primarily for operationalizing congressional mandates. This is not surprising in that Medicare's scale and scope mean that any payment changes inevitably affect a range of powerful political constituencies, including large numbers of health care professionals and provider groups, health insurance plans that participate in Medicare, medical suppliers, and Medicare beneficiaries. Therefore, national payment changes stemming from Medicare demonstrations—such as P4P—prompt debates in a highly politicized forum and must consider the political environment as well as the research findings. Demonstrations offer an opportunity to test constituency and political responses to programmatic changes.

Turning these lessons learned into national policies requires clearing several formidable hurdles. Given the number of P4P projects under Medicare's experience with demonstrations, it is curious that policy makers have considered only limited and nonspecific moves toward national implementation of any existing P4P model. In this chapter, we are interested in understanding the following:

- Why have so few demonstrations been evaluated as "successful" (i.e., met goals for generated savings and quality improvement)?
- Why has Congress failed to incorporate under national payment reform the lessons learned from Medicare P4P demonstrations?
- What will it take operationally and politically to apply the lessons of successful P4P demonstrations to a national payment system?

The rest of this chapter comprises four broad sections to answer these three important questions. First, we describe the ground rules that Congress and CMS impose on the design of Medicare demonstrations, considering the impact of these rules on national implementation. Next, we discuss common threats to successful evaluation findings that limit the generalizability of Medicare demonstrations. Next we lay out the operational challenges of taking a small, geographically constrained demonstration to the national stage. Each section cites specific Medicare demonstrations described in Chapter 10 (refer to that chapter for details of these demonstrations). The chapter concludes with an analysis of the political challenges that Congress and CMS face in incorporating successful demonstrations into a national payment system.

Demonstration Ground Rules and Practical Limitations

Medicare demonstration projects serve to test and evaluate policy innovation within specific boundaries. Section 402 of Public Law 92-603 grants CMS specific demonstration "waiver" authority for variations in the established payment regulations so long as these variations do not result in increased costs to the Medicare program. Known as demonstration payment waiver authority, this provision allows the DHHS Secretary to try alternative payment methods in small demonstrations prior to implementation in the full program. Although the Medicare demonstration statute permits the DHHS Secretary to waive certain Medicare requirements (such as cost-based or charge-based reimbursement) in conducting demonstrations, the statute's language focuses on program efficiency and cost reduction rather than on quality enhancement. Some subsections, however, authorize demonstration projects to examine impacts of various provider payment methods on quality of care.

In addition to this general demonstration authority, over the years Congress has also authorized projects to explore specific policy options, such as payment for case management for chronic illness, cancer prevention for ethnic and racial minorities, and telemedicine.[1] Sections of the law appear to authorize alternative provider payment methods (such as negotiated or discounted fees, bonuses, or withholds) whose objective it is to save program funds,[2] but Congress did not draft the demonstration authority explicitly to permit these payment methods as incentives for meeting quality goals. Finally, legal interpretations of Medicare's demonstration waiver authority typically limit projects to those that increase program efficiency (and generate savings) or that are at least budget neutral. Congress can specifically authorize additional spending for pilot projects through specific legislation. The Affordable Care Act contains many specific mandates for Medicare demonstrations and pilot projects.

Despite the DHHS Secretary's statutory authority to conduct them, Medicare demonstration projects that might disadvantage certain clinicians or provider organizations or beneficiaries relative to the status quo often

---

[1] These laws are printed in the pocket part following 42 USC section 1395b-1 in US Code Annotated and include specific standards for program design features, types of Medicare standards that can be waived, evaluation, and funding.

[2] A court upheld the Secretary's authority to test paying a single negotiated fee for outpatient cataract surgery under this statute in *American Academy of Ophthalmology, Inc. v. Sullivan*, 998 F. 2d 377 (6th Cir. 1993).

faced legal challenges. For example, the American Association of Health Plans in 1997 (US General Accounting Office, 1997) challenged the DHHS Secretary's authority to test a bidding approach for Medicare managed care plans in Colorado (the competitive pricing demonstration, proposed before Congress enacted the Medicare+Choice Medicare managed care program). After the federal district court issued a temporary restraining order that raised questions about the Secretary's authority to undertake the project, CMS did not implement it.[3] Few courts have decided cases involving the Secretary's authority to waive Medicare requirements as part of a demonstration project.[4] Courts generally accord great discretion to administrative agencies in interpreting and implementing federal law, especially complex programs like Medicare.[5] Similarly, legal challenges to CMS's authority to conduct competitive bidding demonstrations for laboratory and durable medical equipment have led both of these potentially promising, competitively based pricing projects to be delayed, with little realistic hope of their being implemented.

The fate of attempts to establish competitive bidding for Medicare managed care is an illustration of the "not in my backyard" (NIMBY) syndrome. Three times CMS attempted to demonstrate competitive bidding bids in markets around the country, and all three attempts failed because of local political opposition (Nichols & Reischauer, 2000). In two of the three attempts to implement this controversial demonstration, congressional representatives quashed the effort once a city was targeted for fear of potential negative impacts on local clinicians and health care organizations and beneficiaries. By the third attempt, the project had been delayed so long that congressional and policy interest waned and was insufficient to counteract continued local opposition; the project was never implemented.

Aside from the political challenges inherent in implementing demonstrations, the legal foundations for Medicare's demonstration

---

3 *AAHP v. Shalala* (D. Colo. Civ. Action No. 97-M-977, May 20, 1997). The case was dismissed when the Secretary agreed not to pursue the proposed demonstration project.
4 Several cases unsuccessfully challenged the Secretary's authority under 42 USC 1395b-1(F) to choose fiscal intermediaries and carriers based on competitive bidding that are not nominated by providers or carriers already in the program: *Health Care Service Corp v. Califano*, 601 F. 2d 934 (7th Cir. 1979); *Blue Cross Assoc. v. Harris*, 622 F. 2d 972 (8th Cir. 1980); *Blue Cross Assoc. v. Harris*, 664 F. 2d 806 (10th Cir. 1981).
5 See cases cited in note 4.

program also result in three practical limitations that nearly all Medicare demonstrations share:

- *Geographic and participant constraints:* specific legislative mandates often dictate where a demonstration takes place and who is invited to participate. Limiting geographic areas for participation can in turn limit the national generalizability of the demonstrations.
- *Voluntary participation by both clinicians and providers and by beneficiaries:* only willing groups and beneficiaries choose to participate. This condition also limits the generalizability of demonstrations because only selected organizations with a narrow range of characteristics participate.
- *Medicare budget neutrality:* the government must at least break even or save money on every Medicare demonstration. The condition limits Medicare from testing a wider range of projects that may have unclear cost impacts or even short-term additional costs, but the potential for longer-term gains.

Geographic and Participant Constraints

As a public program, fee-for-service (FFS) Medicare operates under an "any willing provider" legal requirement. Health care professionals and provider organizations that meet specified Medicare conditions (including certification and acceptance of Medicare payment amounts and balance billing limitations) are welcome to participate in the program. The same is not true in demonstrations. Almost all demonstrations are geographically limited so as to confine the "experimentation" to a manageable number of clinicians and provider organizations or to target the demonstration to providers who have specific capabilities. CMS issues a solicitation for participating clinicians and providers and then selects from among what appear to be the most qualified. Congress, Office of Management and Budget (OMB) and CMS staff impose project expenditure caps that usually constrain the number of providers and beneficiaries an agency can take. Sometimes CMS fails to select a qualified applicant, resulting in protests and, in extreme instances, pressure from Congress to expand eligibility. Such politically motivated geographic expansions often dilute the demonstration's focus on the most qualified applicants and the model to be tested, resulting in evaluations with unclear findings.

Voluntary Participation

For research purposes, the law limits demonstrations to *voluntary participation* on the part of both clinicians/providers and beneficiaries. CMS cannot require hospitals, physicians, and other providers to participate in demonstrations. Similarly, the agency must notify Medicare beneficiaries of a demonstration if it will affect them, and it must offer them the opportunity to drop out of the demonstration at any time; clinicians and providers have nearly the same rights (often at the end of a demonstration year). An agency cannot place limitations on the range of legally entitled Medicare services available to beneficiaries.

The extent of selection bias that this voluntary participation introduces varies by the nature of the intervention and by whether both clinicians and providers and beneficiaries must be recruited. This selection bias is potentially serious when clinicians and providers have to recruit beneficiaries with specific characteristics into the demonstration, which was the case in the Medicare Health Support (MHS) and Care Management for High-Cost Beneficiaries (CMHCB) demonstrations (Centers for Medicare & Medicaid Services [CMS], 2009b), in which only higher-cost beneficiaries with specific diseases were eligible to participate. Voluntary participation was less an issue, however, in the Medicare Physician Group Practice (PGP) and Hospital Gainsharing demonstrations, which required recruiting providers but not beneficiaries with any specific characteristics (Sebelius, 2009).

Still, some important differences might exist between demonstration beneficiaries and regular FFS beneficiaries treated in nondemonstration settings. To preserve its neutral role, the government almost never promotes its own demonstrations through the media or other sources. Provider groups are left to market "weak" imprimaturs to beneficiaries with strict oversight by CMS to ensure fair and accurate information is given to potential beneficiaries.[6] Randomization is almost never applied under Medicare demonstration projects because, although this approach would result in stronger evaluation results, excluding potentially eligible beneficiaries from participation in their local markets has been considered politically unpalatable.

---

[6] One example has been the CMS reticence to refer to some demonstration providers as Centers of Excellence (CoEs), approving instead titles such as Participating Medicare Heart Bypass Center. Providers complain that such titles have little value in gaining market share, even when an expressed goal of the demonstration was to regionalize care in higher-quality institutions.

Budget Neutrality

Unless specifically authorized by Congress, Medicare demonstrations operating under waiver authority must be at least budget neutral, meaning that the total costs under the demonstration cannot exceed those predicted under the existing statutory program. OMB must review and approve the budget neutrality of the demonstrations. OMB usually requires some savings to compensate for additional operational costs incurred during the demonstration (e.g., extra CMS monitoring staff, independent contractors who help set up and evaluate each demonstration). The MHS disease management (DM) pilot originally had a 5 percent savings minimum or organizations had to give back all of their specific DM fees (McCall et al, 2008). CMS staff inserted this requirement in response to the federal legislation that groups must be able to demonstrate that they can bear "financial risk." After the first 6 months, CMS waived the 5 percent requirement and changed to the budget neutrality standard because DM organizations' initial savings predictions were unrealistically high. The PGP Demonstration, by contrast, has no upfront management fees but gives Medicare the first 2 percent of savings while sharing with providers any additional savings above 2 percent (Sebelius, 2009). The Medicare Hospital Gainsharing and Acute Care Episode (ACE) P4P demonstrations do not invoke budget neutrality per se because no additional payments were made to providers under this demonstration, but OMB did put limits on the amount of profit-sharing that physicians can receive from their hospital partners (CMS, 2006, 2009a).

Because clinicians and provider organizations must apply (or otherwise actively volunteer) to participate in a demonstration while demonstrating budget neutrality at a minimum, demonstrations have a distinct "carrot" bias toward those clinicians and provider organizations who have the necessary resources and believe the proposed changes will favor their organization. Thus, under most conditions, testing provider organizations' and clinicians' behavioral responses to CMS's simply paying less rather than more is not possible. CMS would get few, if any, physician or hospital groups to apply if the intervention were to test responses simply to lower physician conversion factors or DRG payment rates. Nor would provider groups volunteer for a DM demonstration if the intervention simply reduced payments for poor quality of care. A win-win, silver-bullet philosophy has therefore pervaded Medicare's demonstration authority simply as a practical effect of these combined requirements for both budget neutrality and voluntary participation.

Given these inherent limitations, CMS has tended toward payment carrots in demonstration projects in four different ways. First, one of the strongest incentives that the government can offer demonstration applicants is *up-front fees* to cover any administrative costs associated with the intervention. In several CMS DM demonstrations (e.g., the Medicare Coordinated Care Demonstration, MHS Demonstration, and the CMHCB Demonstration), CMS pays up-front monthly management fees to cover extra management resources of commercial vendors and provider groups. Costs are substantial for DM interventions that require sophisticated electronic medical records and support staff staying in close touch with high-risk beneficiaries. On the downside, because OMB generally requires demonstrations to be budget neutral, sites failing to generate Medicare savings are at risk of needing to return all or most of their up-front fees. This process has been contentious, even though applicants sign contracts with the explicit acknowledgment that retaining fees is contingent on savings. Prolonged legal negotiations often ensue, with arguments over technical design and implementation issues.

Second, *shared savings* is another way of encouraging participation, but it entails considerably more financial risk for applicants who have to make initial investments on their own. In the PGP Demonstration, physician groups are encouraged to reduce overall billings (from themselves and other health care providers) on Medicare patients for whom they provide most of that patient's primary care (Sebelius, 2009). In return, they share in resulting program savings. By design, Medicare payments must decline more than the savings bonuses paid out.

Third, CMS also uses nonfinancial carrots along with required savings to attract applicants to some demonstrations. In Medicare's Participating Heart Bypass Center Demonstration, 10 hospitals originally applied, and 7 eventually participated by offering Medicare up-front reductions on DRG payments for bypass and valve surgery (Cromwell et al., 1998). In return, they were given the right to market a form of Centers of Excellence (CoE) imprimatur. A major incentive to participate in CMS demonstrations affecting payments has been competitive pressures at the local market level. If Medicare were to designate one hospital a CoE for cardiovascular or orthopedic care, other local hospitals' volumes for these lucrative services may be threatened.

The fourth reason for offering payment discounts with no financial carrot is *physician gainsharing* in any hospital cost savings, generally disallowed under Medicare rules and regulations. Both the Participating Heart Bypass

Center and the Hospital Gainsharing demonstrations offer this incentive to the clinician staff in order to align their incentives with the bundled payment incentives the hospital faces in caring for Medicare patients (CMS, 2006; Cromwell et al., 1998).

Threats to Evaluation Findings

The previous sections describe how Medicare's demonstration waiver authority faces both political and legal constraints that can limit the ability of CMS and policy makers to implement promising innovative program concepts. Once a demonstration can overcome these hurdles of authority and design limitations, it must then be implemented and evaluated to assess its effectiveness in accomplishing its goals. In evaluating demonstration success, at a minimum policy makers such as members of Congress, DHHS, and other stakeholder groups need to know whether evaluation findings are a valid indicator of an intervention's impact on health care costs, quality of care, or both. If the answer is no, or even maybe, then it would be premature and potentially both financially and politically risky to promote the intervention to a national program. Unfortunately, Medicare demonstrations face a wide range of threats to robust evaluations. These difficulties in fully evaluating demonstration outcomes in turn undermine support for national implementation. Before a demonstrated intervention can be promoted to a national level, it must first be deemed a success, at least according to the available evaluations. Many Medicare P4P projects have been subject to formal evaluations.

Each demonstration defines success differently, but to be successful, demonstration interventions must do the following:

- reduce Medicare costs, holding quality of care constant;
- improve quality of care, holding costs constant; or
- both improve quality of care and reduce costs.

Policy makers often use both actuarial and research evaluation methods to consider demonstration success. Actuarial tests usually focus on a narrower definition of cost savings by determining whether the intervention cost less for enrollees than for a matched control group. Actuaries do not tend to consider broader questions of statistical reliability, and they apply their results only to the performance of demonstration participants (e.g., participating hospitals, DM organizations). Participants failing the actuarial test usually are required to pay back any fees to the government under the budget neutrality clause in their CMS contract.

By contrast, research evaluation tests do consider the statistical reliability of the results, using standard confidence intervals. Evaluators test whether savings are statistically greater than zero. Because of the substantial variance in beneficiary monthly and annual costs, actuarial savings can be 5 percent or more yet not statistically different from zero.

In recommending expansion of an intervention to a national program, CMS relies on the evaluator's findings because the government must be fairly certain that the intervention will succeed in other environments entailing greater overall financial risk. To demonstrate 20 percent savings on just 100 patients in one county in one state would not justify a large national program because these results may not be replicable on a larger scale for a variety of reasons. Success on the national stage can be considered the product of expected savings per beneficiary and the number of beneficiaries enrolled in the national program. Both necessary components can be jeopardized by numerous internal and external threats to the validity of the evaluation inherent in the demonstration's design and implementation.

Quasi-Experimental Design

Rarely can CMS conduct a trial that randomly assigns beneficiaries to intervention and control groups; the MHS pilot, with 240,000 beneficiaries, is a notable exception. Nearly always the demonstration entails a quasi-experimental design with hierarchical, or nested, assignment of beneficiaries. Under these designs, beneficiaries are assigned to evaluation groups by categories and subcategories according to characteristics that are relevant and hypothesized to affect the outcomes of the demonstration. Random assignment, although preferable from a research evaluation standpoint, is either administratively impractical or problematic because, under this approach, some otherwise eligible beneficiaries are excluded from the additional benefits provided under the demonstration; this is often considered politically unpalatable in a public program like Medicare. Even under demonstration provisions, Medicare does not have the authority to limit a beneficiary's freedom to choose a Medicare participating provider. Because most applicants to demonstrations are groups of providers (e.g., hospitals, physician practices) rather than Medicare beneficiaries, random assignment of beneficiaries to intervention and control groups has not been possible. Patients are naturally loyal to their clinicians and providers; Medicare cannot require them to switch to another.

Nor is it acceptable to randomize beneficiaries to an intervention arm within a provider group because of likely spillover effects onto control beneficiaries in the same group. Spillover effects are essentially unintended consequences that may affect behavior of others not directly involved in an intervention. An example of a spillover effect would be a physician's changing the way he or she treats all patients (i.e., those participating in the demonstration and those not participating) as a result of what the physician learns or is exposed to through the demonstration intervention.

The best that demonstration evaluators can do is to match the comparison group as closely as possible with loyal intervention beneficiaries. (See discussion in Chapter 10 on strategies for matching intervention and control groups.) Rigorous matching can filter out most of the threats to the internal validity of an evaluation that are associated with history, regression to the mean, and experimental mortality, but some level of threat remains. The following discussion summarizes the threats that arise from the "loyal patient" structure of most P4P demonstrations.

Willingness to Take Risks

Clinician or provider demonstration applicants, by virtue of applying to a P4P demonstration, are more willing to take risk than those who do not. That quality may stem from their internal culture of innovation and passion for improving the delivery of health services. It may also be a result of their having already invested heavily in the intervention's infrastructure (e.g., medical homes with extensive information technology [IT] medical record systems), their already being efficient and able to offer deeper payment discounts, or their having a specially trained and experienced staff and a charismatic leader familiar with the intervention. Their patients may be particularly healthy (or unhealthy), less costly, and more compliant with the intervention requirements than are those in the general population. Clinicians and provider groups may be particularly good at targeting beneficiaries most in need, or they may be part of a larger network of providers with more control over where patients go for care. Being larger, they can spread intervention fixed costs across more participating beneficiaries than can other practices. Groups that apply may be in particularly competitive markets and seek an advantage from marketing a Medicare imprimatur that may increase their market share. They may work in markets with greater health needs that the intervention addresses, or where patients have greater access to lower-cost alternatives to expensive hospitals.

The demonstration's design can influence an evaluation's conclusion of success. Clinicians and health care organizations may not be responsible for certain services (e.g., post-acute care) that encourage early discharges and lower hospital costs. They may receive considerable government support in terms of claims and administrative data that demonstration sponsors and participants use to help monitor progress and refine the intervention. The very fact that the government is closely monitoring the intervention is likely to redouble practitioners' and providers' efforts and keep members of the group in the intervention. Medicare may have chosen a particularly unusual, high-cost population for the demonstration that will likely regress to the mean during the demonstration period. Even if the comparison group similarly regresses to the mean, the churning of patients will add to the statistical "noise" and reduce the likelihood of significant results. Beneficiary exposure to the intervention may be short, either because the demonstration period is too short to capture longer-term success or because something delays patient recruitment into the intervention. These threats undermine the replicability of demonstration results in a much larger program. The biases that voluntary participation and geography introduce suggest less success per beneficiary for other provider groups elsewhere in the country. They also suggest less interest in other groups that are not in the same position to take advantage of the incentives and support associated with the demonstration.

Given the number and range of demonstrations that Medicare has undertaken, it is puzzling that so few have succeeded and become eligible for expansion to a national program. Although Medicare has tested and implemented a range of demonstrations over its almost 50-year history, very few demonstration projects become incorporated into the national program. Occasionally, failure of demonstrations to become national programs results from a lack of applicants or from early dropouts—not from factors affecting the few that remain.[7] Demonstrations may fail to attract a large enough group of voluntary participants initially, sometimes because of long and complex application and approval processes. Others may begin implementation with a robust group of participants but lose some as the demonstration proceeds as a result of operational difficulties, costs, or other reasons. More often, however,

---

[7] Medicare's Residency Reduction Demonstration in New York had 49 participating hospitals early on but only 6 completed the full 3 years. Five of six completers successfully reduced their resident counts, although they may have intended to do so anyway (Cromwell et al., 2005).

participants are unable to show cost savings, although most show modest gains in quality indicators.

One P4P initiative that did show significant success in cost savings was the Medicare Participating Hospital Heart Bypass Demonstration. The demonstration realized substantial savings from discounts on hospital payments, with some regionalization of surgery performed at greater frequency by providers with lower mortality rates (Sebelius, 2009). Yet, as successful as this initiative was, CMS never pursued a national program. The next two sections consider why successfully demonstrated and evaluated Medicare program innovations do not seem to be extrapolated to national programs, particularly for P4P models.

Operational Challenges to National Implementation

Many significant obstacles impede P4P models' implementation as national programs. Some barriers stem from practical operational problems inherent in the way these promising projects begin as demonstrations or other pilot projects. These challenges arise from many of the models' dependence on achieved savings for financing, their operational complexity, and their high operational and data requirement costs.

Paying for Innovation

Current Medicare policy often focuses on finding ways to improve the program's efficiency and to lower its costs while maintaining or improving quality of care. Therefore, most new initiatives—including P4P—aim at either achieving savings for Medicare or, at a minimum, funding the new quality improvement programs from efficiencies gained (termed *budget neutrality*). The fiscal realities of the Medicare program and the political climate in Congress seem to suggest little interest in a major programmatic change that would significantly increase program costs. The promise of savings and increased efficiency accounts for the appeal of various P4P models under a variety of provisions in Affordable Care Act health care reform legislation passed in early 2010.

The need to operate a successful clinical model that is funded on achieved savings creates a challenging obstacle to both implementation and evaluation of success. Some P4P projects have achieved sufficient efficiencies to cover operational costs and still net additional savings to the Medicare program; most notably, these are the original Participating Heart Bypass Center Demonstration and the top-performing Premier Hospital Quality Incentive

Demonstration. Others, once they factor in operational costs, find net savings difficult to achieve. For example, under the MHS pilot, none of the sites achieved the target 5 percent net savings and hence could not keep the upfront management fees they had already received from Medicare. CMS is still evaluating performance for many of the other current Medicare demonstration projects. (See Chapter 2 for detailed discussions of relevant Medicare demonstration projects and available evaluation findings.)

The practical necessity that demonstrations rely on achieved savings to fund P4P initiatives entails what can amount to significant financial risk to both participating sites and the Medicare program. If these projects do not achieve savings, the Medicare program faces the often difficult task of negotiating close-out of operating sites. Because they can measure savings only retrospectively, lack of achieved savings can also represent potential additional costs to Medicare. For participating sites, focus on achieved savings often means that providers must bear the financial risk of the operating costs of the P4P intervention. For example, clinicians and provider organizations that invest in care models, additional staff, and/or upgraded data collection and health IT systems may or may not receive the expected performance-related payments. All these factors may make P4P models that are funded by achieved savings too risky for some groups and, on a large scale, for the Medicare program.

Start-Up and Implementation Operational Complexity

Many P4P models that include carefully defined performance metrics can entail significant operational complexity, both in the process of designing demonstrations and throughout implementation. Negotiating the specific terms and conditions of the measures, payments, and other operational specifics may be enormously time-consuming and thus expensive both for the Medicare program and for participating clinicians and provider groups. Experience from the Medicare demonstrations suggests that the parties make these decisions based on detailed negotiations that attempt to address very specific facility/practice small-scale concerns. For example, despite initial interest from a range of hospital-based organizations, only a subset of the original applicants to the Medicare Hospital Gainsharing, Medicare Physician–Hospital Collaboration, and ACE demonstrations actually participated in the project because sites found it difficult to reconcile their internal goals for participation with the CMS requirements for savings generated, evaluation reporting, and/or other mandated guidelines. In several recent cases (e.g., Medicare Hospital Gainsharing and ACE demonstrations), years have passed between the selection of potential demonstration participants and the official start of the demonstration projects. It is not unusual for sites to withdraw during this period.

Negotiated issues, such as specifics of payment mechanisms, risk responsibilities, and other terms and conditions, are extremely important to both Medicare and potential sites. Still, the negotiation period is costly for both—a factor that adds to these projects' overall operational complexity. This approach of individually negotiating performance metrics and payment terms may not ever be feasible at a national level from either a timeliness or a cost perspective. Reaching agreement on these important specifics on a national scale would only increase this complexity. The difficulty of gaining agreement on details such as which is the appropriate entity to be monitored and "paid" for performance (e.g., the group practice versus the individual physician) will be magnified at the national level. Geographic differences in practice patterns may also complicate a nationally agreed-upon standard.

Once policy makers and purchasers set the performance standards, payment amounts and conditions, and other operational details, implementation is very data intensive and therefore costly. Who would bear this cost? The Medicare program, which is under persistent pressure to reduce costs? Where would these additional resources come from? Experience from several of the current Medicare demonstrations suggests that reconciliations necessary to finalize payments for each initiative can sometimes be arduous and contentious. Because CMS commonly assesses performance of these initiatives relative to comparison groups, determining whether demonstration participants have achieved performance targets—and, consequently, whether they can be paid—requires a significant amount of data processing and analysis.

Historically, the reconciliations—even for a limited number of demonstration sites—have sometimes taken more than a year following queries and questions on methodology from affected sites. The processing of site-specific reconciliations at the national level would likely be time-consuming and expensive at best—and, at worst, potentially unworkable if actual performance payments lag so far behind interventions as to have little behavior-changing incentive value. Policy makers considering nationalization of similar demonstrations would need to identify a method for streamlining final payment reconciliations that is at the same time clearly tied to individual performance. Thus far, this kind of streamlining has been elusive.

Data Demands

One factor related to the high operational costs of P4P initiatives is their data intensity. The data necessary to set and evaluate standards are usually significant. For example, some demonstrations require additional clinical diagnostic and outcome information beyond what is available on Medicare claims. Other demonstrations require reporting of internal provider micro-cost data, necessary to understand how and where the demonstration achieved savings and whether savings are likely to be generated by other similar health care provider organizations should the demonstration be expanded or nationalized. To the extent that some standards require data that are unavailable from administrative sources (such as Medicare claims), a high degree of variability is likely in terms of either provider ability or willingness to collect and report accurate data. Many P4P models require an analysis of costs and/or other performance metrics for each individual patient followed by an analysis of comparison group patients—a resource-intensive activity.

Although advances in health IT have made the necessary data collection and analysis more feasible than we could have imagined even 10 years ago, these costs for participating providers can be substantial and often difficult to justify in an era of shrinking Medicare and private reimbursement. Initiatives to improve electronic health records and overall health IT systems may make these data requirements more feasible in the future, though currently these costs can create a barrier to participation.

In addition to the necessary collection and analysis of requisite data to measure performance, clinicians and providers must contend with regulatory requirements to protect the privacy of these data. Meeting data privacy and protection requirements, set forth in the Health Insurance Portability and Accountability Act (HIPAA) and subsequent regulatory requirements, increases the complexity and price of collecting much of the data necessary for P4P. Detailed clinical and health status information not available through administrative claims sources is an example of HIPAA sensitive data often necessary for P4P initiatives. Therefore, national implementation of P4P models, which rely on data that exceed typical administrative data collection, will raise the costs of participation for both Medicare and provider organizations and clinicians. As a result, these additional data needs may, as a practical matter, limit participation either to those initiatives with the greatest potential savings or to participating clinicians and provider organizations that can afford the additional expense.

Implications for the Future: Political Challenges to National Implementation

The previous sections focused on the programmatic, evaluation, and other analytic challenges to national implementation of Medicare P4P demonstrations. An additional obstacle further accounts for the dearth of demonstration projects that actually transition to a national program: politics. As noted earlier, Medicare is the largest insurer in the United States and as such has enormous market influence. Changes to the Medicare program have a substantial impact on a large proportion of the US economy, affecting a wide range of direct medical care clinicians and provider organizations, insurers, medical device/supply manufacturers and distributors, the pharmaceutical industry, beneficiaries, and other stakeholder pocketbooks. This makes change a highly visible and potentially politically dangerous activity for those who hold this responsibility. Demonstrations, particularly the majority that are designed with voluntary participation, and ones that are crafted to offer primarily positive rewards and incentives, are much more palatable politically than national implementation that would in many cases remove such impact-limiting features.

In addition, because legislative action determines virtually all central provisions of Medicare program eligibility, program payments, and benefits offered, the authority for significant change rests mostly with elected officials (i.e., Congress) rather than with political or career executive branch staff at DHHS or CMS. This is not to imply that DHHS and CMS staff have no impact or influence on the program; CMS staff are in fact responsible for the myriad of details that govern the program and operationalize day-to-day policy. Still, elected officials with accountability to a wide range of interests and organizations focused on self-interest rather than improved performance of the Medicare program are the ones making major programmatic changes such as national implementation of P4P models. The Affordable Care Act health care reform legislation includes specific language to create a Medicare Center for Innovation within CMS, likely to create a forum for reform more removed from the congressional political arena.

The fact that most major policy changes to Medicare occur through federal legislation significantly hampers significant and innovative change to the program. Recognizing this principle, some early health care reform proposals considered shifting cost-cutting policy implementation to the Medicare Payment Advisory Commission (MedPAC) or other nonelected entities. Other proposals would extend to the DHHS Secretary the authority to expand successful demonstrations on a national level (Weaver & Steadman, 2009). The Affordable Care Act ultimately tasked MedPAC with several studies related to Medicare payment reform. The legislation also tasks the DHHS Secretary and, by extension, the DHHS agencies with literally hundreds of health care reform-related projects aimed at improving quality of care and expanding access, in addition to dozens of P4P-related demonstrations and pilot initiatives. These wide-ranging reform initiatives may introduce more examples of demonstration projects that may improve quality and lower costs at some level. Still, the fragmented nature of this "thousand points of light" approach to policy making may not address the core question: *Why* is it so politically difficult to enact large-scale Medicare policy innovations?

Theoretical Explanations

Various theories of political decision making may hold some answers. One classic theory describes policy making along the lines of a cost/benefit analysis (Wilson, 1973). Proposed policies have certain constituencies or supporters, and these groups can be either distributed (such as the tax-paying public) or concentrated (such as a special interest group). As a balance to the support gained from different types of constituencies, costs associated with certain policies can be borne either by a broad or distributed group (such as a general tax increase) or by a concentrated group (e.g., the cost associated with a regulation on a specific industry). Policies that have both distributed constituencies and costs can succeed through political strategies that advocate majoritarian politics, essentially on the logic that a lot of good can be achieved for a lot of people with only limited costs per person (Wilson, 1986). In contrast, the process of entrepreneurial politics refers to distributed constituencies but concentrated costs. In this case, policies with these characteristics can be advocated by arguing for large benefits to large numbers and with costs borne only by a limited group. A third strategy focuses on policies with concentrated benefits but distributed costs: client politics. Policies with client political strategies can face an uphill battle because they argue for limited benefits for the few and a cost burden on many. Finally, policies with both concentrated costs and benefits are commonly interest-group politics (Wilson, 1986).

The purpose of this theoretical model is to describe the most common and successful ways for politicians to approach prospective policies, weighing both their potential benefits and their real costs. The perceived distributions of costs versus benefits can predict the kind of political coalitions that are likely to form successfully around policies that fall into each category. Policies that can be driven by majoritarian politics will likely have the largest supportive constituencies; those supported by interest group politics have the smallest (Wilson, 1986). Unfortunately, Medicare program policy, particularly any aspect that affects payments, does not fit neatly into Wilson's political constituency model. This may explain partly why policy change and innovation within Medicare are relatively rare: essentially, significant Medicare policy change requires a unique political strategy.

Described within the Wilson framework, additional programmatic costs for Medicare are often widely distributed in that tax revenues frequently finance them. However, specific providers affected by payment changes— particularly those that cut payments and generate any kinds of savings—bear these concentrated costs. When payment changes increase reimbursement, we commonly see disagreement and competition for resources among different provider groups and medical specialties—hence, a lack of consensus on policy direction is the norm.

On the benefit side, policy makers see Medicare beneficiaries as a large and powerful political constituency around which a majoritarian political consensus might form. Current Medicare policy options such as P4P, however, rarely grant additional benefits to large groups without additional costs. Moreover, like provider organizations and clinicians, Medicare beneficiaries rarely speak as a group, leading to lack of agreement concerning the most desired benefits or the appropriate costs to support them. This conflict with theory on building political constituencies to support policy making suggests that significant policy changes to Medicare have difficulty creating viable groups of political supporters. Finding the ideal win-win situation in making major Medicare changes is difficult. Strong and united coalitions fail to form, which results in an absence of innovations in policy making.

Punctuated Equilibrium

Although Wilson's classic theory may explain in part why successful coalitions for major policy change can be difficult to achieve, other political theories may suggest hope for major policy changes within Medicare of the sort that might be suggested from the Medicare P4P demonstration and pilot projects. True and colleagues (2007) describes a policy model of "punctuated equilibrium": long periods of equilibrium, during which small incremental change is the norm. According to this theory, policy stability rather than drastic change typifies American policy making. Instances of major change sometimes disrupt, or punctuate, these periods of equilibrium, however. Punctuated equilibrium theory suggests that, under most circumstances, political discourse that generally reinforces existing policies with only small marginal changes drives stable policies. Wildavsky (1964) has also cited the tendency for policy driven by small incremental change to describe federal budgeting.

True and colleagues (2007) note that although maintenance of the policy status quo and general lack of policy change are the norm, simple observation suggests that in some instances—albeit infrequent—major change does occur. This occurrence is more likely when a particular issue gains increased prominence on the overall political agenda because of political newcomers, a crisis, or both. As media attention or other external pressure raises an issue's visibility, the likelihood of a major change increases significantly.

The actions of these newcomers and the extra attention also tend to remove certain issues from their typical forums for debate, such as within congressional committees. Status quo forums, in which many issues are considered simultaneously, have been described as "parallel processing." When certain issues rise to higher-level political institutions, however, such as the interest of a new president, they move to a policy forum of serial processing by macropolitical institutions (Jones, 1994). It is under these circumstances that major change is most likely (True et al., 2007).

Passage in 2010 of the Affordable Care Act health reform legislation, championed as a key priority of then-popular President Obama, is consistent with this theory (i.e., attention from a political newcomer and the news media or other organizations outside the normal political institutions make change possible—but they cannot guarantee it). Original versions of health care reform supported by the Obama Administration called for more substantial policy changes, including development and implementation of a public health care option. As a compromise to accomplish enactment of some measure of health care reform, more modest initiatives including dozens of P4P-related demonstration initiatives were mandated. Inclusion of these models based on the Medicare demonstrations, referred to generally as models of accountable care organizations, does suggest hope for applications of the lessons learned. Unfortunately, the current debates also underscore the serious difficulties surrounding policy change driven by Congress (the primary political organization responsible for Medicare change).

Political scientists often refer to Congress as "the broken branch" because of persistent shortcomings "in the ethical process, the failure to improve the quality of deliberation in committees, and the many moves to restrict the role of the minority" (Mann et al., 2008, p. x). Essentially, the common view is that Congress, driven by partisan politics and the pressures of a "permanent election," has great difficulty enacting policy of any type, including the annual mandated federal government appropriation bills. It is hardly surprising that any policy making that requires difficult choices for the Medicare program will face great barriers in a largely dysfunctional legislative body. No matter the policy change advocated, it will harm some likely powerful constituency in some way, and major costs at a minimum will be concentrated and sometimes distributed through large increases in taxes. Given that significant Medicare policy changes are often lose-lose rather than win-win, that such changes are infrequent should not be much of a surprise.

To illustrate these political dilemmas, consider a theoretical, modest P4P model that would pay a bonus, on a national level, to provider organizations and clinicians who meet specific improved quality performance metrics. Funding for this bonus would come from an overall lowering of base payment rates for all similar providers. The primary political landmine for elected officials would be the outcome that some clinicians and providers would be paid more and others *less* than the status quo, creating winners but also losers. Because most providers participate in Medicare, it is the largest US insurer, and because these clinicians and provider organizations depend on this steady stream of revenue, this modification would potentially create a large number of losers, who may in turn pressure Congress to hold them harmless to policy change.

Such political pressure may then put a strong emphasis on the use of only carrots, or win-win, P4P scenarios. More politically appealing proposals include the use of lower fee updates or fee freezes (as opposed to actual reductions) and payment for higher quality or process improvements (such as data reporting) but no penalties for relatively inefficient providers. Carrot-only approaches may be feasible on a small demonstration scale. Their potential cost implications for Medicare program spending if they do not achieve (at least) budget neutrality, however, make such methods untenable and impossible to implement.

Using Carrots Rather Than Sticks

One way in which Congress had attempted to shift the burden of politically difficult payment and improved-efficiency models was to rely on incentives for clinicians and providers to make simultaneous price reduction and quality improvement changes themselves, using internal mechanisms. These approaches, including competitive bidding, bundled payment, and the CoE models, give participants some type of reward (e.g., access to bonus payments, competitive advantage in Medicare markets, and/or use of an imprimatur for marketing purposes). Still, even these indirect models can encounter significant problems in building successful political constituencies. Congress overturned competitive bidding for Medicare laboratory services based on lower pricing and minimum quality standards, for example, because of political pressure from the laboratory industry—despite specific authorization initially by legislative mandate. In this case, large national laboratory firms launched a campaign that convinced Congress that any limits on laboratory access would be potentially detrimental to beneficiary choice, and CMS (under significant pressure from Congress) halted the demonstration just as implementation was set to begin in 2009. Similarly, in the mid-1990s, Congress specifically mandated, then canceled, competitive bidding for Medicare managed care after local lawmakers raised objections in multiple designated demonstration sites.

For competitive bidding, selection of starting demonstration sites has invoked strong NIMBY responses from lawmakers, despite their professed support for the general concept of market-driven competition as a mechanism for improved quality and lower cost. Medicare Advantage payments feature competitive bidding. However, given that bidding under Medicare Advantage payment rules is pegged against a known, administratively set benchmark and that final payment rates include *minimum* payment rates, it poses little price-reducing risk to bidding insurers. In this case, Congress is able to take credit for implementing "competitive bidding" though under such constrained regulations that the impact—political or otherwise—is limited.

Although pressure to reduce—or at least not increase—Medicare program expenditures is a constant factor in congressional political deliberation, this pressure clearly is not sufficiently strong to force specific action. Congress, unlike most states and localities, is under no legal obligation to pass fiscally balanced budgets. This situation has allowed Congress annually to overturn requirements to cut Medicare physician payments in adherence with sustainable growth limits. Therefore, although Medicare demonstrations have suggested numerous policy innovations that might cut programmatic costs, Congress has likely little political motivation and certainly no legal requirement to enact them.

Awarding marketing imprimaturs as rewards for quality standards and Medicare savings has also faced political opposition. Follow-ons to the original successful CoE demonstrations for cardiac care encountered political issues from the perceived impact of selective designation of the valuable CoE title. Competing local provider groups and clinicians argued that using this imprimatur in marketing gave awardees an unfair advantage. Designers of the demonstration considered this imprimatur simply an objective assessment of participating groups' outcomes and performance, as well as a reward for giving Medicare discounts. Partly in response to this issue in the earlier project, the current implementation of this model (the ACE Demonstration, implemented in five sites in 2009) is permitted to market itself as a "Value-Based Care Center" instead, a potentially less valuable term than the original CoE label.

Another politically unpalatable feature of P4P models implemented on a national scale may be the additional required administrative and operational costs. Experience from the Medicare demonstrations suggests that terms and conditions of demonstration participation and payment are based on detailed negotiations that attempt to address very specific facility/practice small-scale concerns. Gaining agreement on thorny details such as the appropriate entity to be monitored and "paid" for performance (e.g., small rather large group practices) will be magnified at the national level. Geographic differences in practice patterns may also complicate a nationally agreed-upon standard. Once CMS sets the performance standards, payment amounts and conditions, and other operational details, implementation is very data-intensive and therefore costly. Experience from several of the current Medicare demonstrations suggests that reconciliations necessary at payment points for each initiative can be arduous and contentious. All these challenging aspects provide Congress and other policy makers ready fodder for discussion and study—rather than forward momentum and national implementation.

Summary

Change is complicated for a program like Medicare, which has enormous market power, is a critical source of revenues for most US providers of care, and provides essential benefits to a large and vulnerable beneficiary population. It should not be surprising that members of Congress support concepts such as P4P, but only insofar as the effects do not negatively affect segments of their local constituencies. The political status quo of making incremental rather than major policy change certainly applies to Medicare. As a result, despite the long history of policy experimentation through Medicare demonstrations and pilot tests, few if any of these projects result in national program changes. Such is the case with P4P models.

Whether the current focus on implementation of health care reform, and the legislation's numerous calls for new and expanded P4P demonstrations, can change these political realities remains to be seen. Several of the current Medicare P4P demonstrations are highlighted in health care reform efforts even though many have yet to be evaluated—and not one of the demonstrations has been converted to national implementation. That said, the lessons learned from demonstrations can be a road map to continued health policy reform. The good news from Medicare's extensive demonstration experience in P4P is that the problems and challenges in many of these models are generally well known and, as such, can be addressed and accounted for—*if* the nation sees either a political constituency for real change or a rare confluence of events that opens a policy window enabling real progress to occur.

References

Centers for Medicare & Medicaid Services (CMS). (2006). *DRA 5007 Medicare Hospital Gainsharing Demonstration solicitation.* Retrieved February 24, 2010, from http://www.cms.hhs.gov/DemoProjectsEvalRpts/downloads/DRA5007_Solicitation.pdf

Centers for Medicare & Medicaid Services (CMS). (2009a). *Acute Care Episode Demonstration.* Retrieved February 24, 2010, from http://www.cms.hhs.gov/DemoProjectsEvalRpts/downloads/ACEFactSheet.pdf

Centers for Medicare & Medicaid Services (CMS). (2009b, January 13). *CMS press release: Medicare extends demonstration to improve care of high cost patients and create savings.* Retrieved February 24, 2010, from http://www.cms.hhs.gov/DemoProjectsEvalRpts/downloads/CMHCB_ExtensionPressRelease.pdf.

Cromwell, J., Dayhoff, D. A., McCall, N., Subramanian, S., Freitas, R., Hart, R., et al. (1998). *Medicare Participating Heart Bypass Center Demonstration: Final report.* HCFA Contract No. 500-92-0013. Available from http://www.cms.hhs.gov/reports/downloads/CromwellExecSum.pdf; http://www.cms.hhs.gov/reports/downloads/CromwellVol1.pdf; http://www.cms.hhs.gov/reports/downloads/CromwellVol2.pdf; http://www.cms.hhs.gov/reports/downloads/CromwellVol3.pdf

Cromwell, J., Drozd, E. M., Adamache, W., Maier, J., Mitchell, J. B., Pilkauskas, N., et al. (2005). *Evaluation of the New York State & 1997 Balanced Budget Act (BBA) graduate medical education (GME) demonstration and payment reforms. Final report.* Waltham, MA: RTI International.

Hsiao, W. C., Sapolsky, H. M., Dunn, D. L., & Weiner, S. L. (1986). Lessons of the New Jersey DRG payment system. *Health Affairs (Millwood), 5*(2), 32–45.

Jones, B. D. (1994). *Reconceiving decision-making in democratic politics: Attention, choice, and public policy.* Chicago: University of Chicago Press.

Mann, T. E., Ornstein, N. J., Annenberg Foundation Trust at Sunnylands, & Annenberg Public Policy Center. (2008). *The broken branch: How Congress is failing America and how to get it back on track.* New York: Oxford University Press, p. x.

McCall, N. T., Cromwell, J., Urato, C., & Rabiner, D. (2008). *Evaluation of Phase I of the Medicare Health Support Pilot Program under traditional fee-for-service Medicare: 18-month interim analysis. Report to Congress.* Available from http://www.cms.hhs.gov/reports/downloads/MHS_Second_Report_to_Congress_October_2008.pdf

Nichols, L. M., & Reischauer, R. D. (2000). Who really wants price competition in Medicare managed care? *Health Affairs (Millwood), 19*(5), 30–43.

Sebelius, K. (2009). *Report to Congress: Physician Group Practice Demonstration evaluation report.* Available from http://www.cms.hhs.gov/reports/downloads/RTC_Sebelius_09_2009.pdf

True, J. L., Jones, B. D., & Baumgartner, F. R. (2007). Punctuated-equilibrium theory: Explaining stability and change in public policy making. In P. A. Sabatier (Ed.), *Theories of policy process* (2nd ed., pp. 155–188). Boulder, CO: Westview Press.

US General Accounting Office. (1997). *Communication from William Scanlon, Director, Health Financing and Systems Issues, to the Honorable John B. Breaux, June 12, 1997.* Retrieved February 24, 2010, from http://archive.gao.gov/paprpdf1/158828.pdf

Weaver, C., & Steadman, K. (2009, November 3). *Medicare experiments to curb costs seldom implemented on a broad scale.* Retrieved April 26, 2010, from http://www.kaiserhealthnews.org/Stories/2009/November/03/medicare-pilot-projects.aspx

Wildavsky, A. B. (1964). *The politics of the budgetary process.* Boston: Little, Brown.

Wilson, J. Q. (1973). *Political organizations.* New York: Basic Books.

Wilson, J. Q. (1986). *American government: Institutions and policies.* Lexington, MA: D.C. Heath.

CHAPTER 12

Conclusions: Planning for Second-Generation Pay for Performance

Michael G. Trisolini, Jerry Cromwell, and Gregory C. Pope

As discussed in earlier chapters, pay for performance (P4P) is intended to provide a way of responding to several major deficiencies in the fee-for-service (FFS) reimbursement system that prevails in US health care. Of particular concern are the lack of (1) accountability for the range of different types of care that patients may receive, (2) incentives for coordinating care across clinicians and providers or over time, (3) incentives for improving quality or reducing costs, and (4) incentives for constraining the volume of care. P4P programs attempt to remedy these deficiencies by assigning responsibility for overall quality of care and efficiency results, measuring performance, and rewarding documented improvements. P4P programs can work, in theory, because they closely link financial incentives with measurable performance results.

In practice, unfortunately, in recent years P4P has not lived up to the enthusiasm it initially generated in health policy circles. P4P literature reviews have shown mixed results (Chen et al., 2010; Christianson et al., 2008; Integrated Healthcare Association, 2009; Lindenauer et al., 2007; McDonald et al., 2009). As often occurs with health policy innovations, advocates who embraced the core logic but underestimated the complexities of the health care system in which it had to operate clearly oversold the P4P concept (McDonald et al.). In the national debates over health reform legislation in 2009 and 2010, some politicians touted P4P as a win-win silver bullet for curing the problems of the health care system. Yet if P4P were as easy to implement and as effective as they claimed, it would already have been implemented much more broadly.

The institutional, economic, and clinical complexities of the health care system demand more sophisticated approaches; we term these "second-generation" P4P initiatives. The challenge is to design these second-generation P4P programs in ways that achieve significant improvements in quality and cost outcomes, are acceptable to legislative bodies, and are acceptable to physicians and patients.

This concluding chapter draws on the analysis and lessons from earlier chapters and provides recommendations for improving P4P programs in the future. We first review the main problems with private markets and incentives in health care that motivated the development of P4P programs. We next review the challenges in developing effective P4P programs that led to the major shortcomings of the first generation of P4P programs. A set of policy and implementation recommendations—to improve on current initiatives and develop more effective second-generation P4P programs—follows that discussion. We conclude with an analysis of the P4P provisions in the 2010 health care reform law, the Affordable Care Act, suggesting ways that the Secretary of the Department of Health and Human Services (DHHS) could best implement these provisions. Congress grants the DHHS Secretary fairly wide latitude for implementing the law's P4P provisions.

The Challenges for Private Markets and Financial Incentives in Health Care

When considering ways to design and implement new P4P arrangements in health care, policy makers must take into account the three main reasons that public and private health care payers need to carefully design financial incentives. First, consumers are relatively uninformed about the optimal methods for diagnosing and treating their diseases. Consequently, they must rely on physicians and other clinical experts for advice and decisions regarding their care. This asymmetry in information between patients and clinicians gives the latter opportunities to act as "imperfect agents" and sometimes overprescribe services or provide substandard quality.

Second, insurance coverage, designed to spread risk and improve access by reducing financial barriers to necessary care, desensitizes patients to the cost implications of physicians' testing and treatment recommendations and professional referrals. This means that patients feel free to utilize health care services that may sometimes be unneeded or of marginal benefit.

Third, the health care industry suffers from a serious lack of vertical integration for patient care. Too often, no one person or group is responsible for coordinating all of a patient's care (Guterman & Drake, 2010). Rather, incentives abound to bounce patients among primary care physicians, different types of specialist physicians, clinics, hospitals, and post-acute care. Rapid technical change has exacerbated this problem. Physicians have increasingly turned to providing specialty or subspecialty care, leading to more referrals and greater fragmentation of care. Outdated piecework physician payment systems have perpetuated fragmentation of care and discouraged vertical integration of services, leaving the nation with provider-centered instead of patient-centered medicine. Private and public payers reimburse physicians, hospitals, and other provider organizations according to what they provide and not necessarily according to what patients need. Hereafter we use the term *provider* to encompass all health care professionals and provider organizations.

For these and other reasons, paying for performance, or value, makes sense if it can effectively address these problems. The question is how payers can best design and implement P4P programs to motivate providers and insurers to improve quality and reduce costs.

Challenges in Developing Effective Pay for Performance Programs

A major limitation of P4P is that despite the appeal of its basic logic, it is often difficult to implement well in practice. First-generation P4P programs failed to fully consider the complex nature of the health care sector—with its often multiple layers of policies, institutions, and stakeholders—that can mitigate P4P incentives or redirect them toward unintended consequences.

Herzlinger (2006) reminds us of six broad forces that operate in the health care sector: (1) *players*, who represent the broad range of often-feuding stakeholders in health care, including physicians, hospitals, insurance companies, pharmaceutical companies, government agencies, and others; (2) *funding*, characterized by third-party reimbursement of medical services with little cost sensitivity among consumers and confusing payment arrangements for providers; (3) *policy*, reflecting the broad range of government regulations for ensuring the quality of care, improving access, and containing costs; (4) *technology*, characterized by rapid development, diffusion, and often high costs; (5) *patients*, who increasingly want a more active role in decisions about their health care and who sometimes represent a threat to providers, given the prevalence of malpractice litigation and the often high price of malpractice insurance; and (6) *accountability*, which some stakeholder groups seek as a way to rein in costs and ensure quality. To be more effective, the second generation of P4P programs needs to better account for the complexity these six forces represent.

In this book, we have identified five broad issues for designing and implementing successful P4P programs, which policy makers must address in the context of this complexity. We need to consider all of these issues in new and innovative ways to develop more effective second-generation P4P programs:

1. Whom to Pay

Although P4P can potentially realign incentives in health care to counteract the trend toward increased fragmentation and encourage quality and efficiency improvement, finding provider groups who are willing to take responsibility for all of a patient's care is challenging. Payers have sometimes implemented P4P for solo practitioners or small group providers in such a way that the incentives affect only the small percentage of the patient's overall care provided in their office visits. This arrangement will not mitigate—and may even exacerbate—the fragmentation of the health care system. P4P needs to be better targeted to enable its incentives to motivate more global viewpoints on a patient's overall care, from all providers treating the patient. This issue has posed a particular challenge in rural areas, where small practices may be the only kind that small local populations can economically support and geographic distances make coordination of care more difficult.

Larger organizations can manage a broader range of care, take more financial risk, and make performance measurement more reliable. However, incentives for individual clinicians may be diluted in larger organizations facing only a group incentive.

2. How to Measure Performance

Achieving reliable, valid, and comprehensive measurement of quality and cost performance in a field as complex as health care is challenging. Quality measures depend on broadly accepted clinical guidelines for diagnosis and treatment of individual diseases and for preventive care. However, strong evidence bases for clinical guidelines are often lacking, such as for the increasing numbers of aging patients who suffer from multiple chronic diseases. Moreover, quality measures based on structure (input) or process of care guidelines may have only limited direct influence on outcomes of care such as morbidity, mortality, and quality of life. Focusing quality measurement only on clinical areas that have guidelines available runs the risk of providing disincentives for equal attention to other types of care. Outcomes may be too rare (e.g., mortality) to be useful in evaluating routine quality, and many factors outside the providers' control may influence them.

Cost measures usually need detailed case-mix adjustment or risk adjustment for performance to be measured in ways that are fair to providers who are treating sicker patients. The available risk-adjustment models work better for some diseases than others, and all of the models have some limitations in their ability to explain the statistical variation in health care costs among sicker and less sick patients. In which situations do these models provide statistical adjustments that are sufficient for measures of cost performance that are fair to providers treating different patient populations? The answer is still a topic of debate. Moreover, even with risk adjustment, the underlying (random) variation in medical costs is substantial. Without very large sample sizes (tens of thousands of patients), it may not be possible to reliably distinguish small to moderate cost-control or efficiency gains from the normal variation in costs.

3. How, and How Much, to Pay

Financial incentives that are too small to have a significant impact on provider behavior have been a common problem in P4P programs to date. In part, this stems from the typical P4P model of voluntary programs with only positive incentives (i.e., viewed as experimental, P4P cannot be too aggressive so as to keep volunteer provider organizations from dropping out of the program). Few P4P programs have been mandatory or included negative incentives or penalties, although the Affordable Care Act does include some mandatory up-front reimbursement reductions in its new hospital value-based purchasing initiative. Striking a balance between positive and negative incentives, between incentives that may be too small or too large, and between voluntary and mandatory P4P programs are all ongoing concerns in the design of P4P programs.

Structuring P4P financial incentives to achieve the intended goals while avoiding unintended consequences can also be difficult. Unintended consequences of P4P can be unfortunate: for example, an overemphasis by providers on the types of care measured for incentive payments at the expense of other types, a focus on patients considered likely to be more adherent to prescribed care and thus to boost performance scores at the expense of patients believed to be more difficult to treat, and increased competition among physicians and other health care professionals to earn financial incentives at the expense of clinical teamwork.

4. How to Evaluate Success in Pay for Performance Programs

Because of the diversity of P4P programs, generalizing from the success or failure of individual programs can be tricky. How do we determine whether evaluation results from one P4P program are relevant to programs implemented in other institutional settings, with other types of providers, and with other patient populations?

For large-scale P4P programs, identifying randomized control groups or matched comparison groups to evaluate observed changes in quality and cost outcomes is often difficult. The challenge of controlling all of the variables at work while comparing health sector organizations may always limit the goal of rigorous evaluations. Organizations that are intended to serve as control or comparison groups must provide ongoing care to their patients at the same time and respond over time to their own—often varying—policy and institutional contexts by revising their programs, staffing, care patterns, and technologies.

Balancing quality and cost outcomes is another important issue. Many P4P programs have focused on quality outcomes, some have included performance measures for both quality and cost, and some have required simultaneous improvements in cost and quality outcomes to achieve improvements in value. The Affordable Care Act has taken the last approach, by focusing on value-based purchasing programs. These programs require, for example, mandatory reductions in hospital reimbursement, which can then be earned back through P4P bonus payments linked to performance on quality measures. Determining the appropriate combinations of quality and cost performance incentives is a topic of ongoing debate for design of P4P programs.

5. How to Tailor Pay for Performance Programs to Varying Institutional Settings and Provider Cultures

As discussed in Chapter 2, myriad P4P schemes can emerge from various combinations of the key elements of accountable providers, targeted services, types of care processes and outcomes, performance measures, and bonus payment incentives and methods. Given the lack of compelling evidence for particular approaches, payers have experimented with many different P4P models in different institutional settings and types of provider organizations. For example, P4P incentives often run the risk of being mitigated or misdirected to unintended consequences in situations where there are multiple layers of health system institutions, including payers, managed care organizations, physician-hospital organizations, physician groups, and physician practice settings.

Organizational or regional provider cultures, which range widely on a continuum from competitive and fragmented to collaborative and coordinated, can also influence approaches to the design of P4P programs. For example, at the more fragmented end of the spectrum stands McAllen, Texas, "the country's most expensive place for health care," according to Gawande (2009). "In 2005 and 2006, patients in McAllen received 20 percent more abdominal ultrasounds than in nearby El Paso, 30 percent more bone-density studies, 60 percent more stress tests with echocardiography, ... one-fifth to two-thirds more gallbladder operations, knee replacements, breast biopsies, and bladder scopes [and] two to three times as many pacemakers, implantable defibrillators, cardiac-bypass operations, carotid endarterectomies, and coronary-artery stents." Yet Gawande (2009) found no evidence that physicians in McAllen were trained differently from those in El Paso. In this situation, small P4P incentives would likely have little impact, given the focus of providers on the much larger FFS payment streams.

Gawande (2009) contrasted the situation in McAllen, Texas, with the collaborative behavior evidenced at the Mayo Clinic in Rochester, Minnesota, where physicians are salaried and the focus is more on the needs of the patient, quality of care, and teamwork than on financial goals. Other examples of more collaborative and coordinated care include the Wisconsin Collaborative for Healthcare Quality and Kaiser Permanente (Greer, 2008; Tompkins et al., 1999). Collaborative providers may facilitate second-generation P4P in several ways: (1) by serving as accountable providers representing a broader range of a patient's care, (2) by implementing quality of care incentives that are consistent with the provider culture, and (3) by facilitating linkage of P4P incentives with other types of care coordination and quality improvement programs that may benefit from funding that is available in P4P programs but not under FFS. In this situation, small P4P incentives may have more impact in that they are consistent with the provider culture and complement other quality improvement programs.

Policy Recommendations for Second-Generation Pay for Performance Programs

To better address the five broad issues identified in the last section regarding effectiveness of P4P programs, this section presents 10 policy recommendations to guide development of second-generation P4P. They are intended to guide policy makers and stakeholders in future efforts to make P4P programs more effective for improving quality and efficiency in health care.

1. Make Providers More Accountable for Reducing Fragmentation of Care

To integrate vertically and coordinate the care that patients receive in multiple venues, one provider organization must be accountable. To date, the most notable accountable groups are managed care organizations that are insurer-based, not provider-based. The piecework FFS payment incentives, coupled with rapid technical change, unfortunately reinforce a lack of comprehensive patient care accountability among hospitals and physicians. The lack of accountable provider organizations has contributed to a lack of teamwork in medical care—both between physician groups and institutional caregivers and between physicians and other advanced practice clinicians (e.g., nurse practitioners, nurse anesthetists, psychologists). A failure to have one provider organization take responsibility for the total care of a patient has also reinforced fragmentation of care, costly care that is sometimes of marginal value, and less-than-desirable attention to quality.

Fragmentation of care reflects the prevalence of small, independent physician practices in the US health care system. Overall, about one-third of physicians practice in solo or two-physician practices, 15 percent are in group practices of three to five physicians, and only 6 percent are in practices of 51 or more physicians (Boukus et al., 2009). A one- or two-physician practice cannot reap the scale economies inherent in an efficient division of labor involving nonphysician clinicians and other support personnel, health information technology (IT) systems, and facilities. To be cost-effective, nurses managing chronically ill patients must have sizable numbers of patients to work with. Electronic health record (EHR) systems that integrate information across multiple provider venues are quite costly unless spread across thousands of patients.

We expect that no one model for integrated health care systems will be universal; rather, different models may work better in different regional and organizational contexts. In some areas, the larger and more collaborative systems like the Mayo Clinic, Kaiser Permanente, and Geisinger can form accountable care organizations (ACOs). ACOs are entities designed to take responsibility for all of a patient's care, improve coordination of care and quality, take financial risk, and share in cost savings relative to external benchmarks in other markets. They are large enough in terms of physicians and patients to enable reliable and accurate quality and cost indicators to be calculated, to share data on performance results in ways acceptable to clinicians in their collaborative context, and to have the financial resources to bear significant risk. Many regions, however, in which physician practices are smaller and more fragmented, will need a different approach.

One appeal of the recent policy movement to develop ACOs is that these organizations will simultaneously apply both P4P financial incentives and health care delivery system reforms to reduce fragmentation of care across providers (Devers & Berenson, 2009). The ACO concept has meant different things to different people at times but usually includes several core elements. The first is a focus on the provider delivery system rather than private health insurance companies or public payers. The idea is that physicians or provider organizations manage themselves; "outsiders" are not doing the managing. The second element is the development of positive financial incentives to improve quality and reduce costs across services and providers, moving away from the prevailing FFS incentives for increasing quantity of care. Third, the ACO concept allows organizations to receive financial incentives for improving coordination of care across providers and sites of care, and enables more explicit team-based rewards to be provided using P4P. For example, P4P incentives could be directed toward a diabetes disease management program that requires teamwork among endocrinologists, primary care physicians, clinic nurses, diabetes educators, and home health agency nurses.

The medical home concept is another promising approach to improving provider accountability for patients opting to select a particular physician practice as their medical home. This concept also has varying definitions, but it usually entails patients' choosing personal physicians or medical practices to serve as their "home" for managing and coordinating their full range of care. This arrangement is intended to improve coordination of care across providers, expand access to care, improve care management and quality of care, enhance use of EHRs and other health IT interventions, and enhance reimbursement to support care coordination and other medical home services, including P4P incentives for improving quality and reducing episode costs (Backer, 2009; Barr et al., 2006; Carrier et al., 2009).

Increasing accountability can also be linked to efforts to use P4P incentives to reinforce medical professionalism. That perspective is consistent with taking responsibility for the overall care of a patient. Physicians often view P4P programs in a negative light when implemented by distant, for-profit health insurance companies, but P4P could be viewed by physicians as an ally when implemented by local, physician-led ACOs. For example, P4P can provide additional revenue that gives physicians more time to establish stronger partnerships with patients, promote competent practice based on the best available evidence, improve chronic care management, and improve patient satisfaction (Mechanic, 2008). Supportive administrative systems, health IT systems, and medical culture are also needed to achieve these goals, but the financial incentives are an important foundation for other quality improvement systems. Physicians are actively seeking ways to make primary care more viable and professionally rewarding. Longer patient encounters are more financially viable using teams with nonphysician providers when results achieved by teams yield extra P4P reimbursement. P4P revenue can thus open up ways of practicing that may enable primary care physicians to escape the "tyranny of the visit," which is often their only way to gain adequate reimbursement under FFS (Trisolini et al., 2008).

2. Focus Pay for Performance on Larger Provider Organizations, Not on Individual Physicians

Financial risk is a significant problem related to accountability for patients. Managed care organizations, when constituted as risk-bearing insurance companies, are required to maintain adequate financial reserves to ensure that providers will be paid for services rendered to their members. Some proposed P4P strategies would push P4P incentives down to the individual-physician level. Expecting small physician practices to bear the financial risk of cost performance measures is unreasonable, given small patient populations at the individual-physician level and often wide variation in costs for individual patients. Small patient populations for individual diseases also make reliable quality measurement difficult at the individual-physician level. Multispecialty physician group practices, hospitals, physician-hospital organizations, independent practice associations, integrated delivery systems, and managed care organizations are generally large enough in terms of patient sample sizes to justify P4P payments based on robust quality and cost performance measures.

We believe that P4P for individual physicians would also undermine the current trend toward team-based care that is important to improve coordination and quality of care. Individual incentives can increase competition between providers and promote individualistic gaming behaviors such as hoarding of information and skills. P4P should encourage group rewards that promote teamwork by targeting larger physician groups, hospitals, or ACOs.

P4P programs should include incentives that encourage formation of larger, multispecialty groups that can provide larger patient populations for improved performance measurement and facilitate coordination of care across specialties. Although most physicians in the United States remain in smaller practices focusing on a single specialty, many multispecialty groups also exist around the country. For example, insurers might reduce physician payment updates for solo or small group practitioners or make ACO membership mandatory for providers.

The Medicare Payment Advisory Commission (2009) report outlined several advantages and disadvantages for both voluntary and mandatory ACO approaches. Encouraging voluntary ACOs would require positive financial incentives to entice physicians to join the ACOs, and thus may hamper their ability to enforce significant performance improvements because physicians could always drop out of voluntary ACOs. Payers could take a more forceful approach, however, and deny P4P incentive payments to physicians who are not members of a defined ACO.

Mandatory ACOs would likely run into provider opposition but could be formed on a virtual basis among previously unaffiliated physician groups and hospitals. For instance, where a formal ACO does not exist, Medicare might hold providers located in defined geographic areas or those who admit to the same hospital to a fixed level of expenditures per beneficiary. Medicare might also impose mandatory cuts in reimbursement and allow providers in a defined geographic area to earn back the lost revenue through P4P bonus payments based on quality and efficiency performance. Either approach would put all providers in the area at financial risk for fragmented, uncoordinated care and encourage them to form an ACO to coordinate with one another and negotiate with Medicare. The mandatory approach may be needed in high-cost regions, such as McAllen, Texas, where providers would likely see few financial benefits from joining voluntary ACOs compared with continuing to pursue their existing high levels of FFS revenue.

Virtual ACOs may be necessary to promote broader teamwork incentives in rural areas in which low population densities may not economically support larger physician groups and where physicians may be widely dispersed in solo or small practices. Community-wide incentives are one way to develop P4P programs in such settings, either through ACO programs or other types of regional coalitions that providers may organize. Technical assistance programs could support virtual ACOs to help small physician practices and hospitals in rural areas redesign their IT and clinical systems to improve coordination of care and develop more advanced care management systems.

The Office of the National Coordinator for Health Information Technology has set up a nationwide network of Regional Extension Centers to help small practices adopt EHRs to improve the quality and efficiency of their care. National consultants support the Regional Extension Centers through a Health Information Technology Research Center that develops tools and resources the Regional Extension Centers use in their work with small physician practices. Similar technical assistance efforts could be initiated to support development of ACOs. Staff from provider groups that have experience with improving care in rural areas, such as the Wisconsin Collaborative for Healthcare Quality, could be brought in to provide this technical assistance to support development of ACOs in rural areas.

3. Adopt a More Bundled Unit of Payment for Pay for Performance

Bundled payment methods can complement P4P programs by providing incentives for better management of resources within the package of bundled services (Guterman & Drake, 2010). Since the 1980s, when Medicare adopted per-case prospective payment through diagnosis-related groups (DRGs) for hospital payment, hospital and physician incentives have been misaligned. Hospitals have been at financial risk for excess services provided during an admission, whereas the physicians who order all of the hospital services are paid on a FFS basis.

Medicare's Hospital-Acquired Condition (HAC) and Present on Admission (POA) Reporting program exemplifies the current misalignment: the program penalizes hospitals for a list of complications occurring during a hospitalization, including a foreign body left after surgery and surgical infections. Medicare does not hold at any financial risk the physicians treating the patient—only the hospital. From a payment perspective, it is as if the physicians were not part of the team caring for the patient who experienced the medical errors. If P4P is to be successful, physicians must be considered an integral part of the care process, which means they must share in both the rewards for positive performance and in the penalties for poor care, along with the hospitals and other institutions. This can be accomplished by using a bundled episode payment unit for all physician and hospital inpatient services.

Inside hospitals, the government is now exploring bundling Medicare payment for physician services with payment for hospital services into acute inpatient episodes—usually based on the DRG system. Episode-based P4P programs could expand even further the span of accountability of hospitals and physicians by bundling post-discharge medical services into the inpatient global payment. Broader episode bundling including post-acute care may be slower in developing, however, given its increased complexity in comparison with the simpler episode-grouping approaches that include only inpatient hospital and physician services.

4. Involve Patients, Not Just Providers, in Pay for Performance

A major concern of physicians under P4P programs is that patient adherence to their prescriptions and recommendations for tests, treatments, pharmaceutical regimens, diet, and exercise can affect physicians' measured performance on quality indicators. We view patients as a part of the process of producing health care services, and not merely as passive recipients of care provided by physicians and other health care professionals. As a result, patients also have some accountability for the outcomes that result from health care services.

Under the current system, patient accountability begins when people purchase health insurance. Poor health habits (e.g., smoking) may sometimes result in higher insurance premiums. However, the connection between them is often tenuous. Most workers, regardless of their health status, enjoy uniform commercial health insurance premiums paid on their behalf by their employers. Medicare does not account for patients' health habits or adherence in setting the levels of copayments or deductibles that its beneficiaries must pay. It is only in the individual insurance market where poor health habits can result in higher premiums—if the individual can purchase coverage at all. As a result, physicians who may be at financial risk under P4P for patients' poor health outcomes usually do not believe that patients will have any similar financial incentive for adherence to prescribed care or for improving their lifestyles or health-related habits in response to physicians' recommendations.

In this situation, we recommend that P4P financial incentives be provided to patients as well as to provider organizations. These incentives could be implemented in several ways: through lower copayments, lower deductibles, or rebates from private insurance or Part B Medicare premiums based on quality and cost outcomes. The incentives could be based on performance measures similar to those used for providers.

A related approach would be to provide broader incentives for consumers to seek care from providers found to have lower-cost or higher-quality performance. Some employer-based, private-sector health insurance markets have implemented tiering of providers based on cost and quality, with lower patient cost sharing or premiums when patients choose higher-ranked providers. This approach has not yet been widely tested in public insurance P4P programs, but we view it as a logical extension of such initiatives because patients themselves are a part of the health care production process, and their behavior affects quality and cost outcomes. This is another way to expand P4P into its second generation, with broader packages of interventions that move beyond financial incentives targeted only at physicians and provider organizations, to also include financial incentives for patients.

5. Quality Performance Payments Should Be Self-Financing Under Pay for Performance

Our belief, which the US Congress and the majority of Americans presumably share, is that we spend enough money on health care to insist that it be of high quality. As is well-known, the US per capita spending on health care is much higher than that of any other country; moreover, we have no clear evidence of better population health outcomes as a result (Institute of Medicine Board on Health Care Services, 2007). Thus, it is reasonable to expect P4P financial incentive payments for quality improvement to be self-financing.

Self-financing of better quality can be achieved in P4P programs in at least five different ways; we use examples from Medicare to illustrate these methods. First, Medicare could reduce payments in cases where poor quality or medical errors are documented. The Deficit Reduction Act of 2005 HAC legislation prevents higher DRG payments to hospitals for any of 10 hospital-acquired conditions. The Affordable Care Act expands these DRG payment penalties for selected hospitals with high HAC rates.

Second, quality incentive payments could be contingent on demonstrated cost savings. This is the approach that Medicare's Physician Group Practice Demonstration takes. Participating physician groups receive quality incentive payments in this demonstration only if they demonstrate significant cost savings in the same time period used for quality performance assessment.

Third, a group of providers could be put in direct competition with one another, so that reimbursement for the lower-quality providers could be reduced to fund P4P incentive payments for the higher-quality providers. Medicare's Premier Hospital Quality Improvement Demonstration provides bonus payments for participating hospitals in the top two performance deciles and imposes financial penalties for hospitals in the bottom two performance deciles.

Fourth, reimbursement for all providers in a given region or nationwide could be reduced to fund P4P incentive payments for selected high-quality providers. This is the approach taken in the Affordable Care Act for its Hospital Value-Based Purchasing Program (HVBPP). Payments to hospitals nationwide will be reduced by a set percentage to fund P4P incentive payments for hospitals demonstrating high-quality performance. However, the total amount of the incentive payments cannot exceed the funding generated by the nationwide payment reduction. The Affordable Care Act uses a similar approach for private Medicare health plans. The Affordable Care Act cuts overall payments to these Medicare Advantage plans but specifies payment rate bonuses of 1.5 percent in 2012, 3.0 percent in 2013, and 5 percent in 2014 and beyond for plans with four or more stars on a five-star rating scale for quality. The quality-adjusted payment rate is capped at what the rate would have been under payment methodology before passage of the Affordable Care Act.

Fifth, payers could offer providers the right to market themselves as a payer-designated Center of Excellence if they could demonstrate high levels of high-quality performance, and at the same time offer reimbursement discounts for the right to use the Center of Excellence imprimatur for marketing. Medicare's Participating Heart Bypass Center Demonstration allowed highly qualified major heart hospitals to market a similar imprimatur. Participating hospitals were willing to offer Medicare substantial DRG payment discounts for the privilege.

When self-financing entails payment reductions to fund P4P incentive payments for high-quality performance, mandatory legislation such as the Affordable Care Act is usually required. Medicare's Participating Heart Bypass Center Demonstration was an exception because hospitals valued highly the competitive advantage of marketing a Medicare Center of Excellence imprimatur and thus were willing to volunteer for the demonstration and pay for the marketing benefits through discounted DRG payments. Individual provider-funded P4P programs may remain mostly voluntary because these strategies require providers to take financial risks for the chance of receiving quality and cost performance bonuses.

6. Increase the Size of Financial Risks and Rewards in Pay for Performance

As noted in Chapter 2, most P4P systems began with incentives of limited size. Reasons for these size limits included concerns about the validity and reliability of quality measurement and data collection, controversy created by payment disparities among providers, and provider market power to resist P4P programs. Many P4P systems in the United States provide incentives of less than 5 percent of a provider's total FFS income, percentages often viewed as insufficient to motivate major changes in care practices. In this situation, P4P incentives are usually inadequate to counteract the much larger payments and volume incentives from the prevailing FFS reimbursement system.

P4P incentives will probably need to be increased substantially (i.e., up to 10 percent or more) to have a material impact on provider care patterns, innovation in care systems, and ultimately on quality and cost outcomes. The growing concerns over steadily rising costs of health care generally, and the anticipated impact of the baby boom generation on Medicare, may enable public and private payers to experiment more boldly with P4P incentives. Although the United Kingdom enacted sizable P4P quality incentives, it was in the context of a very different health care system, with a much lower percentage of gross domestic product spent on health care. The UK implemented its P4P program with a goal of increasing incomes for general practitioners, with the understanding that this would increase costs for the National Health Service while also providing incentives for improved quality of care (Roland, 2004).

In addition to increasing the positive incentives in P4P, payers should also expand their use of negative ones such as financial penalties or risks for providers, broader use of which would balance the P4P programs' prevailing focus on positive incentives. Though it is important that providers not view P4P as including only negative incentives—and thus being punitive—including only positive incentives has not motivated significant levels of quality improvement or cost reduction. A likely reason for the limited impact of P4P programs to date is the limited downside risk to provider organizations for poor performance. If they perform poorly and fail to get P4P bonus payments, under most P4P programs they still receive their regular FFS revenue, which dwarfs the dollar amounts of the P4P bonuses. Thus provider groups do not lose much by failing to earn the P4P payments. However, if these penalties put some of the regular FFS revenue at risk, that may motivate much stronger efforts by providers to meet the P4P performance targets.

Using negative incentives would likely mean that P4P programs would need to be mandatory, because voluntary programs may not be able to enforce negative incentives and still convince provider organizations to join. We can view requiring P4P programs to be self-financing as a first step in this direction, such as by requiring mandatory reductions in reimbursement that providers can then earn back through improved quality performance. However, negative incentives could go further to put larger portions of FFS reimbursement at risk.

A historical example of successful financial incentives that put regular FFS revenue at risk, while balancing negative and positive incentives, was Medicare's introduction of the DRG-based Inpatient Prospective Payment System for hospitals in the 1980s. Prospective payment had a potential for negative incentives because hospitals could lose money if costs were higher than the fixed level of reimbursement per admission under the DRG payment schedule. However, positive incentives balanced these negative incentives in that if costs were lower than the fixed level of reimbursement, the hospitals could gain money. Moreover, the Inpatient Prospective Payment System was a mandatory program for most hospitals receiving Medicare reimbursement; thus, they could not opt out or fail to volunteer for this new program. This prompted a wave of innovation in hospitals that reduced lengths of stay for most admissions and shifted to outpatient settings many surgical procedures that had previously been done on an inpatient basis.

A benefit of negative incentives is that they can be structured to avoid upfront investments by Medicare or private payers to fund P4P bonus payment pools. They also do not require self-financing mechanisms, such as "shared savings" models, in which P4P programs must document cost reductions before positive bonus payments can be made for quality improvement.

7. Make Quality Improvement Goals More Ambitious Under Pay for Performance

Applying more ambitious quality measures in P4P programs is another approach for achieving larger improvements in quality. In this book, we have explored several methods for determining quality improvements: (1) setting an individual quality target and paying only if it is achieved; (2) setting a target as the percentage difference the between current and an ideal performance rate; (3) a composite score based on weighted average performance across several different quality indicators; and (4) a composite score based on simultaneous achievement required across several different quality indicators (often referred to as "all-or-nothing" or "all-or-none"). Generally viewed as the most challenging, the all-or-nothing approach is our recommendation for making quality improvement goals more ambitious under P4P (see also Nolan & Berwick, 2006).

It is clear that any composite score that allows above-target performance to offset below-target performance will minimize the provider's financial risk and almost guarantee bonus payments—even without much effort. A better idea is to set a single target for the percentage of a physician group's patients who achieve goals for all of the quality measures for a given disease such as diabetes. Percentage scores on all-or-nothing measures could result in baseline provider performance levels of 30 percent of patients or fewer achieving the goals for all measures. One recent study of 7,333 diabetic patients found that 34 percent received care reflecting all eight process measures studied, whereas only 16 percent achieved targets for all three intermediate outcome measures studied (Shubrook et al., 2010). A P4P program benchmark based on these data could provide the low initial levels of performance to enable more ambitious quality improvement goals to be set. To improve from 30 to 60 percent of patients on these all-or-nothing quality measures taken together, providers would likely have to make sizeable investments in quality improvement interventions, such as point-of-care quality monitoring, improved clinical reporting for feedback to physicians, and active exploration of new patient management methods.

Making quality improvements more difficult to achieve is challenging if P4P incentives—whether positive or negative—are voluntary. Ambitious quality improvement goals have sometimes been difficult for Medicare to negotiate with providers in voluntary P4P programs because providers have been able to opt out. If a P4P program is voluntary, with small positive incentives for improved quality, then most providers may simply choose to remain in the much larger FFS revenue system and bill for services without financial risk. Payers must realize that if they want substantial quality improvements in a voluntary P4P system, they will have to make the rewards greater. They must decide how much it is worth to have, for example, a 10 percentage point increase across a range of quality measures. For providers to voluntarily make the effort for such across-the-board gains will require substantial rewards. Naturally, invoking penalties for failing to achieve a broad 10 percentage point gain cannot be enforced under a voluntary arrangement.

In designing second-generation P4P programs, sponsors could also combine patients across chronic diseases to make quality improvement goals more ambitious. For example, the percentage of diabetics achieving goals for 10 different quality measures could be combined with the percentage of heart failure patients achieving all 10 of their disease-specific process of care and intermediate outcome goals. We would not expect most providers to achieve perfect scores on these ambitious measures, but the idea is to combine a number of desirable indicators to show increased room for improvement compared with reviewing each indicator individually.

Another potential benefit of this all-or-nothing approach could be to achieve a tighter linkage between process of care measures and final outcomes such as reduced morbidity, mortality, and complication rates. Individual process of care quality measures, such as annual HbA1c testing, may have limited direct (short-term) impact on outcomes of care. However, patients who achieve clinical goals for a broader set of 8 or 10 process of care quality measures may be more likely to experience positive outcomes. When process measures are combined with intermediate outcome measures, such as HbA1c levels and blood pressure levels, then the linkage to final outcomes is expected to become even stronger. Studies can be conducted to determine which groups of process and intermediate outcome measures are more closely linked to outcomes. This could also improve the credibility of P4P programs among physicians, who sometimes view individual process of care measures as of limited importance in isolation from other aspects of a patient's care.

Expanding direct use of outcome measures in provider performance assessment is another general approach in improving the quality measures used in P4P programs. Most P4P programs have focused primarily on process measures of quality because these are usually more acceptable to providers. As discussed in Chapter 4, many outcome measures currently available do have several limitations, such as low frequency, long time horizons, and the need for detailed risk adjustment to account for factors outside providers' control that may affect outcomes. However, private-sector P4P programs have applied patient satisfaction outcome measures in P4P; and other types of outcomes, such as complications of chronic diseases and hospital-acquired conditions (HACs), are now being more broadly measured and reported for quality improvement programs and public reporting of quality. If ACOs can be implemented for specified geographic regions, they may provide sufficient numbers of patients to enable broader application of outcome measures as well.

Another potential approach to making process quality targets more ambitious is to pay bonuses for the percentage of recommended care provided. In this approach, no specific target for performance is established, other than that 100 percent of patients should receive recommended care. Thus, even providers who are currently performing well will have an incentive to improve their performance until all of their patients are receiving care recommended by clinical guidelines.

8. Utilize Electronic Health Record Systems to Implement Patient-Specific Quality Targets

Even more sophisticated approaches to quality measurement may be on the horizon if the United States can surmount the hurdle of broad—nationwide—implementation of EHRs in the next several years. In the future, EHRs could apply clinical decision support tools to link an individual patient's own clinical data to the results of available clinical trials and clinical guidelines (Pawlson & Lee, 2010). This could enable quality measures to reflect more patient-specific assessments of appropriate care and not just population averages for people diagnosed with a given disease. For some diabetics, for example, controlling HbA1c to a level below 7 percent may not provide additional clinical benefits, and 8 percent may be sufficient. Evidence-based quality measures are currently limited to targets for one clinical level because of the administrative burden and high cost of designing and implementing measures that consider a patient's other comorbid diseases.

EHRs could make patient-specific quality targets a reality for second-generation P4P programs. Payers could use P4P incentives to encourage providers with EHRs to implement and apply clinical decision support tools for patient-specific quality measures by providing higher levels of bonus payments for these more sophisticated measures.

9. Acknowledge That Clinical Uncertainty Will Limit the Scope of Pay for Performance

Medicine is far more complex than it was 50 years ago, as evidenced by the vast array of drugs, medical specialties, treatments, devices, surgical procedures, and sites of care. At the same time, despite medical advances, uncertainty continues to be a prominent characteristic of medical practice. For example, given a range of presenting symptoms of patients with multiple chronic diseases, what is the right test to do, drug to prescribe, specialist to refer the patient to, or procedure to perform? Will the intervention lead to a successful outcome despite the possibilities of interactions between different diseases and different pharmaceutical treatments?

Payers are now promoting evidence-based medicine to develop clinical diagnosis and treatment guidelines that are grounded in systematic reviews of the available scientific studies of diagnosis methodologies, drugs, and other treatments. At one level, this appeals to physicians' scientific training. Others in the medical profession, however, have complained that it leads to "cookie cutter" medicine—there is no such thing as "the average patient"—and limits their autonomy. Moreover, evidence-based clinical guidelines are not currently available for many of the more complex medical situations that lack detailed scientific research, such as treating elderly patients who have multiple chronic diseases, which may include diabetes, heart failure, coronary artery disease, hypertension, chronic kidney disease, cancer, chronic obstructive pulmonary disease, and others.

Cognitive problems are also more prevalent in the elderly than in other population groups. These deficits can inhibit physicians' efforts to promote patients' adherence to prescribed medications, testing, and other interventions needed to treat multiple chronic diseases. The clinical uncertainty that persists in these and other areas of medicine may limit the ability to accurately measure cost and quality performance of providers and thus also limit the size and scope of P4P programs and the bonuses or penalties that they provide for physicians and provider organizations.

Although we believe that a larger role for P4P is needed in health care reimbursement systems to counterbalance the weaknesses of FFS discussed above, policy makers also need to acknowledge the long-term need for a balance between P4P and FFS. At present the role of FFS is too large, but P4P's role will need to focus on clinical areas where programs can establish broadly accepted guidelines so that they can measure performance with confidence.

10. Acknowledge That Pay for Performance Is Necessary but Not Sufficient for Improving Quality

Some advocates have billed P4P as sufficient on its own to improve quality of care, but we view P4P as necessary but not sufficient for quality improvement. Applying P4P programs where possible can be an improvement over relying solely on FFS reimbursement systems, because P4P can provide the direct incentives for quality improvement that FFS lacks. However, it is doubtful that P4P programs alone can achieve a high-quality and cost-effective health care system. Financial incentives are only one lever and can only go so far in ensuring that the health care system provides the highest-quality care. A better perspective on P4P is that it can promote health care system benefits with incentives that increase pressure for improving quality, provided that other policy, health care delivery system, health IT, and organizational factors are also aligned toward those same goals. We can use this perspective to consider ways to tailor P4P programs to varying institutional settings and provider cultures.

For example, P4P programs could be jointly implemented with other quality-improvement programs such as the Wisconsin Collaborative for Healthcare Quality, which was explicitly designed as a physician-led intervention: it comprises mainly medium and large physician groups but also includes hospitals and health plans (Greer, 2008). The Wisconsin Collaborative for Healthcare Quality pursued goals of reinforcing the medical profession's norms of peer support, sharing treatment ideas and knowledge so that all could improve together, and vesting ownership of quality improvement systems in the medical profession rather than external organizations. Physician participation and leadership in the Collaborative led to "… 1) acceptance of measures as valid indicators of performance; 2) "apples-to-apples" comparisons with colleagues practicing in similar settings for reliable performance benchmarks; and 3) opportunities to meet with peers to share strategies and practices employed by the high-performing organizations" (Greer, 2008). The Wisconsin Collaborative for Healthcare Quality represents one model, collaborative and locally organized, of a package of quality-improvement interventions that could be integrated with second-generation P4P programs. These programs could target the financial incentives to reinforce the goals and culture of the Wisconsin Collaborative for Healthcare Quality, so that together they could provide a more sufficient package of interventions for quality improvement.

Assessment of the Pay for Performance Components of the Affordable Care Act

P4P played a fairly prominent role in the debate over the 2010 health care reform bill (HR 3590; the Affordable Care Act), which President Obama signed into law on March 23, 2010. (The reconciliation bill, HR 4872, subsequently amended the act in minor ways.) Members of Congress often cited P4P as a way to improve on the current FFS system during the debate over this bill. In this section, we provide an assessment of the main P4P provisions of the Affordable Care Act.

In most cases the Affordable Care Act specifies P4P programs in only general terms; as mentioned, it delegates a substantial portion of the design and implementation details to the Secretary of DHHS. As a result, we anticipate that DHHS and, especially, Centers for Medicare & Medicaid Services (CMS) staff will play major roles in developing the final specifications for these P4P programs, in shaping how they are implemented, and in working to make them as effective as possible for improving quality and reducing costs. Our objective in this section is to assess the provisions of that legislation that provided for implementation or planning of P4P programs in light of our 10 policy recommendations presented in the last section.

We discuss six sections of the Affordable Care Act below, in order of their presentation in the legislation. In general, the Affordable Care Act refers to P4P programs using the rubric of "value-based purchasing," indicating that they are pursuing both quality and cost goals simultaneously. However, we can also view these programs as P4P given their focus on measuring performance and providing financial incentives for improved quality and cost performance.

Section 3001: Hospital Value-Based Purchasing Program

The Hospital Value-Based Purchasing Program (HVBPP), which will be implemented in 2012, provides incentive payments to hospitals that meet performance standards for a range of quality measures, to be determined by the Secretary of DHHS. These measures will include, at least, those focused on acute myocardial infarction, heart failure, pneumonia, surgeries, health-care associated infections, and the Hospital Consumer Assessment of Healthcare Providers and Systems patient survey. Outcome measures will be risk adjusted to ensure that hospitals have incentives to treat severely ill patients. Efficiency measures will be added during 2014, including, at least, Medicare spending per beneficiary with case-mix adjustments for age, sex, race, severity of illness, and other factors to be determined.

Congress explicitly linked this P4P effort to public reporting of the performance data. Hospitals must have reported the quality and efficiency measures used to calculate the incentive payments on the CMS Hospital Compare Web site for at least 1 year prior to the beginning of the first P4P performance period.

Reducing routine DRG payments for all hospitals nationwide will fund the HVBPP incentives. CMS will reduce payments for all hospitals by a percentage of the base DRG payment for each discharge. The percentage reduction is 1.0 percent in fiscal year 2013, 1.25 percent in 2014, 1.50 percent in 2015, 1.75 percent in 2016, and 2.0 percent in 2017 and subsequent years.

CMS will make bonus payments to high-performance hospitals by adding a percentage increase to the base operating DRG payment amount for a hospital for each discharge occurring in the fiscal year by a value-based incentive payment amount. However, the total bonus payments must not exceed the dollar amount funded for the HVBPP by the associated reduction in DRG payments to all hospitals.

We believe that it is a positive step that the HVBPP balances negative and positive P4P incentives by reducing all hospital reimbursement and then requiring hospitals to earn back at least a portion of the lost revenue through quality measure performance. The HVBPP is mandatory and thus avoids the problems of voluntary P4P programs, which need to provide extra financial inducements to provider organizations to join the program, and which make negative incentives difficult to implement given that providers can opt out.

However, perhaps because of the HVBPP's mandatory nature, the size of the P4P incentives appears low, starting at just 1 percent and rising only to 2 percent of hospital revenues. The relatively small size of these financial incentives may limit the impact of this program. One way to concentrate the effect of these incentives would be to target the bonus payments to the higher-performing hospitals. In 2008, Medicare paid $110 billion to short-stay hospitals, for which DRG reimbursement mostly applies (CMS, 2009). Therefore, a 1.0 percent reduction in reimbursement under the HVBPP would generate about $1.1 billion that would be available for bonus payments. If the bonus payments could be concentrated on fewer hospitals, such as the top 25 percent, then the payments could represent substantial additional revenue for those hospitals earning the bonuses.

If initial experience with the HVBPP is positive, and as performance measures are refined and extended to more conditions and domains, we recommend that Congress increase the payment amounts that are withheld initially but that hospitals can earn back through the P4P bonus payments. This step would provide both stronger negative incentives and stronger positive incentives for hospitals, both of which may result in larger impacts on quality and cost outcomes. DHHS has the responsibility to design the quality measures and the performance targets to be used, so it also has an opportunity to implement more ambitious quality measures and targets for measuring quality of care performance. Those can also lead to larger impacts on the quality of hospital services.

Section 3006: Plans for a Value-Based Purchasing Program for Skilled Nursing Facilities and Home Health Agencies

Section 3006 requires that the Secretary of DHHS submit to Congress two reports by October 1, 2011, containing plans for value-based purchasing for skilled nursing facilities and home health agencies. This provision provides latitude in designing these programs. It gives DHHS the opportunity to again include more powerful P4P design features, such as larger negative financial incentives, larger potential P4P bonus payments, and more ambitious quality measures and targets.

Section 3007: Value-Based Payment Modifier Under the Physician Fee Schedule

Section 3007 requires CMS to establish a payment modifier to the physician fee schedule, based on performance on quality-of-care measures; it recommends but does not require use of composite measures and indicates that risk adjustment will be needed if outcome measures of quality are used. Cost performance will also be assessed and used in the payment modifier.

Implementation for some physicians will occur in 2015 (those physicians deemed appropriate for earlier implementation of this payment modifier by the Secretary of DHHS) and for all physicians in 2017.

This provision of the health reform law also provides latitude in program design. This is a mandatory P4P program, but the legislation does not specify the size of the incentives. This flexibility will give DHHS and CMS the opportunity to include more powerful design features, such as larger negative and positive financial incentives, larger potential P4P bonus payments, and more ambitious quality measures and targets. The challenge will be in implementing these design features; as noted in our recommendations above, we believe that P4P incentives should not focus on individual physicians, but rather on physician groups or other larger provider organizations. Moreover, this P4P incentive faces the challenge of working with a fee schedule defined by thousands of procedure codes. As a result, we recommend that DHHS and CMS use the latitude the legislation gives them to focus these incentives on larger physician groups and provide a creative set of second-generation P4P design features to make this program more effective.

Section 3008: Payment Adjustment for Conditions Acquired in Hospitals

Section 3008 reduces hospital DRG payments for admissions in which patients experienced HACs to 99 percent of the regular DRG payment the hospital would otherwise have received. However, this provision specifies that the hospitals for which this is applicable include only those in the top quartile of HAC rates, relative to the national average (after risk adjustment). Public reporting of the data is required; HAC results for each hospital are to be publicly reported on the CMS Hospital Compare Web site. Hospitals will have an opportunity to review the data and submit corrections.

Section 3008 provides negative incentives by reducing reimbursement for some hospitals when patients experience HACs. Because it is mandatory, it provides stronger incentives than voluntary P4P programs. However, the size of the incentives appears low, with just a 1 percent reduction in revenue—and limiting even this to just the 25 percent of hospitals with the highest rates of HACs. These features may limit the impact of this program. This is an improvement on the prior HAC legislation from the Deficit Reduction Act, which did not impose penalties but only prevented higher DRG payments that would have resulted solely from HACs, but this new Affordable Care Act approach could be made still stronger.

We recommend that Congress increase the size of the percentage cut in hospital reimbursement for admissions associated with HACs to at least 5 percent and make it applicable to at least 75 percent of hospitals. With this approach, only the best-performing hospitals would escape these negative incentives and the program would likely have a much greater impact.

Section 3021: Establishment of a Center for Medicare and Medicaid Innovation Within the Centers for Medicare & Medicaid Services

The Center for Medicare and Medicaid Innovation (the Innovation Center) is intended to test a broad range of innovative payment and service delivery models to improve quality and reduce costs. The legislation recommends testing 18 specific models, and it provides latitude for DHHS and CMS staff to fund pilot programs to test other models that they deem promising. Of the 18 models recommended, at least 3 are P4P programs:

- varying payment to physicians who order advanced diagnostic imaging services according to appropriateness criteria for ordering these services;
- paying providers for using patient decision-support tools; and
- providing payment incentives for cancer care based on evidence-based guidelines.

The Innovation Center holds promise for testing a broader range of P4P models than have been implemented to date, and it can provide a vehicle for implementing the second-generation P4P design features we have recommended in this chapter. A risk is that the Innovation Center's focus on pilot programs may mean that its programs consist mostly of voluntary participation by provider organizations. This approach (as we have argued above) may limit the size of negative P4P incentives that can be implemented, and it may also limit how ambitious the quality measures and performance targets can be. Some mandatory programs may be needed to test the more aggressive second-generation P4P program designs.

Section 3022: Medicare Shared Savings Program

The Medicare Shared Savings Program (MSSP) promotes the development of ACOs by groups of providers serving Medicare beneficiaries. ACOs that meet quality performance standards will receive P4P payments if they can also demonstrate savings in costs for the Medicare program. ACOs must have at least 5,000 beneficiaries assigned to participate in the ACO program; assignment of patients to an ACO is based on beneficiaries' use of ACO-provided primary care. ACOs will continue to receive regular FFS payment under Medicare Parts A and B, but they will also receive P4P bonus payments based on shared savings if they meet quality performance standards.

The legislation defines Medicare savings as occurring when estimated average per capita Medicare expenditures for beneficiaries assigned to the ACO are at least a prespecified percentage amount below a benchmark spending level. CMS is to estimate this benchmark from 3 years of per-beneficiary expenditures for beneficiaries assigned to the ACO. CMS will then adjust this benchmark for beneficiary characteristics and other factors it deems appropriate; the benchmark is then updated (increased) each year by the projected national per capita absolute expenditure growth rate for Parts A and B services. ACOs can earn through P4P payments a percentage of the difference between the actual per capita expenditures of their patients in a given year and the benchmark level, subject to a requirement that a percentage of the savings goes to Medicare and an overall limit on the shared savings that the ACO can earn.

The MSSP includes several positive features we have recommended for P4P programs. It is self-financing because it requires that cost savings be demonstrated before P4P bonus payments are made (i.e., cost saving incentives are built into the program). CMS calculates cost savings based on all Medicare Parts A and B services, so that ACOs have an incentive to become accountable for the full range of care provided to their patients. The MSSP also allows for a range of different types of quality measures, so the P4P performance measures could include some of the more ambitious "all or nothing" quality indicators recommended earlier in this chapter.

One concern is that the use of a national absolute benchmark update, coupled with the voluntary nature of the MSSP, may inhibit participation by provider groups in regions of the country with higher baseline costs and costs that are growing faster than the national average. In those regions, provider groups may consider that demonstrating Medicare savings against a national absolute cost-increase benchmark update is too challenging; as a result, they may be hesitant to join this ACO program voluntarily. For example, McAllen, Texas, may be considered a region in which ACOs could provide significant benefits by containing or even reducing the high rates of FFS reimbursement. It is unlikely, however, that providers in that city would join an ACO voluntarily while they are doing so well financially under FFS.

After initial experience with the ACO program, we recommend that Congress consider a mandatory ACO program, as discussed earlier in this chapter, to enable more ambitious P4P incentives to be implemented more broadly across the country. A mandatory program would also allow the imposition of penalties (negative incentives).

Summing Up

This concluding chapter has identified numerous ways in which second-generation P4P programs could improve and expand P4P. The first generation enabled exploration of the P4P approach and began to shift incentives away from the FFS focus on increasing volume of services to a new focus on quality improvement and cost containment. However, the first-generation P4P programs did not fully account for the complexities of the US health care system, so its impact did not live up to the initial enthusiasm it had generated in health policy circles.

Second-generation P4P programs will offer a broader range of design features to address the fragmentation of care, the institutional complexities of the health care sector, and the need for stronger incentives. They will integrate P4P with complementary quality improvement and cost containment interventions, harmonize its focus with the norms of medical professionalism, and involve both patients and providers. They will also include more ambitious performance measures, balance negative financial incentives with positive ones, and explore how mandatory programs can be implemented to enable more significant improvements in quality and cost outcomes. In these and other ways, second-generation P4P programs will be better designed to achieve larger impacts in the health care sector.

References

Backer, L. A. (2009). Building the case for the patient-centered medical home. *Family Practice Management, 16*(1), 14-18.

Barr, M., American College of Physicians, et al. (2006). *The advanced medical home: A patient-centered, physician-guided model of health care.* Washington, DC: American College of Physicians.

Boukus, E., Cassil, A., & O'Malley, A. (2009). *A snapshot of US physicians: Key findings from the 2008 Health Tracking Study* (Data Bulletin No. 35). Washington, DC: Center for Studying Health System Change. Available from http://www.hschange.com/CONTENT/1078/.

Carrier, E., Gourevitch, M. N., & Shah N. R. (2009). Medical homes: Challenges in translating theory into practice. *Medical Care, 47*(7): 714-22.

Centers for Medicare & Medicaid Services (CMS). (2009). *Medicare & Medicaid statistical supplement*, 2009 Edition. Available from http://www.cms.gov/MedicareMedicaidStatSupp/10_2009.asp

Chen, J. Y., Tian, H., Taira Juarez, D., Hodges, K. A., Jr., Brand, J. C., Chung, R. S., et al. (2010). The effect of a PPO pay-for-performance program on patients with diabetes. *American Journal of Managed Care, 16*(1), e11–e19.

Christianson, J. B., Leatherman, S., & Sutherland, K. (2008). Lessons from evaluations of purchaser pay-for-performance programs: A review of the evidence. *Medical Care Research and Review, 65*(6 Suppl), 5S–35S.

Devers, K., & Berenson, R. (2009). *Can accountable care organizations improve the value of health care by solving the cost and quality quandaries?* Washington, DC: The Urban Institute. Available from the Robert Wood Johnson Foundation Web site: http://rwjf.org/qualityequality/product.jsp?id=50609

Gawande, A. (2009, June 1). Annals of Medicine: The cost conundrum. The New Yorker, p. 36. Retrieved from http://www.newyorker.com/reporting/2009/06/01/090601fa_fact_gawande

Greer, A. (2008). *Embracing accountability: Physician leadership, public reporting, and teamwork in the Wisconsin Collaborative for Healthcare Quality.* Available from The Commonwealth Fund Web site: http://www.commonwealthfund.org/Content/Publications/Fund-Reports/2008/Jun/Embracing-Accountability--Physician-Leadership--Public-Reporting--and-Teamwork-in-the-Wisconsin-Coll.aspx

Guterman, S., & Drake, H. (2010). *Developing innovative payment approaches: Finding the path to high performance.* Available from The Commonwealth Fund Web site: http://www.commonwealthfund.org/Content/Publications/Issue-Briefs/2010/Jun/Developing-Innovative-Payment-Approaches.aspx

Herzlinger, R. (2006). Why innovation in health care is so hard. *Harvard Business Review, 84*(5), 58–66.

Institute of Medicine, Board on Health Care Services (2007). *Rewarding provider performance: Aligning incentives in Medicare.* Washington, DC: National Academies Press.

Integrated Healthcare Association (2009). *The California Pay for Performance Program, the second chapter: Measurement years 2006–2009.* Available from http://www.iha.org/pdfs_documents/home/FINAL%20White%20Paper%20June%202009.pdf

Lindenauer, P. K., Remus, D., Roman, S., Rothberg, M. B., Benjamin, E. M., Ma, A., et al. (2007). Public reporting and pay for performance in hospital quality improvement. *New England Journal of Medicine, 356*(5), 486–496.

McDonald, R., White, J., & Marmor, T. R. (2009). Paying for performance in primary medical care: Learning about and learning from "success" and "failure" in England and California. *Journal of Health Politics, Policy and Law, 34*(5), 747–776.

Mechanic, D. (2008). Rethinking medical professionalism: The role of information technology and practice innovations. *Milbank Quarterly, 86*(2), 327–358.

Medicare Payment Advisory Commission (2009). *Report to the Congress: Improving incentives in the Medicare program.* Available from http://www.medpac.gov/documents/Jun09_EntireReport.pdf

Nolan, T. & Berwick, D. M. (2006). All-or-none measurement raises the bar on performance. JAMA, 295(10), 1168–1170.

Pawlson, L. G., & Lee, T. H., Jr. (2010). Clinical guidelines and performance measures. *American Journal of Managed Care, 16*(1), 16–17.

Roland, M. (2004). Linking physicians' pay to quality of care—a major experiment in the United Kingdom. *New England Journal of Medicine, 351*(14), 1448–1454.

Shubrook, J. H., Jr., Snow, R. J., McGill, S. L., & Brannan, G. D. (2010). "All-or-none" (bundled) process and outcome indicators of diabetes care. *American Journal of Managed Care, 16*(1), 25–32.

Tompkins, C. P., Bhalotra, S., Trisolini, M., Wallack, S. S., Rasgon, S., & Yeoh, H. (1999). Applying disease management strategies to Medicare. *Milbank Quarterly, 77*(4), 461–484.

Trisolini, M., Aggarwal, J., Leung, M. Y., Pope, G., & Kautter, J. (2008). The Medicare Physician Group Practice Demonstration: Lessons learned on improving quality and efficiency in health care. *Commonwealth Fund, 84*, 1–46.

Index

A

accountability, 151, 343, 347–350, 353
accountable care organizations (ACOs), 348–349, 351, 366–367
accreditation, 66–67
ACE Demonstration. *see* Medicare Acute Care Episode (ACE) Demonstration
ACOs. *see* accountable care organizations
ACSCs. *see* ambulatory care sensitive conditions
active engagement, 233
activities of daily living (ADLs), 112, 115
Acute Care Episode (ACE) Demonstration. *see* Medicare Acute Care Episode (ACE) Demonstration
administrative claims data, 112, 118–119
administrative efficiency and compliance, as performance indicator, 36
Affordable Care Act (Patient Protection and Affordable Care Act)
    about, 1–2, 7, 22, 152
    Hospital Value-Based Purchasing Program, 14
    incentive payments, 354–355
    Medicare demonstration projects, 318, 332
    pay for performance components of, 362–368
    Section 3001: Hospital Value-Based Purchasing Program (HVBPP), 14, 26, 363–364
    Section 3006: value-based purchasing program for skilled nursing facilities and home health agencies, 364
    Section 3007: value-based payment modifier under physician fee schedule, 364–365
    Section 3008: payment adjustment for hospital-acquired conditions, 365–366
    Section 3021: Center for Innovation, 366
    Section 3022: Medicare Shared Savings Program (MSSP), 366–368
    value-based purchasing, 7, 346, 362
Agency for Healthcare Research and Quality (AHRQ), 10, 112, 114
agency theory, 79
AHCPR. *see* US Agency for Health Care Policy and Research
AHRQ. *see* Agency for Healthcare Research and Quality
allocative efficiency, 142
all-or-nothing payment algorithm, 206
ALOS. *see* average length of stay
ambulatory care, structure measures and, 104
ambulatory care sensitive conditions (ACSCs), 124
AQA Alliance, 150
assignment of patients. *see* patient assignment
asymmetry of information, 79
attribution
    exclusion of patients from, 189
    importance of, 181–182
    to individual physicians vs. physician organizations, 188–189
    patient attribution, 181–199
    of responsibility, 164–166, 197
attribution rules
    "default" rule, 195
    "first contact" rule, 192
    majority rule, 193, 194
    minimum-share rule, 193, 194
    multiple physician assignment rule, 194
    nonexclusive assignment algorithm, 193
    one-touch rule, 41, 192, 193
    plurality rule, 192, 194
    share rules, 193
    simulations of alternative rules, 195
    type of physician and, 197
automatic assignment of patients, 56
average length of stay (ALOS), 151

B

Balanced Budget Act of 1997, 227
benchmarks, 46–48, 49, 206, 208
blood pressure, 110, 111, 125
Blue Cross Blue Shield of Hawaii, 16–17, 22

Blue Cross Blue Shield of Michigan Rewarding Results, 16–17, 22, 37, 50–51, 169
bonus pool, 50
bonuses, 44, 50, 51, 127. *see also* financial incentives
breast cancer patients, 34, 107
Bridges to Excellence, 16–17, 22, 105, 148
British National Health Service, 17, 20–21, 25, 106, 114. *see also* United Kingdom
budget neutrality, 322–324, 328
bundled episodes, 169, 170, 352
bundled payment, 24, 95, 352

C

California Cooperative Healthcare Reporting Initiative, 186
California Physician Performance Initiative (CPPI), 185–186
CAMC. *see* Charleston Area Medical Center
capitation, 7, 9, 10, 167
cardiovascular disease patients
 annual tests for, 13
 episode of care, 167
 heart failure, 123–124, 151
 mortality report cards for, 125
 myocardial infarction, quality measures for, 123, 125, 151
 outcome measures and, 109
 process measures and, 110
Care Management for High-Cost Beneficiaries Demonstration (Medicare). *see* Medicare Care Management for High-Cost Beneficiaries Demonstration
care management organizations, 176–177, 222–223, 226–240
Centers for Medicare & Medicaid Services (CMS)
 Cancer Prevention and Treatment Demonstration for Ethnic and Racial Minorities, 20–21, 24, 63, 224
 Innovation Center, 152, 332, 366
 Inpatient Prospective Payment System rule, 167
 political challenges to national implementation of Medicare demonstration projects, 332–338
Centers of Excellence (CoE) model
 Expanded Medicare Heart and Orthopedics Centers of Excellence Demonstration, 224–225, 249–251
 as incentive, 57–58, 95–96, 174, 355

Medicare Acute Care Episode (ACE) Demonstration, 18, 24, 226–227, 251–252
 overview, 24, 58–59, 247–248
 validity of findings, 290
change management, 91–93
Charleston Area Medical Center (CAMC), 258, 308–311
claims data, 43, 101, 112, 118–119
clinical outcomes, 34, 110–113, 165–166. *see also* health care outcomes; outcome measures
clinical process quality, 34–35, 37
clinical process-of-care guidelines, 46
clinical providers, rewarding, 162–164
clinical quality-of-care domain, 43
clinical uncertainty, 360
clinicians, performance measures and, 40–42
CMHCB (Care Management for High-Cost Beneficiaries Demonstration). *see* Medicare Care Management for High-Cost Beneficiaries Demonstration
CMS. *see* Centers for Medicare & Medicaid Services
coding
 CPT-II system, 126
 diagnosis code data, 118
 POA codes, 109, 119
CoE model. *see* Centers of Excellence (CoE) model
cohort-based longitudinal patient-level indicators, 144
collaborative providers, 347
comparison groups, 47, 277, 292–293
competition, 65, 68, 90, 290, 337, 350
competitive bidding, for Medicare demonstration projects, 337
composite measures, 121–122
composite payment algorithm, 206, 207
computerized physician order entry (CPOE), 102–103, 105
constrained payment algorithm, 206
Consumer Assessment of Healthcare Providers and Systems, 114
contingency theory, 93–96
continuous quality improvement (CQI) programs, 131
continuous rewards, 49
continuous unconstrained payment algorithm, 206
Coordinated Care Demonstration (Medicare). *see* Medicare Coordinated Care Demonstration

coordination of care, 92
cost containment, 9, 123–124
cost efficiency
    as performance indicator, 35–36, 141–142
    of physicians, 195
cost measures, 344–345
cost of care
    changes in how success is measured, 274
    counterfactuals, 276–277
    as performance indicator, 35–36, 150
cost utilization measures, pay for performance, 15
cost-effectiveness
    as performance indicator, 36, 142–143
    of performance indicators, 39–40, 61–62
counterfactual, determining, 275–281
covariance regression, 272
CPOE. *see* computerized physician order entry
CPPI. *see* California Physician Performance Initiative
CPT-II system, 126
CQI programs. *see* continuous quality improvement (CQI) programs
critical care specialists, 105
Current Procedural Terminology (CPT-II), 112

D

data collection for quality measures
    administrative claims data, 112, 118–119
    efficiency of, 39
    health-information exchanges (HIEs), 116
    medical records, 43, 101, 115–116
    patient surveys, 44, 110, 116–118
    process measures, 107
data demands of Medicare demonstration projects, 331
data envelopment analysis (DEA), 144
decubitus ulcers, 112
"default" rule, 195
Deficit Reduction Act of 2005 (DRA), 252, 354, 365
demonstration payment waiver authority, 318
demonstration projects. *see* Medicare demonstration projects
deprofessionalization, 84, 85
design of P4P systems, 33–70, 163–179, 270n2, 325–326
diabetes patients
    annual tests for, 13, 15–20, 43, 49, 100, 110, 127
    evidence-based clinical guidelines for, 360
    HbA1c test for, 110, 111, 125, 289

    HEDIS quality measures for, 121
    incentives to patients, 68
    neuropathy in, 111
    per-member bonuses, 51
    quality improvement goals for, 357
    reducing complications, 123
    teamwork in treating, 95, 349
    United Kingdom P4P organizations and, 106
diagnosis code data, 118
diagnosis-related groups (DRGs), 9, 352
Dialysis Facility Compare (Medicare Web site), 125, 131
differential cost sharing, 58
differential growth rate method, 280
differential mortality, 273–274, 302–303, 306–307
differential patient selection, 269, 300, 302–303, 306, 308–309
differential payment update, 51
"difficult" patients, 62
disability scales, 115
disease management organizations
    differential patient selection, 269
    P4P programs overview, 42, 176–177, 222–223, 226–240
    validity of results, 274
disease reporting, inconsistencies in, 275
disincentives (negative incentives), 50, 55, 80–81, 356–357
Donabedian's model, 100, 102–103
DRA. *see* Deficit Reduction Act of 2005
DRGs. *see* diagnosis-related groups

E

economic agency theory, 88
economics, of health care, 79–82
efficiency
    allocative efficiency, 142
    defining, 141–143
    evaluation of efficiency measurement, 148–149
    hospital efficiency measurement, 144–145, 149, 151
    measurement of, 143–149
    motivation for inclusion in P4P, 140–141, 328
    physician efficiency measurement, 145–147, 150, 151
    productive efficiency, 142
    risk adjustment, 149–150
    technical efficiency, 142

efficiency index, 145
efficiency of care, 150
efficiency-based programs, 4
EHRs. *see* electronic medical records/electronic health records (EMRs/EHRs)
elderly patients, cognitive problems of, 360
electronic medical records/electronic health records (EMRs/EHRs), 43–44, 101, 105, 116, 351, 359–360
eligibility, Medicare demonstration projects, 320
EMRs. *see* electronic medical records/electronic health records (EMRs/EHRs)
enrollment data, 118
episodes of care
   about, 95
   annual episodes, 167
   attribution of responsibility, 197
   bundled episodes, 169, 170, 352
   defined, 167–168
   episode length, 168–170
   fixed-length episodes, 167
   health care outputs and, 141
   as hospital efficiency measure, 144
   longer episodes, 168–169
   patient vs. episode assignment, 189
   as physician efficiency measure, 146, 151
   post-discharge episode, 168
   shorter episodes, 169–170
   variable-length episodes, 168
evidence-based medical practice
   adherence to, 36
   health outcomes and, 203
   hospital referrals, 105
ex post regression matching, 267, 281–287
Excellus/Rochester (New York) Individual Practice Association Rewarding Results demonstration project, 50, 169
Expanded Medicare Heart and Orthopedics Centers of Excellence Demonstration, 224–225, 249–251
extended hospital medical staff, 188–189

F

fee schedule adjustment, 50–51
fee-for-service (FFS) systems
   capitation as alternative to, 9, 10, 65–66
   health care quality, 12, 65, 92
   patient attribution in, 182, 183, 191–197
   pay for performance systems contrasted with, 7–8, 33, 65–66
   piece-rate compensation, 80
   preventive care and, 9
   value-based purchasing contrasted with, 11
FFS payment systems. *see* fee-for-service (FFS) systems
FIM. *see* Functional Independence Measure
final outcomes, 112–113
financial incentives
   Affordable Care Act value-based payment modifier under physician fee schedule, 364–365
   benchmarks for, 45–49
   bonuses, 44, 50, 51, 127
   for collaboration, 95
   combined benchmark approaches, 48
   comparison of programs, 16–17
   continuous rewards, 49
   differential cost sharing, 58
   differential payment update, 51
   differential premiums, 58
   fee schedule adjustment, 50–51
   funding of, 45, 124
   gainsharing, 174, 255–259, 308–312, 323
   graduated or tiered rewards, 49
   to groups of physicians, 172
   to health maintenance organizations (HMOs) and other managed care organizations, 163, 175–176
   health outcomes and, 203
   implementing, 50–52
   for improved performance, 46, 80
   to individual physicians, 171–173
   to integrated delivery systems (IDSs), 174–175
   lack of payment for poor performance, 51–52
   limitations of, 80–82
   magnitude of incentives, 52–54
   need to make larger, 355–357
   negative incentives (disincentives), 50, 55, 80–81, 356–357
   for participation or reporting, 51
   to patients, 353
   for patients to use high-performing providers, 57, 58–59, 68–69
   payment frequency, 56
   penalties, 50
   per-member payment, 51
   to physician-hospital organizations (PHOs), 174–175
   provider tiering, 58
   for provision of a service, 51
   quality grants or loans, 52
   quality measures and, 120

for relative performance, 46–48, 206, 208
risk-taking in, 54–56, 62–63, 326–328
shared savings, 52, 323
simulations of, 210–216
single vs. multiple reward pools, 52
size of, 345
for target attainment, 46
unintended consequences of, 62–64, 70
in United Kingdom, 25, 53
up-front fees, 323
who gets the payment, 162–179, 344
"first contact" rule, 192
frontier modeling, 144
Functional Independence Measure (FIM), 112
functional outcomes, 112–113
funding
of financial incentives, 45, 124
of second-generation P4P programs, 343

G

gainsharing, 174, 255–259, 308–312, 323
gatekeeper, 182
generated savings, 45
graduated or tiered rewards, 49
group practices. *see* physician group practices (PGPs)

H

HAC (Hospital-Acquired Condition) program (Medicare). *see* Medicare Hospital-Acquired Condition (HAC) program
HACs. *see* hospital-acquired conditions
Hawaii Medical Service Association, 16–17, 22, 105, 114
Hawthorne effect, 109, 286, 292
HbA1c test, 110, 111, 125, 289
health care
cost of services, 140
economics of, 79–82
fragmentation of, 348
institutional layers, 89–90
IOM goals for improving health care, 102
organization theory and, 88–93
organizational culture, 90–91
per capita health care expenditures in U.S., 140
psychology of, 85–88
sociology of, 82–85
health care costs. *see also* cost containment
economics of, 79–82
efficiency and, 140, 141
health care outcomes and, 12
per capita spending in U.S., 140

health care organizations. *see* providers
health care outcomes. *see also* outcome measures
as compared with other countries, 11–12
and health care costs, 12
pay for performance, 13, 15, 34, 203
health care quality. *see also* health care outcomes; performance indicators; performance measures
access and availability of care, 35
accreditation, 66–67
adherence to evidence-based medical practice, 36
administrative efficiency and compliance, 36
adoption of information technology, 36, 37
clinical outcomes, 34
cost efficiency, 35–36, 141
cost of care, 35–36
cost-effectiveness, 36, 142–143
efficiency and, 141–143
fee for service, 12
history of, 10–11
participation in performance-enhancing activities, 37
patient experience or satisfaction, 35, 37, 44, 113–114
patient safety, 35
pay for performance, 12
process measures, 13, 34–35, 39, 61, 100–101, 104, 107–108
productivity, 36
professionalism and, 66, 84, 85, 94–95, 349
public reporting of quality measures, 26–27, 57, 68, 131–132, 151
quality assurance, 12–13
quality improvement, 91–93
quality regulation, 66–67
reporting of performance indicators, 37
service quality, 35
health economics, 79–83
health insurance plans
bonus pools in, 50
differential premiums as patient incentives, 58
performance measures and, 42
provider network designation, 59
Health Insurance Portability and Accountability Act (HIPAA), 331
health maintenance organizations (HMOs), 59, 175
Health Outcomes Survey, 114

376   Index

health policy, history, 8–12
Health Support Pilot Program (Medicare). *see* Medicare Health Support Pilot Program
Healthcare Effectiveness Data and Information Set (HEDIS), 104, 121, 126, 131, 163
health-information exchanges (HIEs), 116
Heart Bypass Center Demonstration (Medicare). *see* Medicare Participating Heart Bypass Center Demonstration
heart disease patients. *see* cardiovascular disease patients
heart failure patients, 123–124, 151. *see also* cardiovascular disease patients
HEDIS. *see* Healthcare Effectiveness Data and Information Set
hemoglobin HbA1c test, 110, 111, 125, 289
HIEs. *see* health-information exchanges
high blood pressure patients, annual tests for, 13
high-performing providers
  financial incentives for patients to use, 57, 58–59, 68–69
  identifying, 126–130
HIPAA. *see* Health Insurance Portability and Accountability Act
HMOs. *see* health maintenance organizations
home health agencies, Affordable Care Act plans for a value-based purchasing program for, 364
Hospital Compare (Medicare Web site), 13, 26, 119, 131, 151
Hospital Gainsharing Demonstration (Medicare). *see* Medicare Hospital Gainsharing Demonstration
Hospital Value-Based Purchasing Program (HVBPP), 14, 26, 363–364
Hospital-Acquired Condition (HAC) program (Medicare). *see* Medicare Hospital-Acquired Condition (HAC) program
hospital-acquired conditions (HACs), payment adjustment for, 365–366
hospital-focused pay for performance, 224–225, 247–259
hospitals
  average length of stay (ALOS), 151
  efficiency measurement for, 144–145, 149, 151
  evidence-based hospital referrals, 105
  extended hospital medical staff, 188–189
  incentive payments to, 162–163
  intensive care unit staffing, 105
  P4Ps, 23, 88, 174, 223
  performance payments to, 174

physician-hospital organizations (PHOs), 174–175
post-discharge episodes, 168
rehospitalization, 168, 169
HQID. *see* Medicare Premier Hospital Quality Incentive Demonstration
HVBPP. *see* Hospital Value-Based Purchasing Program
hypertension patients, process measures and, 110

I

IADLs. *see* instrumental activities of daily living
ICER. *see* incremental cost-effectiveness ratio
IDSs. *see* integrated delivery systems
IHA. *see* Integrated Healthcare Association
IHIE. *see* Indiana Health Information Exchange Demonstration
improvement
  improvement-over-time targets, 128–129
  P4P not sufficient for improving quality, 361–362
  rate of improvement payment algorithms, 46–48, 206, 208
  room for improvement in performance, 123
  statistical analysis of, 130
improvement rate simulations, 216
improvement-over-time targets, 128–129
incentives. *see also* bonuses; financial incentives; nonfinancial incentives; performance payments
  for collaboration, 95
  gainsharing, 174, 255–259, 308–312, 323
  negative incentives (disincentives), 50, 55, 80–81, 356–357
  patient assignment and, 182
  for patients to use high-performing providers, 57–59, 68–69
  risk-taking in, 54–56, 62–63, 326–328
  unintended consequences of, 62–64, 70
incremental cost-effectiveness ratio (ICER), 142–143
Indiana Health Information Exchange Demonstration (IHIE), 297–301
indicators of performance. *see* performance indicators
indirect providers, performance payments to, 163
individual service (unit), 167
information, asymmetry of, 79
information technology (IT), as performance indicator, 36, 37

infrastructure sunk costs, 291
inpatient prospective payment system (IPPS), 247
Institute of Medicine (IOM), 100–103, 121
institutional providers, performance measures and, 40
instrumental activities of daily living (IADLs), 112, 115
instrumentation problems, as validity of findings, 274–275, 296, 303
insurance premiums, differential premiums as patient incentives, 58
integrated delivery systems (IDSs), 42, 88, 174–175
Integrated Healthcare Association (IHA)
  bonus payments, 124
  coordinating multiple programs, 64
  efficiency and, 151
  information technology and, 105
  overview, 16–17
  patient satisfaction, 113–114
  public reporting of quality performance, 26, 131
  quality measures, 22, 37
intensivists, 105, 197
intermediate outcomes, 110–112, 121
internal threats to validity
  about, 267, 268–289
  changes in how success is measured, 274–275, 327
  changes over time, 268
  determining the counterfactual, 275–281
  differential mortality, 273–274, 302–303, 306–307
  differential patient selection, 269, 300, 302–303, 306, 308–309
  ex post regression matching, 267, 281–287
  instrumentation, 274–275, 296, 303
  levels of confidence, 272–273
  propensity score matching, 267, 287–289
  statistical regression, 270–272
  statistical significance, 272–273, 300–301, 304, 307, 309
intrinsic motivation, of physicians, 87
investments, performance-enhancing, 55–56
IOM. *see* Institute of Medicine
IPPS. *see* inpatient prospective payment system

J

joint attribution algorithm, 187
Joint Commission, 16, 66, 104

K

Kaiser Permanente, 9–10, 347
kidney disease patients, 114, 147
Kidney Disease Quality of Life Scale, 114

L

lawsuits, 67
Leapfrog Group, 15, 104–105, 120, 148
levels-based simulations, 215–216
Local Initiative Rewarding Results Demonstration, 20–21, 25
lock-in, 190

M

majority rule, 193, 194
malpractice insurance, 67
managed care, 7
managed care systems
  attribution of responsibility, 165
  performance measures and, 41
  performance payments to, 163, 175–176
managers, organizational culture and, 90–91
mandatory ACOs, 351
mandatory pay for performance programs, 44
Massachusetts Health Quality Partners (MHQP), 169, 185
Mayo Clinic (Rochester, MN), 347
McAllen (TX), collaborative behavior in, 346–347
MCCD. *see* Medicare Coordinated Care Demonstration
MCS. *see* Mental Component Summary
Medicaid
  about, 8
  disparities in health care, 63
  Michigan Medicaid Health Plan Bonus/Withhold system, 203
  nonfinancial incentives in, 56
  P4P payment models, 204
  performance-based systems, 163
medical education, 82–83
medical health support organizations (MHSOs), 232
Medical Home Demonstration (Medicare). *see* Medicare Medical Home Demonstration
medical home systems, 41, 106–107, 183, 186, 199, 349
medical professionalism, 66, 84, 85, 94–95, 349

medical records
 abstracts of, 115, 116
 as data source for quality measures, 43, 115–116
 disadvantage of, 116
 electronic medical records (EMRs), 43–44, 101, 105, 116, 351, 359–360
Medicare
 about, 2, 8, 22–23, 221–222
 converting successful Medicare demonstrations into national programs, 315–339
 cost of services, 140
 efficiency of, 140, 149
 FFS demonstrations, 47
 lack of payment for poor performance, 51–52
 P4P programs, 16–21, 22–23, 221–260
 patient attribution and, 198–199
 prospective payment system, 9, 316
 public reporting of hospital outcome measures, 151
 public reporting of quality measures, 26–27
 severity diagnosis-related groups (MS-DRGs), 149
Medicare Acute Care Episode (ACE) Demonstration, 18, 24, 226–227, 251–252, 322, 329
Medicare Advantage program, 199, 337, 354–355
Medicare Cancer Prevention and Treatment Demonstration for Ethnic and Racial Minorities
 disparities in health care, 24, 63
 findings, 239–240
 incentives, 24
 overview, 20–21, 24, 224–225, 237–240
 project status, 238–239
Medicare Care Management for High-Cost Beneficiaries Demonstration (CMHCB)
 findings, 237
 overview, 18–19, 23, 224–225, 235–237
 project status, 236–237
 prospective assignment, 187
 selection bias, 321
 up-front fees, 323
 validity of findings, 305–307
Medicare Center for Innovation, 152, 332, 366
Medicare Coordinated Care Demonstration (MCCD)
 findings, 230–231
 overview, 20–21, 25, 222–223, 227–231
 project status, 228–229

 up-front fees, 323
 validity of findings, 273
Medicare demonstration projects. *see also* pay for performance (P4P) systems; *individual demonstration project names*
 budget neutrality, 322–324, 328
 care management and disease management organization projects, 222–223, 226–240
 compared to pilot tests, 315
 comparison groups, 292–293
 competitive bidding for, 337
 Congress and, 335–336
 converting successful demonstrations into national programs, 315–339
 data demands, 331
 demonstration payment waiver authority, 318
 design of, 33–70, 163–179, 270n2, 325–326
 early policy changes, 316–317
 eligibility, 320
 evaluating interventions, 267–312
 failure of, 327
 geographic and participant constraints, 320
 hospital-focused projects, 224–225, 247–259
 legal challenges to, 319, 319nn3, 4
 limitations of, 320
 models, 161–179, 204–217
 paying for innovation, 328–329
 physician-focused projects, 224–225, 240–247
 political challenges to national implementation, 332–338
 quasi-experimental design, 325–326
 rules for, 318–328
 second-generation pay for performance initiatives, 341–368
 selection bias, 321
 shared savings, 323
 start-up complexity, 329–330
 statutory authority, 318–319
 threats to evaluation findings, 267–292, 324–328
 up-front fees, 323
 validity of findings, 267–292
 voluntary vs. mandatory participation, 321
Medicare Health Support Pilot Program (MHS)
 accountability, 42
 budget neutrality, 322
 engagement problems, 294–295
 findings, 230, 233–234

instrumentation problems, 296
overview, 16–17, 23, 222–223, 231–232
participation discontinuities, 295–296
patient assignment, 187
prerandomization selection problems, 294
project status, 232–233
selection bias, 321
selection-experiment interactions, 296–297
statistical problems, 296
up-front fees, 323
validity of findings, 293–297
Medicare Hospital Gainsharing Demonstration
budget neutrality, 322
differential selection, 308–309
findings, 259
operational complexity of, 329
overview, 226–227, 255–259, 308–312
patient selection, 321
physician rewards in, 174
project status, 257–259
selection-experiment interactions, 309, 312
statistical significance, 309
Medicare Hospital-Acquired Condition (HAC) program, 352
Medicare Medical Home Demonstration, 224, 246–247
Medicare Participating Heart Bypass Center Demonstration
bundled payment, 23–24
efficiency, 328
findings, 248–249
incentives, 323
overview, 18–19, 23–24, 224–225, 247–249, 355
project status, 248
Medicare Payment Advisory Commission (MedPAC)
on attribution rules, 188, 194
health care payment reform and, 332, 333
on hospital efficiency, 145
on physician efficiency, 147
Medicare Physician Group Practice (PGP) Demonstration
budget neutrality, 322
differential mortality, 303
differential selection, 302–303
findings, 244–245
funding bonus payments, 124
incentive payments in, 52, 354
instrumentation problems, 303
overview, 16–17, 23, 224–225, 240–243
patient assignment, 42

patient attribution, 184–185, 197
paying groups of physicians, 173
performance payments in, 54
physician efficiency measures, 152
project status, 243–244
quality measures, 23, 119
reaction to experimental arrangements, 304
reduction-in-performance-gaps approach, 128
regression-to-the-mean problems, 304
selection bias, 321
selection-experiment interactions, 304
shared savings, 52
statistical significance, 304
threshold targets, 128, 129
validity of findings, 301–304
Medicare Physician Quality Reporting Initiative, 51, 104
Medicare Physician-Hospital Collaboration Demonstration, 18–19, 24, 226–227, 255–259, 329
Medicare Premier Hospital Quality Incentive Demonstration (HQID)
comparison with other providers, 47, 129
efficiency, 328–329
findings, 254–255
incentive payments in, 354
overview, 18–19, 23, 226–227, 252
penalties for poor performance, 23, 50
project status, 253–254
Medicare Present on Admission (POA) Reporting program, 352
Medicare reform, patient attribution and, 198–199
Medicare Reporting Hospital Quality Data for Annual Payment Update program, 51
Medicare Residency Reduction Demonstration, 327n1
Medicare severity diagnosis-related groups (MS-DRGs), 149
Medicare Shared Savings Program (MSSP), 366–368
Medicare Web sites
Dialysis Facility Compare, 125, 131
Hospital Compare, 13, 26, 119, 131, 151
Nursing Home Compare, 131
MedPAC. see Medicare Payment Advisory Commission
Med-Vantage, 15
Mental Component Summary (MCS), 114
merit-based pay, 81
metropolitan statistical areas (MSAs), 145, 297

MHQP. *see* Massachusetts Health Quality Partners
MHS. *see* Medicare Health Support Pilot Program
MHSOs. *see* medical health support organizations
Michigan Medicaid Health Plan Bonus/Withhold system, 203
minimum-share rule, 193, 194
Modified Fatigue Impact Scale (MFIS), 114
money, as physician's motivator, 86
morbidity measures, 112, 146
mortality measures, 101, 113, 119
    differential mortality, 273–274, 302–303, 306–307
    risk adjustment, 124–125
    validity of findings and, 273–274
motivators, for physicians, 85–88
MS. *see* multiple sclerosis (MS) patients
MSAs. *see* metropolitan statistical areas
MS-DRGs. *see* Medicare severity diagnosis-related groups
MSQLI. *see* Multiple Sclerosis Quality of Life Inventory
MSSP. *see* Medicare Shared Savings Program
multiple physician assignment rule, 194
multiple sclerosis (MS) patients, 108–109, 113, 114, 115, 118–119
Multiple Sclerosis Quality of Life Inventory (MSQLI), 114
myocardial infarction patients, 123, 125, 151. *see also* cardiovascular disease patients

N

National Committee for Quality Assurance (NCQA), 80, 101, 104
national health insurance, 7, 8
National Institute for Health and Clinical Excellence (NICE), 143
National Quality Forum, 38, 80
NCQA. *see* National Committee for Quality Assurance
negative incentives (disincentives), 50, 55, 80–81, 356–357
"never" events, 112, 166–167
NICE. *see* National Institute for Health and Clinical Excellence
nonexclusive assignment algorithm, 193
nonfinancial incentives
    Centers of Excellence, 24, 57, 58–59, 95–96, 174, 226–227, 247–248, 251–252, 290

designation as high-performing provider, 57–58
for patients to use high-performing providers, 57–58
public reporting, 26–27, 57, 68
report cards, 57
Nursing Home Compare (Medicare Web site), 131

O

Obama, Barack, 152, 335
one-touch rules for assignment, 41, 192, 193
online surveys, 117
organization theory, P4P and, 88–93
organizational culture, 90–91
outcome measures
    advantage over process measures, 39
    clinical outcomes, 34, 110–113, 165–166
    expanding use of, 13, 15
    final outcomes, 112–113
    functional outcomes, 112–113
    intermediate outcomes, 110–112
    mortality and morbidity, 34, 101–102
    "never" events, 112
    overview, 108–115
    patient-reported outcomes, 110, 113–115
    process measures and, 109
    structure measures and, 104, 109
out-of-network providers, 59, 175–176

P

P4P systems. *see* pay for performance
P4R. *see* payment for reporting
Participating Heart Bypass Center Demonstration (Medicare). *see* Medicare Participating Heart Bypass Center Demonstration
patient assignment
    about, 182
    "first contact" rule, 192
    geographic assignment, 190–191
    incentives and, 182
    majority rule, 193, 194
    minimum-share rule, 193, 194
    multiple physician assignment rule, 194
    nonexclusive assignment algorithm, 193
    as nonfinancial incentive, 56–57
    one-touch rules, 41, 192, 193
    patient lock-in, 190
    patient notification and lock-in, 190
    plurality rule, 192, 194
    randomization, 321

selection bias in demonstration projects, 321
share rules, 193
two-touch rules, 41
patient attribution
  about, 181–199
  assignment to single or multiple physicians, 187–188
  basic concepts in, 186–191
  California Physician Performance Initiative, 185–186
  challenges in, 64, 183–184
  exclusion of patients from attribution, 189
  in fee-for-service situation, 182, 183, 191–196
  geographic unit of assignment, 190–191
  joint attribution algorithm, 187
  Massachusetts Health Quality Partners, 185
  Medicare Physician Group Practice Demonstration, 184–185, 197
  Medicare reform and, 198–199
  multiple attribution, 188
  patient notification and lock-in, 190
  patient vs. episode assignment, 189
  prospective vs. retrospective, 186–187
  rules for determining responsible physician, 192–197
patient lock-in, 190
Patient Protection and Affordable Care Act (P.L. 111-148). *see* Affordable Care Act
patient safety, 35
patient satisfaction, 35, 37, 44, 113–114
patient selection
  capitation, 10
  Medicare Health Support Pilot Program, 294
  as threat to validity of findings, 269, 270–272, 300, 302–303
patient surveys, 44, 110, 116–118
patient-centered teams, 95
patient-reported outcomes, 110, 113–115
patients. *see also* patient assignment; patient attribution
  accountability of, 353
  attrition of sample, 273–274
  as challenge, 343
  differential patient selection, 269, 300, 302–303, 306, 308–309
  "difficult" patients, 62
  dropouts and drop-ins, 295
  elderly patients, 360
  exclusion from attribution, 189
  financial incentives to, 353
  health status variation of, 150
  incentive to use high-performing providers, 57–59, 68–69
  involvement in P4P, 352–353
  limited patient resources, 165
  patient attribution, 64, 181–199
  satisfaction of, 35, 37, 44, 113–114
  selection of, 10, 270–272
  surveys of, 44, 110, 116–118
pay for performance (P4P) systems
  about, 69–70, 139
  administration of, 61–62
  alternatives and complements to, 65–69
  attribution of responsibility, 164–166, 197
  care management P4P demonstrations, 222–223, 226–240
  clinical uncertainty limiting scope of, 360
  competition and, 68
  contingency theory and, 93–96
  cost containment in, 123–124
  cost or resource utilization measures, 15
  definition of term, 7, 33
  definitions for, 14
  design of, 33–70, 162–179, 270n2, 325–326
  economics of, 79–83
  effectiveness of, 27, 53
  efficiency measures in, 139–153
  evaluating interventions, 267–312
  failure of, 327, 345
  fee-for-service systems compared to, 7–8, 33
  health care outcomes, 13
  health care quality and, 12
  history of, 8–12, 139
  hospital-focused demonstrations, 224–225, 247–259
  limitations of, 59–64, 70
  mandatory vs. voluntary participation in, 44
  models, 161–179, 204–217
  multiple payers with inconsistent programs, 64
  organization theory and, 88–93
  organizational culture and, 90–91
  outcome measures, 13, 15
  patient attribution, 64
  pay for quality, 13, 14
  payment models, 204–217
  performance benchmarks for, 45–49
  physician-focused demonstrations, 224–225, 240–247

(continued)

pay for performance (P4P) systems (*continued*)
  private sector programs, 16–17, 22, 119, 199, 359
  providers' role in, 25–26
  psychology of, 85–88
  public sector programs, 16–20, 22–25, 119
  quality regulation and accreditation, 66–67
  second-generation pay for performance initiatives, 341–368
  simulations of bonus payments, 210–216
  sociology of, 82–85
  theoretical perspectives, 77–96
  types of, 14–25
pay for performance (P4P) systems, United Kingdom
  financial incentives in, 25, 53
  National Institute for Health and Clinical Excellence (NICE), 143
  P4P performance measures, 41, 106, 127
  patient-reported outcomes, 114
  pay for performance program, 17, 20–21, 25, 41, 53, 161
  quality measures in P4Ps, 106, 114, 119
  size of incentives, 356
  unintended consequences of P4P programs, 63–64
payment for efficiency, 14
payment for quality, 13
payment for reporting (P4R), 14
payment for value, 14
PBPM fee. *see* per beneficiary per month (PBPM) fee
PCPs. *see* primary care physicians
PCS. *see* Physical Component Summary
peer recognition, physicians, 87
penalties, as disincentive, 50
per beneficiary per month (PBPM) fee, 226
performance benchmarks. *see* benchmarks
performance domains, 37–38
performance incentives. *see* financial incentives; nonfinancial incentives
performance indicators
  comprehensiveness of, 61
  cost-effectiveness and cost benefits of, 39–40, 61–62
  data collection for, 39, 43–44, 101, 107
  graduated or tiered rewards, 49
  importance and relevance of, 39
  improved performance, 46
  lack of high-quality indicators, 59–61
  number and interdependence of quality indicators, 210
  relative performance, 46–48, 206, 208
  reliability of, 38, 59–61
  target attainment, 46
  validity of, 38
performance measures. *see also* outcome measures; quality measures
  about, 344–345
  comparison of programs, 15–20
  data collection for, 39, 43–44, 101, 107
  defining domains of performance, 34–37, 204–210
  defining units for, 40–43
  how to measure performance, 344–345
  lack of flexibility of, 61
  patient attribution, 64
  selecting performance domains and indicators, 37–40, 204–210
  unintended consequences of, 62–64, 70
performance payments. *see also* incentives
  about, 161–162
  all-or-nothing algorithm, 206
  attribution of responsibility, 164–166, 197
  bundled payment, 24, 95, 352
  composite algorithm, 206, 207
  constrained algorithm, 206
  continuous unconstrained algorithm, 206
  to disease or care management organizations, 176–177
  to groups of physicians, 162–163, 173
  to health maintenance organizations (HMOs) and other managed care organizations, 163, 175–176
  to hospitals, 174
  to individual physicians, 171–173
  to integrated delivery systems (IDSs), 174–175
  options for whom to pay, 171–177, 344
  payment algorithms, 205–208
  payment models, 204–217
  to physician-hospital organizations (PHOs), 174–175
  relative rates of improvement algorithm, 46–48, 206, 208
  rewarding clinical providers vs. other organizations, 162–164
  self-financing, 354–355
  setting targets, 208–210
  simulations, 210–216
  statistical algorithm, 206, 207
  unit of care for, 166–168
performance-enhancing investments, 55–56
per-member payment, as financial incentives, 51

PGP Demonstration. *see* Medicare Physician Group Practice (PGP) Demonstration
PGPs. *see* physician group practices
PHOs. *see* physician-hospital organizations
Physical Component Summary (PCS), 114
Physician Group Practice Demonstration. *see* Medicare Physician Group Practice (PGP) Demonstration
physician group practices (PGPs). *see also* Medicare Physician Group Practice Demonstration
   assignment algorithms and, 196
   incentives for, 53–54
   organization theory and, 88, 89–90
   performance payments to, 162–163, 173
   performance standards, 48
   threshold targets, 129
physician organizations (POs), 145, 188–189
Physician Quality Reporting Initiative (Medicare). *see* Medicare Physician Quality Reporting Initiative
physician-focused pay for performance, 224–225, 240–247
Physician-Hospital Collaboration Demonstration (Medicare). *see* Medicare Physician-Hospital Collaboration Demonstration
physician-hospital organizations (PHOs), performance payments to, 174–175
physicians. *see also* physician group practices (PGPs)
   attribution to individual physicians vs. physician organizations, 188–189
   cost efficiency of, 195
   deprofessionalization, 84, 85
   economics of health care and, 79–80
   efficiency measurement for, 145–147, 150, 151, 194
   extended hospital medical staff, 188–189
   fragmentation of care, 348
   gainsharing, 323
   gatekeeper, 182
   incentive payments to, 162–163, 350
   institutional layers interacted with, 89
   intensive care unit staffing, 105
   interdependency of, 164–165
   medical training, 82–83
   motivators for, 85–88
   organizational culture and, 90–91
   P4P programs, 25–26
   patient assignment to single or multiple physicians, 187–188
   peer recognition, 87
   performance measures and, 40–42, 55
   performance payments to groups of physicians, 172
   performance payments to individual physicians, 171–173
   physician-hospital organizations (PHOs), 174–175
   private practice, 92
   professionalism, 66, 84, 85, 94–95, 349
   salaried, 88, 347
   "sample size" problem, 172
piece-rate compensation, 80
pilot testing, 315
P.L. 111-148 (Patient Protection and Affordable Care Act). *see* Affordable Care Act
plurality assignment rule, 192, 194
POA codes. *see* present on admission (POA) codes
POA program (Medicare). *see* Medicare Present on Admission (POA) Reporting program
population-based measures, 122, 146, 290
POs. *see* physician organizations
post-discharge episodes, 168
PPOs. *see* preferred provider organizations
PQIs. *see* Prevention Quality Indicators
practice sanctions, 56
practice sites, 186
preferred provider organizations (PPOs), 59, 175–176, 182
Premier Hospital Quality Incentive Demonstration (Medicare). *see* Medicare Premier Hospital Quality Incentive Demonstration
present on admission (POA) codes, 109, 119
Present on Admission (POA) program (Medicare). *see* Medicare Present on Admission (POA) Reporting program
Prevention Quality Indicators (PQIs), 124
primary care physicians (PCPs)
   efficiency of, 147
   fee schedule adjustment, 50
   patient attribution and, 187, 197
   performance measures and, 41, 85–86, 182
primary prevention, 9–10
principal-agent problem, 79, 80
private health insurance plans, cost containment, 9

private practice, 92
private sector pay for performance programs, 16–17, 22, 119, 199, 359
process measures
  benefits of, 107–108
  characteristics of, 107
  clinical process quality, 34–35
  defined, 13, 39
  health care outcomes and, 109
  intermediate outcomes substituted for, 121
  overview, 13, 61, 100–101, 107–108
process-of-care indicators, 38
productive efficiency, 142
productivity, as performance indicator, 36
professionalism, 66, 84, 85, 94–95, 349
propensity score matching, 267, 287–289
prospective patient assignment, 186–187
prospective payment system, 9
provider education, 66
provider network designation, 59
provider organizations
  differential patient selection, 269
  performance measures and, 40, 48, 55
  second-generation P4Ps, 348, 350
  validity of findings, 290–291
provider reimbursement, P4P and, 65–66
provider tiering, 58
providers
  institutional layers in, 89–90
  malpractice insurance, 67
  organization theory and, 88–93
  ownership of, 88–89
  P4P programs, 25–26, 40, 58
  patient attribution, 64
  provider education, 66
  provider network designation, 59
  provider tiering, 58
  quality regulation and accreditation, 66–67
  reimbursement systems for, 65–66
  stages of organizational development, 91–92
psychology, of health care, 85–88
public recognition, as nonfinancial incentive, 56
public reporting
  advantages and disadvantages of, 131–132
  of hospital outcome measures, 151
  as nonfinancial incentive, 26–27, 57, 68
  risk adjustment and, 125
public sector pay for performance programs, 16–20, 22–25, 119
punctuated equilibrium, 334–336

Q

QALY. *see* quality-adjusted life years
QOL scales, 114
quality
  assurance of, 12–13, 127–128, 129
  of care (*see* health care quality)
  making quality improvement goals more ambitious, 357–359
  P4P not sufficient for improving quality, 361–362
quality bonuses. *see* financial incentives; performance payments
quality indicator weights, 210
quality measures. *see also* outcome measures; performance measures
  about, 344
  accreditation organizations, 104
  analytical methods for, 124–130
  background, 100–103
  changes in how success is measured, 274–275
  composite measures, 121–122
  data collection for, 39, 43–44, 101, 107
  Donabedian's vs. IOM model, 100–103
  identifying high-performing providers, 126–130
  improvement-over-time targets, 128–129
  making quality improvement goals more ambitious, 357–359
  mortality data and, 101
  number and interdependence of quality indicators, 210
  number of, 119–121
  outcome measures, 13, 15, 34, 39, 101, 108–115
  overview of, 99–100, 102
  population-based measures, 122, 146
  Prevention Quality Indicators (PQIs), 124
  process measures, 13, 24, 34–35, 39, 61, 100–101, 103, 104, 107–108
  public reporting of, 26–27, 57, 68, 131–132, 151
  resource utilization measures, 15
  room for improvement in performance, 123
  sample size for, 130, 172
  selection of, 119–124
  statistical analysis of, 130
  structure measures, 100, 102–103, 104–107
  systems-level measures, 122
  in United Kingdom P4P, 106, 114, 119
  validity of findings and, 274–275
quality-adjusted life years (QALY), 141

R

randomization, risk adjustment and, 125
rate of improvement payment algorithm, 46–48, 206, 208
readmission/rehospitalization, 168, 169
rebasing benchmarks, 49
reduction-in-performance-gaps approach, 128
regression, multivariate regression, 273n1
regression-based risk adjuster, 149
regression-to-the-mean (RtoM) effects, 270–271, 296, 304, 307
relative rates of improvement algorithm, 46–48, 206, 208
relative value scale, 9
reliability, of performance indicators, 38
report cards, as nonfinancial incentives, 57
resource utilization measures, pay for performance, 15
responsibility, attribution of, 164–166, 197
retrospective patient assignment, 187
Rewarding Results Demonstration sites
    Blue Cross Blue Shield of Michigan, 16–17, 22, 37, 50–51, 169
    Excellus/Rochester (New York), 50
risk, in performance incentives, 54–56, 62–63, 326–328
risk adjustment
    of efficiency, 149–150
    of functional outcomes, 113
    of intermediate outcomes, 111
    of mortality measures, 124–125
    of quality measures, 119
    regression-based risk adjuster, 149–150
risk-adjusted mortality rates, 113
room for improvement in performance, 123
RtoM effects. *see* regression-to-the-mean (RtoM) effects
Rubin's causal model, 275

S

salaried physicians, 88, 347
sample size, for quality measures, 130, 172
sampling
    sample size for quality measures, 130, 172
    statistical regression, 270–272
scope of care, 166
secondary prevention, 9–10
second-generation pay for performance initiatives
    about, 341–368
    accountability, 347–350

    challenges in development of, 343–347
    policy recommendations for, 347–362
selection bias, Medicare demonstration projects, 321
selection-intervention interactions
    Indiana Health Information Exchange Demonstration, 301
    Medicare Care Management for High-Cost Beneficiaries Demonstration, 307
    Medicare Health Support Pilot Program, 296–297
    Medicare Hospital Gainsharing Demonstration, 309, 312
    Medicare Physician Group Practice Demonstration, 304
    overview, 290–291
self-financing, 354–355
service quality, as performance indicator, 35
SFR. *see* stochastic frontier regression
share rules, 193
shared-savings payment model, 52, 54, 152, 241
simulations
    of attribution rules, 195
    improvement rate simulations, 216
    levels-based simulations, 215–216
    performance payment algorithms, 210–216
skilled nursing facilities, Affordable Care Act plans for a value-based purchasing program for, 364
sociology, of health care, 82–85
Southern California Evidence-Based Practice Center, 143
specialist physicians, 41, 186, 197
"stair step" model, 127
statistical analysis
    ex post regression matching, 267, 281–287
    multivariate regression, 273n1
    for outcomes measures, 110, 113
    propensity score matching, 267, 287–289
    of quality improvement, 130
    regression-to-the-mean (RtoM) effects, 270–271, 296, 304, 307
    statistical noise, 272, 273n1
    two-sided 95 percent confidence interval, 272
statistical noise, 327
statistical payment algorithm, 206, 207
status quo forums, 335
stochastic frontier regression (SFR), 144
structure measures, 100, 102–103, 104–107, 109

success
    evaluating success, 345–346
    measuring success, 274, 327
surveys. *see* patient surveys
Symmetry Episode Risk Groups, 149–150
systems-level measures, 122

T

Tax Relief and Health Care Act of 2006 (TRHCA), 246
team-based care, 92, 95
technical assistance, as nonfinancial incentive, 56
technical efficiency, 142
third-party care management organizations, performance measures and, 42
threshold targets, 127–128, 129
tiered rewards, 49
total quality management, 91
tournament approach, 47–48
TRHCA. *see* Tax Relief and Health Care Act of 2006
two-sided 95 percent confidence interval, 272
two-touch rules for assignment, 41

U

undertreatment, 10
unit of care, 166–168
United Kingdom
    diabetes patients, 106
    financial incentives in, 25, 53
    National Institute for Health and Clinical Excellence (NICE), 143
    P4P performance measures, 41, 106, 127
    patient-reported outcomes, 114
    pay for performance program, 17, 20–21, 25, 41, 53, 161
    quality measures in P4Ps, 106, 114, 119
    size of incentives, 356
    unintended consequences of P4P programs, 63–64
units of accountability, 151
up-front fees, 323
US Agency for Health Care Policy and Research (AHCPR), 10
usual-care comparison group, 47

V

validity (of findings)
    about, 267–292, 324–328
    changes in how success is measured, 274–275, 327

changes over time, 268
determining the counterfactual, 275–281
differential mortality and, 273–274, 302–303, 306–307
differential patient selection and, 269, 300, 302–303, 306, 308–309
ex post regression matching and, 267, 281–287
instrumentation and, 274–275, 296, 303
internal threats to, 267, 268–289
levels of confidence and, 272–273
propensity score matching and, 267, 287–289
reactive effects to experimental arrangements, 291–292, 297–298, 301, 304, 307
selection-intervention interactions, 290–291, 296–297, 301, 304, 307, 309, 312
statistical regression and, 270–272
statistical significance and, 272–273, 300–301, 304, 307, 309
validity (of performance indicators), 38, 59–61
value, 11
value of care, 151
Value-Based Care Centers, 24, 251, 338
value-based purchasing, 7, 10–11, 346, 362. *see also* Hospital Value-Based Purchasing Program (HVBPP)
variable pay, 81
vendor-based measures of efficiency, 146
virtual ACOs, 351
visit cost, 193
voluntary pay for performance programs, 44, 47

W

Wisconsin Collaborative for Healthcare Quality, 347, 361
World Health Organization (WHO), on cost-effectiveness, 143

Y

yardstick competition, 47–48